Bioactive Compounds

Bioactive Compounds

Health Benefits and Potential Applications

Edited by

Maira Rubi Segura Campos

ELSEVIER

WP
WOODHEAD PUBLISHING
An imprint of Elsevier

Woodhead Publishing is an imprint of Elsevier
The Officers' Mess Business Centre, Royston Road, Duxford, CB22 4QH, United Kingdom
50 Hampshire Street, 5th Floor, Cambridge, MA 02139, United States
The Boulevard, Langford Lane, Kidlington, OX5 1GB, United Kingdom

Library of Congress Cataloging-in-Publication Data
A catalog record for this book is available from the Library of Congress

British Library Cataloguing-in-Publication Data
A catalogue record for this book is available from the British Library

ISBN (print): 978-0-12-814774-0
ISBN (online): 978-0-12-814775-7

For information on all Woodhead Publishing publications visit our website at https://www.elsevier.com/books-and-journals

Publisher: Andre G. Wolff
Acquisition Editor: Megan R. Ball
Editorial Project Manager: Susan Ikeda
Production Project Manager: Debasish Ghosh
Cover Designer: Greg Harris

Typeset by TNQ Technologies

Contents

PART 9: Bioactive Compounds and Functional Foods as Therapeutic Alternative

List of Contributors

Juan José Acevedo Fernández Facultad de Medicina, Universidad Autónoma del Estado de Morelos, Cuernavaca, Mexico

Laís Marinho Aguiar University of Campinas, School of Food Engineering, Department of Food and Nutrition, Campinas, Brazil

Víctor Ermilo Arana Argáez Facultad de Química, Universidad Autónoma de Yucatán, Mérida, Mexico

Betty Jarma Arroyo Department of Domestic Science, Program of Food Science and Technology, Universidade Federal Rural de Pernambuco, Recife, Brazil

Cíntia Reis Ballard Department of Food and Nutrition, Faculty of Food Engineering, University of Campinas, Campinas, Brazil

Ludy Pabón Baquero Researcher of Department of Basic Sciences, Universidad de La Salle, Bogotá, Colombia; Biomigen Research Group, Universidad de La Salle, Bogotá, Colombia

Cínthia Baú Betim Cazarin Department of Food and Nutrition, Faculty of Food Engineering, University of Campinas, Campinas, Brazil

Luis Carlos Boyano-Orozco Department of Biochemical Engineering of the Escuela Nacional de Ciencias Biológicas, Instituto Politécnico Nacional, Mexico City, Mexico

Alexis Buitrago Organic Biomolecular Research group, Research Institute, Faculty of Pharmacy and Bioanalysis, University of Los Andes, Mérida, Venezuela; Analysis and Control Department, Faculty of Pharmacy and Bioanalysis, University of Los Andes, Mérida, Venezuela

Anderson Campos Department of Domestic Science, Program of Food Science and Technology, Universidade Federal Rural de Pernambuco, Recife, Brazil

Roger Cauich-Kumul Facultad de Ingeniería Química, Universidad Autónoma de Yucatán, Mérida, Mexico

Ivan Chan Zapata Facultad de Ingeniería Química, Universidad Autónoma de Yucatán, Mérida, Mexico

Michele Silveira Coelho Laboratory of Food Technology, School of Chemistry and Food, Federal University of Rio Grande, Rio Grande, Brazil

Enayde de Almeida de Melo Department of Domestic Science, Program of Food Science and Technology, Universidade Federal Rural de Pernambuco, Recife, Brazil

Helena Dias de Freitas Queiroz Barros Department of Food and Nutrition, Faculty of Food Engineering, University of Campinas, Campinas, Brazil

Sibele Santos Fernandes Laboratory of Food Technology, School of Chemistry and Food, Federal University of Rio Grande, Rio Grande, Brazil

Marina Vilar Geraldi University of Campinas, School of Food Engineering, Department of Food and Nutrition, Campinas, Brazil

Patricia Hernández-Rodríguez Researcher Professor Biology Program, Department of Basic Sciences. Universidad de La Salle, Bogotá, Colombia; Researcher of Department of Basic Sciences, Universidad de La Salle, Bogotá, Colombia; Biomigen Research Group, Universidad de La Salle, Bogotá, Colombia

Harold Rodríguez Larrota Student Biology Program, SIEZ Research Incubator

Verena Silva Lima Department of Food and Nutrition, School of Food Engineering, University of Campinas, Campinas, Brazil; Health and Biotechnology Institute, Federal University of Amazonas, Coari, Brazil

Lara Lins Department of Domestic Science, Program of Food Science and Technology, Universidade Federal Rural de Pernambuco, Recife, Brazil

Mário Roberto Maróstica Junior Department of Food and Nutrition, Faculty of Food Engineering, University of Campinas, Campinas, Brazil

Armando M. Martín Ortega Facultad de Ingeniería Química, Universidad Autónoma de Yucatán, Mérida, Mexico

Edwin Eenrique Martínez Leo Facultad de Ingeniería Química, Universidad Autónoma de Yucatán, Mérida, Mexico

Janne Rojas Organic Biomolecular Research group, Research Institute, Faculty of Pharmacy and Bioanalysis, University of Los Andes, Mérida, Venezuela

Myriam de las Mercedes Salas-Mellado Laboratory of Food Technology, School of Chemistry and Food, Federal University of Rio Grande, Rio Grande, Brazil

Andrelina Pinheiro Santos Department of Chemical Engineering, Universidade Federal de Pernambuco, Recife, Brazil

Maira Rubi Segura Campos Facultad de Ingeniería Química, Universidad Autónoma de Yucatán, Mérida, Mexico

Milena Morandi Vuolo Department of Food and Nutrition, School of Food Engineering, University of Campinas, Campinas, Brazil

PART 1

Antioxidant Activity of Phenolic Compounds

CHAPTER 1

Antioxidant Activity of Phenolic Compounds Biosynthesized by Plants and Its Relationship With Prevention of Neurodegenerative Diseases

Janne Rojas[1], Alexis Buitrago[1,2]

[1]Organic Biomolecular Research group, Research Institute, Faculty of Pharmacy and Bioanalysis, University of Los Andes, Mérida, Venezuela; [2]Analysis and Control Department, Faculty of Pharmacy and Bioanalysis, University of Los Andes, Mérida, Venezuela

1.1 Introduction

Oxygen is the most predominant element in earth layer and it exists in air as a diatomic molecule, O_2. Almost all living organisms use O_2 for energy production, thus it is essential for life. To produce energy, an organism requires a process called oxidation, which implies the loss of electrons. However, oxidation also involves the inactivation of enzymes that do not have antioxidant mechanisms and do not survive in an O_2 environment (Magder, 2006). Under physiological conditions oxidizing agents and antioxidant defenses are in balance. Enzymes such as catalase, superoxide dismutase, glutathione peroxidase, and nonenzymatic antioxidants such as glutathione, and vitamins C and E are among the antioxidative defense molecules. However, if the production of free radicals exceeds the antioxidant capacity of a living system, these species may react with lipids, proteins, and DNA causing structural and functional damage to enzymes and genetic material (Barreiros et al., 2006). The predominance of oxidants, and their consequent damage is called oxidative stress and is considered as an etiological or pathogenic agent of cardiovascular and neurodegenerative diseases, such as cancer, Alzheimer, diabetes, and aging, among others (Jang et al., 2010; Krishnaiah et al., 2011). These pathologies and the evidence that are promoted by oxidative stress have brought the attention of scientists to find antioxidants for the prevention and treatment of such diseases (Halliwell and Gutteridge, 2007). Interest in naturally occurring antioxidants has increased; the food, cosmetic, and pharmaceutical industries are focused on natural products to replace synthetic antioxidants which are often restricted due to carcinogenic effects (Djeridane et al., 2006; Wannes et al., 1970). Plants are source of numerous secondary metabolites, many of these are natural

Bioactive Compounds. https://doi.org/10.1016/B978-0-12-814774-0.00001-3

antioxidants like polyphenols, flavonoids, essential oils, etc., that may be considered as sources of these substances. This chapter aims to review the basic concepts related to free radicals and their relationship to several ailments, to describe some natural products with proved antioxidant activity, as well as a number of in vivo and in vitro antioxidant activity assessment methods for natural products, developed by different researchers. A selection of reported studies in the last 10 years, related to antioxidant activity carry out in plant extracts, has also been included.

1.2 Free Radicals and Their Relationship to Several Ailments

Free radicals are produced to support life under aerobic conditions, keeping a balance between oxidizing agents and antioxidant defenses (Gupta and Verma, 2010). However, if the production of free radicals exceeds the antioxidant capacity of a living system, these reactive oxygen and nitrogen species, called reactive species, can defy lipids, proteins, and DNA, causing structural and functional damage to enzymes and genetic material (Barreiros et al., 2006). This condition is called oxidative stress.

In order to understand the mechanism of these reactive species it is necessary to define each group. Reactive oxygen species (ROS) and reactive nitrogen species (RNS), are radicals or chemical species that take part in radical type reactions (gain or loss of electrons), such as superoxide anion radicals ($O_2^{\cdot -}$), hydroxyl radical species ($OH^{\cdot}$), nitric oxide (NO), and nitrogen dioxide (NO_2) (Halliwell and Gutteridge, 2007; Magder, 2006).

According to specialized studies, there are at least three mechanisms that the human body uses to combat the excess of ROS, hence, oxidative stress: (1) preventive mechanism, proteins which have a coordinated nucleus of iron or copper with the capacity to bind (albumin, myoglobin, metallothionein, ceruloplasmin, ferritin, transferrin), which prevents the overproduction of $OH^{\cdot}$; (2) repairing mechanism, enzymes which repair or eliminate damaged biomolecules by ROS, like glutathione peroxidase, glutathione reductase, and methionine-sulfoxide reductase; and (3) scavenger mechanism, enzymes with capacity to scavenge excess ROS like superoxide dismutase, glutathione peroxidase, catalase, other metalloenzymes, and chemical entities with scavenging capacity like polyunsaturated fatty acids, vitamins C and E, uric acid, bilirubin, carotenoids, and flavonoids (Martínez Sánchez et al., 2003).

Investigations have stated that under stress, the human body can end up having more reactive oxygen species than antioxidant species, an imbalance that leads to cell damage (Krishnaiah et al., 2011; Gupta and Verma, 2010). If an imbalance between ROS and the antioxidative defense systems take place, cardiovascular and neurodegenerative diseases, such as, cancer, Alzheimer, diabetes, cardiovascular, neurological, endocrine, respiratory, immune, self-immune, ischemia, gastric disorders, tumor progression, and carcinogenesis, etc., may occur (Jang et al., 2010).

A number of investigations have been carried out in order to find antioxidant substances that are able to either prevent or treat these diseases (Halliwell and Gutteridge, 2007) and these will be explained in detail in this chapter.

1.3 Natural Products and Antioxidant Activity

Antioxidants are substances present at low concentrations that significantly delay or inhibit oxidation. According to investigations, natural antioxidants are phenolic compounds (tocopherols, flavonoids, and phenolic acids), nitrogen compounds (alkaloids, chlorophyll derivatives, amino acids, and amines), carotenoids, or ascorbic acid (Velioglu et al., 1998). Some of these groups of natural compounds will be explained separately in the following section.

1.4 Ascorbic Acid and Tocopherols

Ascorbic acid (vitamin C, Fig. 1.1) is widely known for its antioxidant activity; it is therefore used in cosmetics and degenerative disease treatments. Vitamin C has many physiological functions, among them is a highly antioxidant ability to recycle vitamin E in membrane and lipoprotein lipid peroxidation (Haslam, 1996).

On the other hand, tocopherols and tocotrienols are widely distributed in nature. Vitamin E is the common name given to a group of lipid-soluble compounds of which α-tocopherol (Fig. 1.1), is the most commonly known. It is found in lipoproteins and membranes blocking the chain reaction of lipid peroxidation by scavenging intermediate peroxyl radicals being generated. The highly steric α-tocopherol radical is much less reactive in attacking fatty acid side chains and converts back to its parent phenol through ascorbic acid, thus breaking the chain reaction (Haslam, 1996).

1.5 Carotenoids

The antioxidant activity of carotenoids is due to the ability to delocalize unpaired electrons through their structure of conjugated double bonds. These secondary metabolites are not very efficient as quenchers of peroxyl radicals, but are able to quench singlet oxygen, thus aiding in protecting lipids against peroxidative damage. β-Carotene (Fig. 1.1) is the most abundant of the carotenoids and is also highly reactive with electrophiles and oxidants. It is widely used in therapies. Despite this, many studies have shown β-carotene inhibition of lipid auto-oxidation in biological tissues and food; few details of either the kinetics or mechanism of these reactions have been revealed (Alves et al., 2010).

1.6 Phenolic Compounds

Phenolic compounds are a kind of secondary metabolite found commonly in plants and are known to possess multiple biological effects, including antioxidant activity. They have been

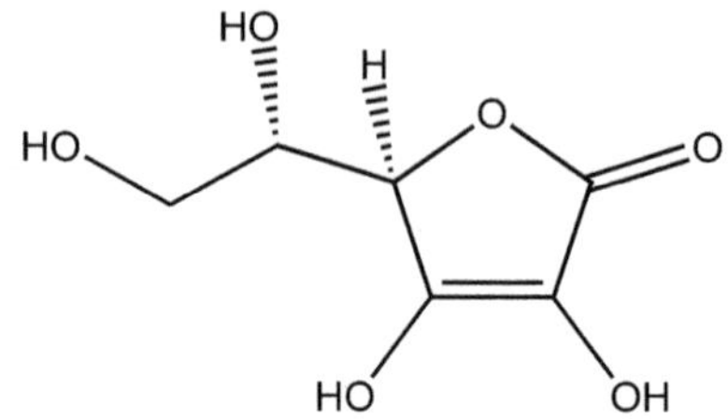

Ascorbic acid

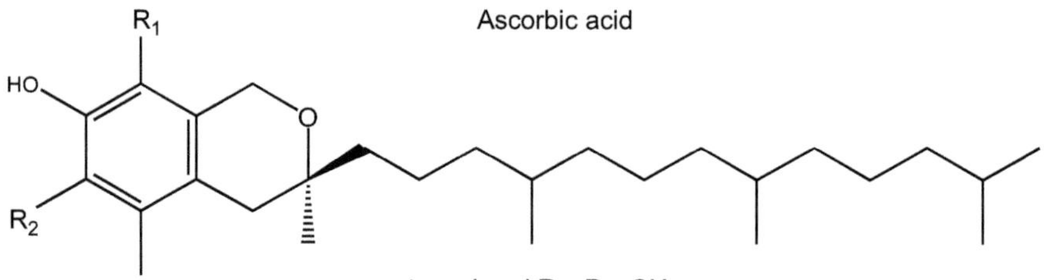

α-tocopherol $R_1=R_2=CH_3$
β-tocopherol $R_1=CH_3$ $R_2=H$
γ-tocopherol $R_1=H$ $R_2=CH_3$
α-tocopherol $R_1=R_2=H$

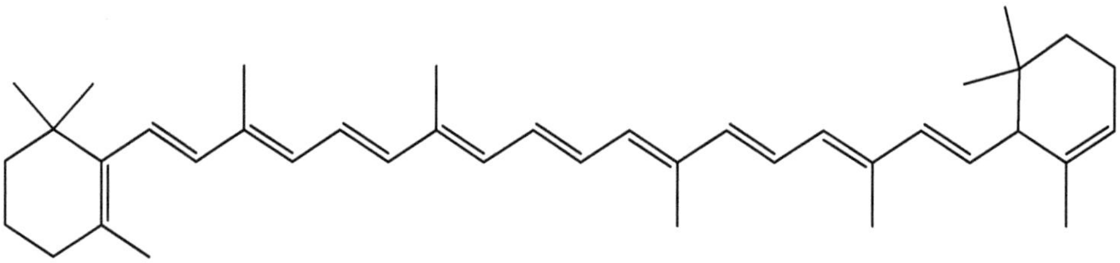

β-carotene

Figure 1.1
Ascorbic acid, tocopherols, and β-carotene as natural products with antioxidant activity.

classified into several categories: simple phenolics, phenolic acids, coumarins, flavonoids, stilbenes, tannins, lignans, and lignins (Fig. 1.2) (Naczk and Shahidi, 2004). From these classifications, flavonoids stands out as they have shown a wide range of antibacterial, antiviral, antiinflammatory, anticancer, and antiallergic activities (Montoro et al., 2005). Furthermore, flavonoids have proved to be highly effective scavengers of most oxidizing molecules, including singlet oxygen, and free radicals implicated in several diseases (Bravo, 1998).

On the other hand, flavonoids can be divided into seven categories: chalcones, anthocyanins, flavones, isoflavones, flavanones, flavononols, and flavanols (Ignat et al., 2011). The chemical structures of some flavonoids are shown in Fig. 1.2, from these, anthocyanins are probably the largest group of phenolic compounds in the human diet, and their strong antioxidant activity suggests their importance in maintaining health, when consumed regularly. These flavonoids have been associated with a reduction in the incidence of cancer and heart diseases (Velioglu et al., 1998).

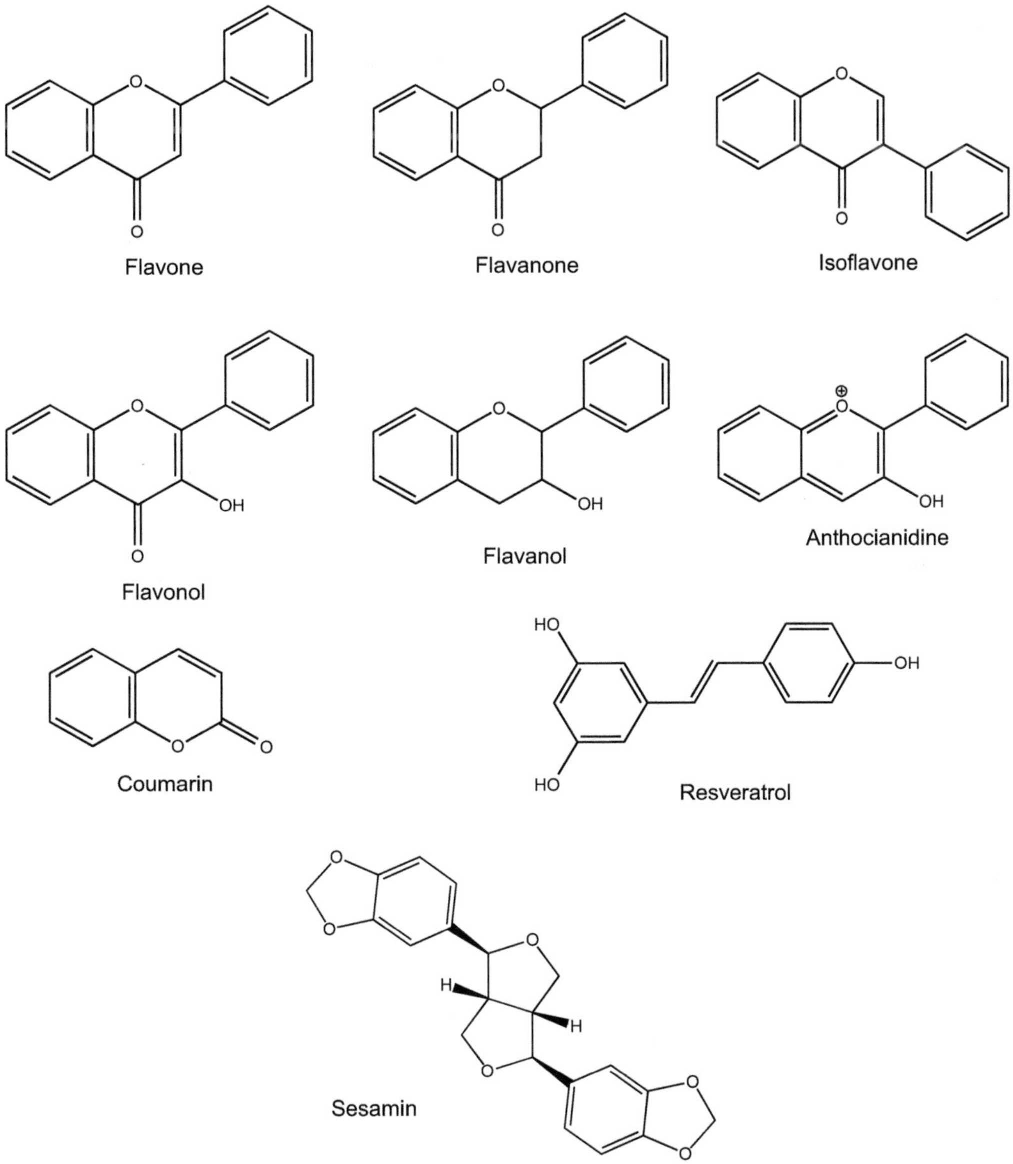

Figure 1.2
Chemical structures of flavonoid nucleus and coumarin, resveratrol and sesamin.

1.7 *In vivo* and *in vitro* Antioxidant Activity Assessment Methods for Natural Products

A large number of methods have been developed and tested for several years as there has been huge interest from researchers to investigate the antioxidant properties of plants, however, the advantages and limitations of these methods are still in discussion. Several researchers have classified these methods as in vivo and in vitro, and have stated that antioxidant activity should not be concluded based on a single antioxidant test model (Nur Alam et al., 2013; Badarinath et al., 2010).

In this regard, in vitro methods provide useful information of antioxidant activity present in either plant extracts or isolated compounds; however, data obtained by these methods are difficult to apply to biological systems, while in vivo assays are difficult to carry out due to difficulties related to cellular uptake and transport processes of these molecules. Any method assessed must be considered only as a preliminary result (Antolovich et al., 2002).

In terms of accessibility, the free radical scavenging DPPH method is rather rapid and simple as it does not involve many steps and reagents, and thus is less expensive in comparison to other test models. ABTS decolorization assay is also convenient as it is applicable for both hydrophilic and lipophilic antioxidants (Karadag et al., 2009; Nur Alam et al., 2013).

Due to the different types of free radicals and their different forms of action in living organisms, it is unlikely that a single, simple, and accurate universal method by which antioxidant activity may be measured will ever be developed (Alves et al., 2010). Therefore, the following section summarizes the most common methods used to assess antioxidant activity.

1.8 *In vitro* Methods

1.8.1 2,2-Diphenyl-1-Picrylhydrazyl Radical (DPPH) Assay

This method was first described by Blois in 1958, and has been modified by numerous researchers. DPPH is a stable free radical that reacts with compounds able to donate a hydrogen atom, thus, this assay is based on scavenging of DPPH through the addition of a radical species or antioxidant capable of discoloring the DPPH solution from deep violet. When a solution of DPPH is mixed with a substrate able to donate a hydrogen atom, this leads to the reduced form with the loss of the violet color (Fig. 1.3). According to the methodology, the degree of discoloration is proportional to the concentration of antioxidant type molecules. The activity is measured by UV spectrophotometry. A low absorbance indicates a high free radical scavenging activity of the compound under investigation (Krishnaiah et al., 2011; Nur Alam et al., 2013).

Considering previously published investigations, this methodology is one of the easiest and most accurate for evaluation of antioxidant activity in plant extracts and pure isolate substances like flavonoids and terpenoids (Alves et al., 2010). Antioxidants like butylated hydroxyanisole (BHA), butylated hydroxytoluene (BHT), and Trolox are commonly used as references in the experiments and the results are expressed as IC_{50} (μg/mL), meaning the

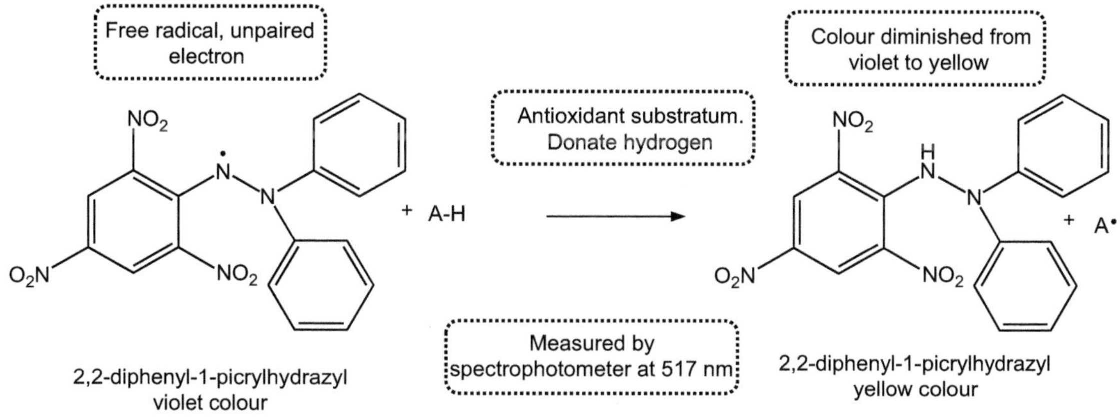

Figure 1.3
2,2-Diphenyl-1-picrylhydrazyl radical reaction.

concentration required to cause a 50% DPPH inhibition. The percentage of DPPH radical scavenging is calculated using the following equation (Eq. 1.1):

Inhibition percentage of DPPH radical

$$\% \text{ inhibition of DPPH radical} = \frac{(A_1 - A_0)}{A_0} \times 100 \tag{1.1}$$

where: A_1, absorbance before reaction ($DPPH^{\cdot +}$ methanol), A_0, absorbance after reaction (DPPH + sample).

1.8.2 2,2′-Azinobis(3-Ethylbenzothiazoline-6-Sulfonic Acid) Method, $ABTS^+$

This method was developed by Rice-Evans et al. (1997) and later modified by Re et al. (1990). The modification was based on the activation of metmyoglobin with hydrogen peroxide in the presence of $ABTS^+$ to produce a radical cation; this is generated by the oxidation of $ABTS^+$ with potassium persulfate. Reagent is prepared by adding solid manganese dioxide (80 mg) to a 5 mM aqueous stock solution of $ABTS^+$ prepared by mixing 75 mM of Na/K at buffer pH 7. Trolox (6-hydroxy-2,5,7,8-tetramethylchroman-2-carboxylic acid) is used as a positive control. A calibration curve is assembled for Trolox at concentrations of 0, 50,100, 150, 200, 250, 300, and 350 mM. Samples are diluted in Na/K buffer pH 7, mixed with 200 μL of $ABTS^+$ solution and placed in 96-well plates. Absorbance is read at 750 nm, after 5 min, in a microplate reader (Seeram et al., 2006). The activity is expressed in terms of Trolox equivalent antioxidant capacity for the extract or sample analyzed (TEAC/mg). Oxidation of $ABTS^+$ radicals is shown in Fig. 1.4 (Krishnaiah et al., 2011).

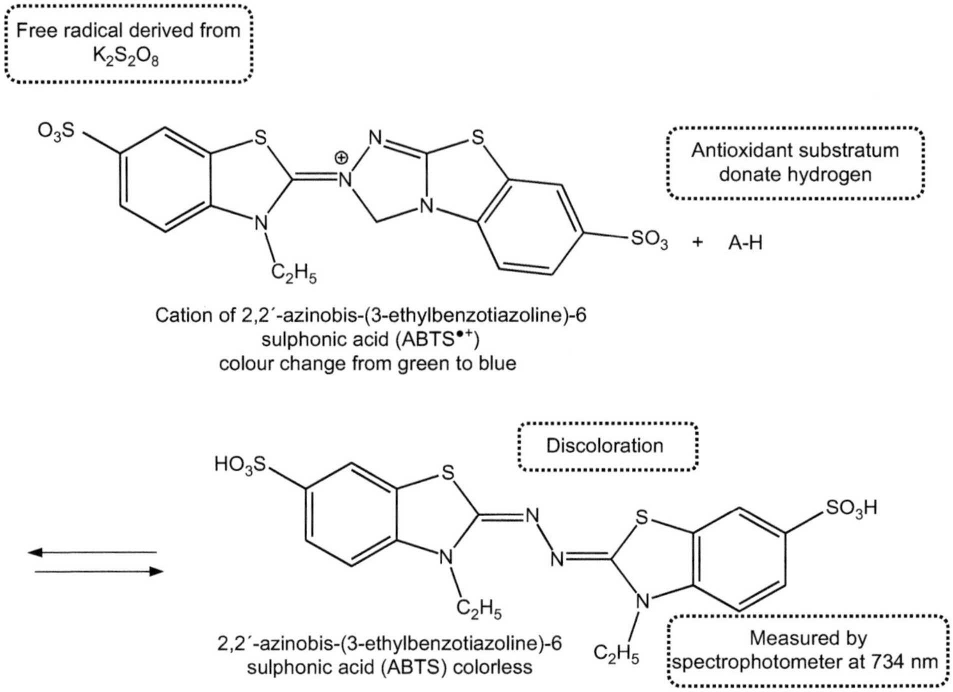

Figure 1.4
Oxidation of $ABTS^+$ radical.

1.8.3 β-Carotene Test

The *β*-carotene method evaluates the inhibitory activity of free radicals generated during the peroxidation of linoleic acid. This method is based on spectrophotometric discoloration measurements of *β*-carotene-induced oxidative degradation products of linoleic acid. A positive response is observed as a loss of *β*-carotene yellow color due to its reaction with radicals formed by linoleic acid oxidation (Kulisic et al., 2004). Results are expressed as IC_{50} (μg/mL), the concentration required to cause a 50% *β*-carotene decolorizing inhibition. BHA, BHT, and Trolox, or natural, such as gallic acid and quercetin, are used as standards (Alves et al., 2010).

1.8.4 Total Radical Trapping Antioxidant Parameter (TRAP) Method

Phycoerythrin (PE) is a red protein pigment complex produced by the light-harvesting phycobiliprotein family. It is present in red algae and cryptophytes as an accessory to the main chlorophyll pigments responsible for photosynthesis (Ficner and Huber, 1993; van der Weij-De Wit et al., 2006). The TRAP method is based on the protection provided by antioxidants on the fluorescence decay of R-phycoerythrin (R-PE) during a controlled peroxidation reaction. The fluorescence of R-phycoerythrin is quenched by ABAP [2,2-azo-bis(2-amidino- propane) hydrochloride] as a radical generator. This quenching reaction is measured in the presence of antioxidants. According to Ghiselli et al. (1995), 120 μL of diluted sample is added to 2.4 mL of phosphate buffer (pH 7.4), 375 μL of bidistilled water, 30 μL of diluted R-PE, and 75 μL of ABAP; the reaction is recorded for 45 min at 38°C by a luminescence spectrometer. A positive reaction is evaluated by measuring the decay in decoloration. TRAP values are calculated from the length of the lag phase due to the sample compared to standard solution (Nur Alam et al., 2013).

1.8.5 Oxygen Radical Absorbance Capacity Assay (ORAC)

This method uses β-phycoerythrin (β-PE) as an oxidizable protein substrate; 2,2′-azobis-(2-amidinopropane)-dihydrochloride (AAPH) as a peroxyl radical generator; and Cu^{2+}/H_2O_2 system as a hydroxyl radical generator. The quantification is carried out using the area under the curve (AUC) and takes the free radical reaction to completion, combining both the inhibition percentage and length of inhibition time for free radical action into a single quantity (Krishnaiah et al., 2011).

On the other hand, according to Prior et al. (2003), in this assay either β-PE or fluorescein is used as target free radical damage, AAPH as a peroxy radical generator, and Trolox as standard control. After addition of AAPH to the sample solution, the fluorescence is recorded and the antioxidant activity is expressed as Trolox equivalent (Cao et al., 1993; Frei et al., 1990). Assays are conducted at pH 7.0 with Trolox (6.25, 12.5, 25, and 50 μmol/L for lipophilic assays; 12.5, 25, 50, and 100 μmol/L hydrophilic assays) as the standard and 75 mM/L phosphate buffer as blank solution. After the addition of AAPH, the plate is placed in a multilabel counter preheated to 37°C. The plate is shaken in an orbital manner for 10 s and fluorescence is read at 1-min intervals for 35 min at a wavelength of 485–520 nm (exiting/emission wavelengths, respectively). The results are expressed in μM of Trolox equivalents (TE) per g dry weight of sample (μM TE/g) (Nur Alam et al., 2013).

1.8.6 Reducing Power Assay

This method is based on the reduction of Fe^{3+}/ferricyanide complex to the ferrous form. Therefore, Fe^{2+} may be monitored by UV spectrophotometry at absorbance of 700 nm. A yellow color of test solution changes to green depending on the reducing power of the sample analyzed (Krishnaiah et al., 2011). An absorbance increase indicates that an antioxidant reaction has taken place (Jayaprakash et al., 2001). According to the methodology described by Oyaizu (1986); 2.5 mL of 0.2 M phosphate buffer (pH 6.6) and 2.5 mL of $K_3Fe(CN)_6$ (1% w/v) are added to 1.0 mL of sample dissolved in distilled water. The resulting mixture is incubated at 50°C for 20 min, followed by the addition of 2.5 mL of trichloroacetic acid (10% w/v). The mixture is centrifuged at 3000 rpm for 10 min, then the upper layer is collected

(2.5 mL) and mixed with distilled water (2.5 mL) and 0.5 mL of $FeCl_3$ (0.1%, w/v). The absorbance is measured at 700 nm against a blank sample.

1.8.7 Hydrogen Peroxide Scavenging (H_2O_2) Assay

Hydrogen peroxide (H_2O_2) is formed in humans and animals as a short-lived product in biochemical processes, however, it is also toxic to cells. The toxicity is due to oxidation of proteins, membrane lipids, and DNA by peroxide ions. Furthermore, human beings are also exposed to H_2O_2 indirectly via the environment and it may enter the body through either inhalation of vapor/mist or eye and skin contact. This chemical component is rapidly decomposed into oxygen and water generating hydroxyl radicals that might initiate lipid peroxidation and cause DNA damage in the body (Nur Alam et al., 2013).

The hydrogen peroxide scavenging (H_2O_2) assay has been established by Ruch et al. (1989). According to this method a solution of hydrogen peroxide (40 mM) is prepared in phosphate buffer (50 mM pH 7.4). The concentration of hydrogen peroxide is determined by absorption at 230 nm using a UV spectrophotometer (Nur Alam et al., 2013). Extract (20–60 μg/mL) solved in distilled water is added to hydrogen peroxide and absorbance at 230 nm is determined after 10 min against a blank solution containing phosphate buffer without hydrogen peroxide. The percentage of hydrogen peroxide scavenging is calculated following Eq. (1.2):

$$\%\ \mathbf{H_2O_2} = \frac{(A_i - A_t)}{A_i} \times 100 \tag{1.2}$$

where, A_i: absorbance of control, A_t: absorbance of sample tested.

1.8.8 Nitric Oxide (NO) Scavenging Activity

Nitric oxide (NO) is a radical gas that it is considered as a genetic messenger since it plays a role in a variety of biological processes (Roszer, 2012). It is biosynthesized endogenously from L-arginine, oxygen, and NADPH by nitric oxide synthase enzymes (Ghafourifar and Cadenas, 2005; Virginia et al., 2003; Nur Alam et al., 2013).

In plants, nitric oxide may be produced by any of these four routes: (1) L-arginine-dependent nitric oxide synthase; (2) plasma membrane-bound nitrate reductase; (3) mitochondrial electron transport chain; or (4) nonenzymatic reactions. It is a signaling molecule, acting mainly against oxidative stress and also plays a role in plant–pathogen interactions (Roszer, 2012).

Two important biological mechanisms of nitric oxide are S-nitrosation of thiols and nitrosylation of transition metal ions. S-nitrosation involves the reversible conversion of thiol groups, including cysteine residues in proteins, to form S-nitrosothiols (RSNOs). On the other hand, nitrosylation involves the binding of NO to a transition metal ion such as iron or copper.

Regarding the scavenging activity, the compound sodium nitroprusside is known to decompose in aqueous solution at physiological pH (7.2) producing NO. Under aerobic conditions NO reacts with oxygen-producing stable products like nitrate and nitrite, quantification of these may be measured through the Griess reagent (Marcocci et al., 1994). This methodology is as follows: 2 mL of 10 mM sodium nitroprusside dissolved in 0.5 mL phosphate buffer saline (pH 7.4) is mixed with 0.5 mL of sample prepared at different concentrations (0.2–0.8 mg/mL). The mixture is incubated at 25°C for 150 min. After incubation, 0.5 mL of this solution is removed and mixed with 0.5 mL of Griess reagent (0.33% in 20% glacial acetic acid at room temperature for 5 min with 1 mL of naphthylethylenediamine dichloride [0.1% w/v]). The mixture is then incubated at room temperature for 30 min and its absorbance is measured at 546 nm. The amount of nitric oxide radical inhibition is calculated following Eq. (1.3):

$$\% \ \mathbf{NO} = \frac{(A_0 - A_1)}{A_0} \times 100 \tag{1.3}$$

where, A_0, absorbance before reaction, A_1, absorbance after reaction with Griess reagent.

1.8.9 Superoxide Anion Scavenging Assay

Xanthine oxidase (XO) is the enzyme responsible for conversion of xanthine into uric acid, leading to the production of hydrogen peroxide and superoxide radicals. These forms are considered to be major biological sources of reactive oxygen species (Alves et al., 2010).

The reaction is carried out by mixing 125 μL of buffer (50 mM KH_2PO_4/KOH, pH 7.4), 20 μL of a 15 mM Na_2EDTA solution in buffer; 30 μL of a 3 mM hypoxanthine solution in buffer; 50 μL of a 0.6 mM nitroblue tetrazolium (NBT) solution in buffer; 50 μL of xanthine oxidase in buffer (1 unit per 10 mL buffer), and 25 μL of plant extract in buffer. The reaction takes place in microplates (96 wells); 5 min after the addition of xanthine oxidase the reaction is read at 450 nm using a microplate reader (Krishnaiah et al., 2011).

1.8.10 Peroxynitrite Radical (ONOO˙) Scavenging Activity

Peroxynitrite (ONOO˙) is a radical with strong oxidizing properties toward cellular constituents, including sulfhydryls, lipids, amino acids, and nucleotides, leading to lipid peroxidation, carcinogenesis, and aging. It is generated in vivo by endothelial cells, Kupffer cells, neutrophils, and macrophages. Its excessive formation may also be involved in several human diseases, such as Alzheimer's disease, rheumatoid arthritis, cancer, and atherosclerosis. In order to perform this assay, researchers follow the method described by Kooy et al. (1994), where a stock solution of dihydroxyrhodamine (DHR 123.5 mM) in dimethylformamide at a final concentration of 5 mM, is used. This stock solution is placed on ice in the dark just before the experiment is carried out. Furthermore, a buffer solution of

50 mM sodium phosphate (pH 7.4), containing 90 mM sodium chloride and 5 mM potassium chloride with 100 mM diethyl-ene-triamine-penta acetic acid (DTPA), is purged with nitrogen and placed on ice before use. Scavenging activity of $ONOO^{\cdot}$ by the oxidation of DHR 123 is measured on a microplate fluorescence spectrophotometer with excitation and emission wavelengths of 485 and 530 nm at room temperature, respectively. The intensities are measured 5 min after treatment without 3-morpholino-sydnonimine (SIN-1) or $ONOO^{\cdot}$. Oxidation of DHR 123 by decomposition of SIN-1 gradually increased, whereas $ONOO^{\cdot}$ rapidly oxidized DHR 123 with its final fluorescent intensity being stable over time (Nur Alam et al., 2013).

1.8.11 Ferric Reducing-Antioxidant Power (FRAP) Assay

This method measures the ability of antioxidants to reduce ferric iron, thus, it is based on the reduction of the complex of ferric iron and 2,3,5-triphenyl-1,3,4-triaza-2-azoniacyclopenta-1,4-diene chloride (TPTZ) to the ferrous form at low pH. This reduction is monitored by measuring the absorption change at 593 nm, using a diode array spectrophotometer (Benzie and Strain, 1999). The procedure is as follows: 3 mL of prepared FRAP reagent is mixed with 100 μL of diluted sample; the absorbance at 593 nm is recorded after 30 min incubation at 37°C. FRAP values are obtained by comparing the absorption change in the test mixture with those obtained from increasing concentrations of Fe^{3+} and expressed as mM of Fe^{2+} equivalents per mg or μg of sample (Nur Alam et al., 2013).

1.8.12 Superoxide Radical Scavenging Activity (SOD)

The superoxide anion itself is a weak oxidant, however, it generates hydroxyl radicals as well as singlet oxygen, both of which contribute to oxidative stress (Meyer and Isaksen, 1995). In this regard, a method to measure superoxide anion scavenging activity has been proposed by Robak and Gryglewski (1988) where 3.0 mL of Tris–HCl buffer (16 mM, pH 8.0), containing 0.5 mL of nitroblue tetrazolium (NBT; 0.3 mM), 0.5 mL NADH (0.936 mM) solution, 1.0 mL sample extract, and 0.5 mL Tris–HCl buffer (16 mM, pH 8.0) is mixed. The reaction is initiated by adding 0.5 mL phenazine methosulfate (PMS) solution (0.12 mM) to the mixture, incubated at 25°C for 5 min. Finally, the absorbance is measured at 560 nm against a blank sample.

1.8.13 Determination of Phenol Content by the Folin–Ciocalteu Method

Folin–Ciocalteu phenol reagent consists of a mixture of phosphomolybdic and phosphotungstic acids in which the molybdenum and tungsten are in the 6^{+} state. When reduction occurs, molybdenum blue and tungsten blue are formed, in which the oxidation state of metals is between 5 and 6. It is known that Folin–Ciocalteu reagent reacts not only

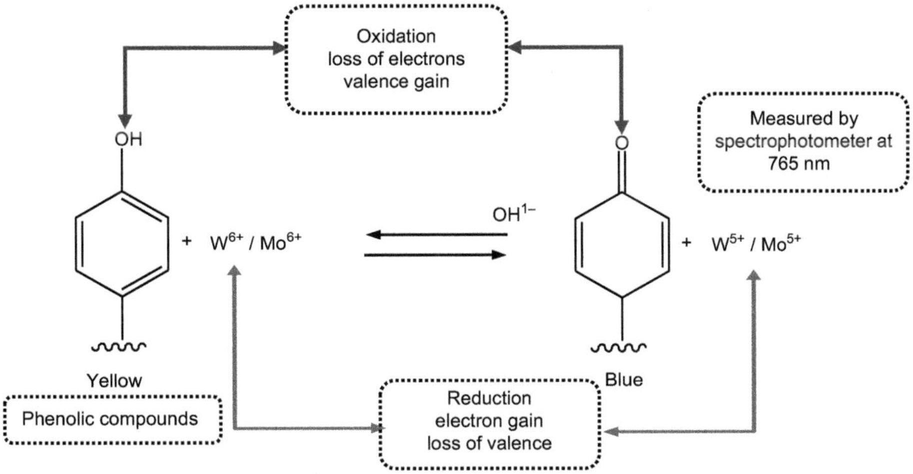

Figure 1.5

Reaction of gallic acid with molybdenum blue (gray in print version) and tungsten blue (gray in print version). *Adapted from Oliveira, A.C., Valentim, I.B., Silva, C.A., Bechara, E.J.H., Barros, M.P., Mano, C.M., Goulart, M.O.F., 2009. Total phenolic content and free radical scavenging activities of methanolic extract powders of tropical fruit residues. Food Chemistry 115(2), 469–475.*

with phenols but also with a variety of other compounds, thus, it does not give a full picture of the quantity or quality of phenolic constituents in the extracts (Singleton and Rossi, 1965). Gallic acid is used as a standard for the calibration curve. Total phenolic content is expressed as mg of gallic acid equivalent (GAE) per gram of extract. Fig. 1.5 show the reaction of gallic acid with molybdenum, a component of the Folin–Ciocalteu reagent.

1.8.14 Total Flavonoid Content

Total flavonoid content is determined using a colorimetric method described by Dewanto et al. (2002). EtOH or AcOEt extracts (0.30 mL) are mixed with 1.50 mL of distilled water in a test tube followed by addition of 90 μL of a 5% $NaNO_2$ solution. After 6 min, 180 μL of a 10% $AlCl_3{\cdot}6H_2O$ solution is added and allowed to stand for another 5 min before 0.6 mL of 1 M NaOH is added. The mixture is brought to 330 μL with distilled water and mixed. The absorbance is measured against a blank at 510 nm using a spectrophotometer (Fig. 1.5). (+)-Catechin is used as standard solution. Results are expressed as mg of catechin equivalents per gram of extract (mg CE/g) (Pereira, 2012; Fig. 1.6).

1.9 In vivo Methods

All in vivo methods explained in this section are carried out using experimental animals, mainly mice and rats. Samples to be analyzed are administered at defined doses for a

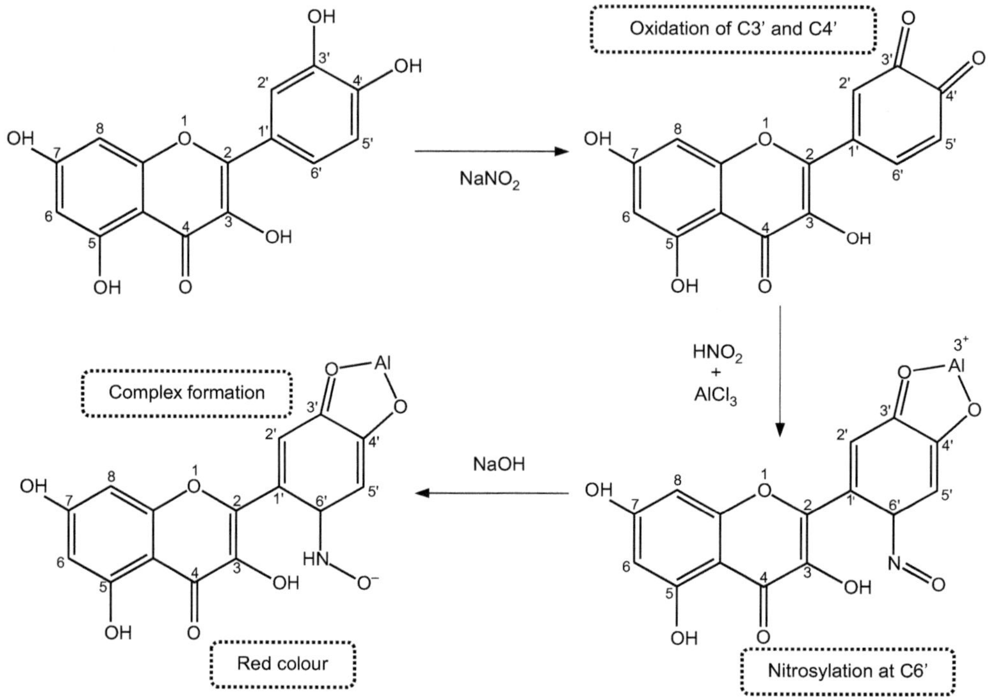

Figure 1.6
Complex formation to determine flavonoid content.

period of time, depending on the method performed. Once the treatment is completed, animals are sacrificed and either blood or tissue is taken from these and used to accomplish the assay.

1.9.1 *Ferric Reducing Ability of Plasma (FRAP) Assay*

This assay has been widely used in nutritional science not only to measure the total antioxidant content of food but also to explore absorption of antioxidants from soya milk, cocoa, and tea, and also to investigate the effect of processing and cooking on the antioxidant content of foods (Benzie and Choi, 2014). The method proposed by Benzie and Strain (1996) involves the use of blood samples that are collected from the rat retro-orbital venous plexus into heparinized glass tubes at 0, 7, and 14 days of treatment. Three mL of freshly prepared and warm (37°C) FRAP reagent [1 mL (10 mM) of 2,4,6 tripyridyl-s-triazine (TPTZ) solution in 40 mM HCl, 1 mL 20 mM $FeCl_2Æ6\text{-}H_2O$, 10 mL of 0.3 M acetate buffer (pH 3.6)] is mixed with 0.375 mL distilled water and 0.025 mL of test samples. The absorbance is measured spectrophotometrically at 593 nm (Nur Alam et al., 2013).

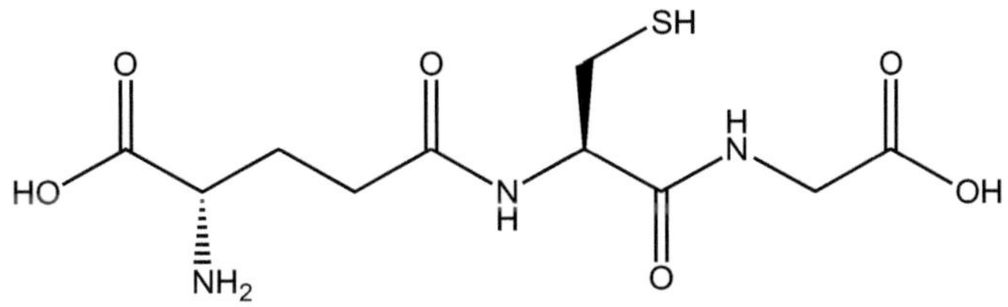

Figure 1.7
Tripeptide glutathione.

1.9.2 Reduced Glutathione (GSH) Assay

Glutathione (GSH, Fig. 1.7) is an intracellular reducer that plays a role as an antioxidant in plants, animals, fungi, and some bacteria, and thus is able to prevent damage to cellular components caused by free radicals, peroxides, and other oxidant species. It is a tripeptide with a gamma peptide linkage between the carboxyl group of the glutamate side chain and the amine group of cysteine. The carboxyl group of cysteine is attached by normal peptide linkage to a glycine. This peptide, once oxidized, may be reduced back by glutathione reductase, using NADPH as an electron donor (Couto et al., 2013).

The ratio of reduced glutathione to oxidized glutathione within cells is often used as a measure of cellular oxidative stress (Lu, 2013). To determine antioxidant activity, researchers follow the method described by Ellman (1959). The tissue homogenate in 0.1 M phosphate buffer pH 7.4 is added with an equal volume of 20% trichloroacetic acid (TCA) containing 1 mM EDTA to precipitate the tissue proteins. The mixture is allowed to stand for 5 min prior to centrifugation for 10 min at 2000 rpm. The supernatant (200 μL) is then transferred to a new set of test tubes and added with 1.8 mL of Ellman's reagent (5,50-dithiobis-2- nitrobenzoic acid [0.1 mM] prepared in 0.3 M phosphate buffer with 1% of sodium citrate solution). Once this reaction is completed, solutions are measured spectrophotometrically at 412 nm against a blank. Absorbance values are compared with a standard curve generated from pure GSH (Nur Alam et al., 2013).

1.9.3 Glutathione Peroxidase (GPx) Assay

Glutathione peroxidase (GPx) is an enzyme with peroxidase activity whose main biological role is to protect the organism from oxidative damage. The biochemical function of glutathione peroxidase is to reduce lipid hydroperoxides to their corresponding alcohols and to reduce free hydrogen peroxide to water (Muller et al., 2007). GPx is found throughout the tissues, being present as four different isoenzymes: cellular glutathione peroxidase, extracellular glutathione peroxidase, phospholipid hydroperoxide glutathione peroxidase, and gastrointestinal glutathione peroxidase. This assay is carried out following the methodology described by Wood (1970). Solutions of 50 μL of 60 mM/L glutathione reductase solution (30 U/mL), 50 μL of 0.12 mM/L NaN_3, 0.10 of 0.15 mM/L Na_2EDTA, 100 μL of 3.0 mM/L NADPH, and 100 μL of cytosolic fraction are mixed in a cuvette (3 mL) containing 2.0 mL of 75 mM/L phosphate buffer, pH 7.0. Water is added to make a total volume of 2.9 mL. Once the mixture is complete, 100 μL of 7.5 mM/L H_2O_2 is added and the reaction takes place. The conversion of NADPH to NADP is monitored by recording, every 1 min for 5 min continuously, the change of absorbance at 340 nm. The results are expressed in terms of mg of proteins (Nur Alam et al., 2013).

1.9.4 Glutathione-S-Transferase (GSt) Assay

GSt is an enzyme involved in detoxification of harmful electrophilic endogenous and exogenous compounds (Ammar et al., 2016). This enzyme catalyzes the reaction of such compounds with the –SH group of glutathione neutralizing the electrophilic sites leading to more water-soluble products. To perform this assay researchers follow the methodology described by Jocelyn (1972). The reaction mixture (1 mL) consisted of 0.1 N potassium phosphate (pH 6.5), 1 nM/L GSt, 1 M/L l-chloro-2,4-dinitrobenzene as substrate, and cytosol (6 mg protein/mL). The reaction mixture is incubated at 37°C for 5 min before the addition of substrate. Results are measured spectrophotometrically at 340 nm (Nur Alam et al., 2013).

1.9.5 Glutathione Reductase (GR) Assay

Glutathione reductase (GR) is a homodimer enzyme that catalyzes the reduction of glutathione disulfide (GSSG) to glutathione (GSH). This enzyme is involved in a wide range of enzymatic reactions where it assists the oxidation/reduction process (Fagan and Palfey et al., 2010).

The method described by Kakkar et al. (1984) is as follows: livers (about 400 g) are obtained from experimental rats (200–250 g). Livers are cut into small pieces and homogenized in 9 mL of 0.25 M ice-cold sucrose/g of rat liver, in a blender. The homogenate solution is centrifuged for 45 min at 14,000 rpm. Separated, the reaction system is composed by 1 mL of a mixture as follow: 1.0 mM GSSG, 0.1 mM NADPH, 0.5 mM EDTA, 0.10 M sodium phosphate buffer (pH 7.6), and a suitable amount of glutathione reductase sample to give a change in absorbance of 0.05–0.03/min; is suspended in a small volume of 0.25 M sucrose and centrifuged. Both centrifuged solutions, as explained before, are combined and adjusted to pH 5.5 with cold 0.2 M acetic acid and centrifuged again for 45 min at 14,000 rpm. The rate of oxidation of NADPH by GSSG at 30°C is used as a standard measure of enzymatic activity. The oxidation of 1 μM of NADPH/min under these conditions is used as a unit of glutathione reductase activity. The specific activity is expressed as units per mg of protein (Nur Alam et al., 2013).

1.9.6 Superoxide Dismutase (SOD) Assay

Superoxide dismutase is an enzyme found in all living cells. It is used in treating pain and inflammation caused by osteoarthritis, sports injuries, and rheumatoid arthritis, among others. Furthermore, this enzyme aids in breaking down potentially harmful oxygen molecules in cells, which might prevent damage to tissues. To measure the antioxidant capacity of this enzyme, a method described by McCord and Fridovich (1969), is commonly used. Seventy-five mM of Tris–HCl buffer (pH 8.2), 30 mM EDTA, and 2 mM of pyrogallol are mixed with 50 μL of erythrocyte lysate. An increase in absorbance is recorded by spectrophotometer at 420 nm for 3 min. One unit of enzyme activity represents 50% inhibition of pyrogallol auto-oxidation rate. The results are expressed as units/mg protein (Nur Alam et al., 2013).

1.9.7 Antioxidant Activity Studies Reported for Extract and Isolated Compounds From Different Plant Species in the Last 10 Years

Antioxidant capacity has been related to chemical compounds capable of protecting a biological system against the potentially harmful effect of processes or reactions involving reactive oxygen and nitrogen species. These protective effects of antioxidants have received increasing attention within biological, medical, nutritional, and agrochemical fields, resulting in the requirement of simple, convenient, and reliable antioxidant capacity determination methods. On the other hand, the search for natural antioxidants that might avoid degenerative diseases has attracted the attention of researchers who have stated that such diseases may be prevented by a proper intake of antioxidants. Since plant extracts have revealed a number of secondary metabolites with antioxidant activity, there has been an increased interest in identifying such compounds from plants with low or no side effects for use in preventive medicine and the food industry. Several investigations carried out in the last 10 years, showing the antioxidant capacities of plant extracts have been summarized in Table 1.1.

1.10 Conclusions

Oxidative stress is involved in the development of several degenerative diseases. Therefore, there has been huge interest from researchers to investigate the antioxidant activity of natural products. A number of in vitro and in vivo methods have also been developed in order to prove such activity. These tests have demonstrated the importance of secondary metabolites such as phenols, flavonoids, carotenes, among others, with verified antioxidant activity for the prevention of degenerative disorders. However, more investigations, especially in vivo tests, are required to confirm the beneficial and harmless effects of these natural components before recommending the consumption of any plant species to prevent oxidative stress signals.

Table 1.1: Recent Studies Related to Antioxidant Activity in Plant Extracts and Isolated Compounds Carried Out by Different Methods

Plant Species	Part Used	Extract/Isolated Compound	Antioxidant Method	Results	References
Vismia baccifera Planch. and Triana *Vismia macrophylla* Kunth (Hypoericaceae)	Leaves	n-Hexane (H) Dichloromethane (D) Ethyl acetate (EA) Butanol (B) Methanol (M)	DPPH Total phenols (TPh) Total flavonoids (TF)	DPPH (IC_{50} μg/mL) M: *V. baccifera* 5.35 μg/mL; *V. macrophylla* 5.87 μg/mL TPh (mg AGE/g Ext) M: *V. baccifera* 306.21 mg; *V. macrophylla* 321.98 mg TF (mg QE/g Ext) M: *V. baccifera* 185.90 mg; *V. macrophylla* 267.07 mg	Buitrago et al. (2016)
Chuquiragua jussieui J.F.Gmel. *Pseudognaphalium elegans* (Kunth) *Kartesz* (Asteraceae) *Gustavia pubescens* Ruiz and Pav. Ex Berg (Lecythidaceae) *Aeghiphila alba* Moldenke (Lamiaceae) *Cleome spinosa* Jacq. (Cleomaceae) *Phyllantus acuminatus* Vahl (Phyllantaceae) *Croton rivinifolius* Kunth (Euphorbiaceae)	Aerial parts	Ethanol	DPPH TPh TF	DPPH (IC_{50} μg/mL) *P. elegans* 41.85 μg/mL (56.4%) *C. jussieui* 41,02 μg/mL (58.2%) TPh (mg QE/100g dry extract) *P. elegans* 130.69 mg *C. jussieui* 244.18 mg TF (mg CatE/100g dry extract) *P. elegans* 1362.08 mg *C. jussieui* 1979.07 mg	Rondón et al. (2015)
Gomphrena celosioides Mart. (Amaranthaceae)	Leaves	Water (W)	TPh TF Lipid peroxidation	Total phenols (mg GAE/g Ext) W: 3.44 mg Total flavonoids (mg QE/g Ext) W: 9.91 mg Lipid peroxidation W: 1000 mg/kg	Konan et al. (2014)

Acorus calamus L. (Acoraceae)	Aerial parts	Ethanol Hydro-ethanol Water	DPPH NO Hydroxyl radical SOD FRAP Estimation of total phenolic and flavonoids Standard ascorbic acid	IC_{50} values of ethanol extract DPPH 54.82 μg/mL NO 118.802 μg/mL Hydroxyl radical 109.85 μg/mL SOD 38.3 μg/mL	Baruah and Barua (2014)
Euphrasia officinalis L (Scrophulariaceae)	Aerial parts	Chloroform (C) Methanol (M)	TPh TF	TPh (mg GAE/g Ext) M: 105.23 mg TF (mM α-tocopherol/g Ext) M: 239.08 mM	Dimitrova et al. (2014)
Senecio anteuphorbium Sch. Bip. (Asteraceae)	Leaves	Methanol (M) Water (W)	DPPH TPh TF	DPPH (IC_{50} μg/mL) M: 3.38 μg/mL W: 39.98 μg/mL TPh (mg GAE/g Ext) M: 20.38 mg W: 2.33 mg TF (mg QE/g Ext) M: 39.47 mg W: 9.68 mg	Azzahra Lahlou et al. (2014)
Dalbergia sissoo Roxb (Fabaceae)	Leaves	n-Hexane (H) Ethyl acetate (EA) Ethanol (E)	Reducing power Assay HP TPh TF	Reducing power assay (50 μg/mL, I %) EA: 76.14% E: 73.89% HP (50 μg/mL, I %) EA: 70.02% E: 62.29%	Muthu Lakshmi et al. (2014)

Continued

Table 1.1: Recent Studies Related to Antioxidant Activity in Plant Extracts and Isolated Compounds Carried Out by Different Methods—cont'd

Plant Species	Part Used	Extract/Isolated Compound	Antioxidant Method	Results	References
Myrtus communis L. (Myrtaceae)	Leaves Stem Berries	Ethyl acetate (EA) Butanol (B)	DPPH TPh TF	DPPH (IC_{50} μg/mL) Leaves: EA: 0.09 μg/mL B: 0.26 μg/mL TPh (mg GAE/g dry Ext) Leaves 119.23 mg Stem 112.96 mg berries 70.26 mg TF (mg CE/g dry Ext) Leaves 6.56 mg Berries 27.20 mg	Kanoun et al. (2014)
Hamelia patens Jacq. (Rubiaceae)	Bark	Acetone (A) Methanol (M)	HP TPh	HP (500 μg/mL, I %) A: 93.07% M: 91.9% TPh (mg GAE/g dry Ext) A: 303.6 mg M: 310.8 mg	Singh et al. (2014)
Asparagus racemosus Willd. (Asparagaceae) *Ocimum sanctum* L. (Lamiaceae) *Cassia fistula* L. (Caesalpiniaceae) *Piper betel* Blanco (Piperaceae) *Citrus aurantifolia* L. (Rutaceae) *Catharanthus roseus* (L.) G.Don (Apocynaceae) *Polyalthia longifolia* (Sonn.) Hook.f. and Thomson (Annonaceae)	Leaves	Methanol (M)	DPPH β-carotene TPh	DPPH (% inhibition, I %) *C. auantifolia* 87.05% *O. sanctum* 81.80% *C. roseus* 71.4% β-Carotene *P. longifolia* 0.65 mg/L TPh (mg GAE/g Ext) *C. roseus* 7.14 mg	Kaur and Mondal (2014)

Mentha arvensis L. (Lamiaceae) *Moringa oleifera* Lam. (Moringaceae) *Spinacia oleracea* L. (Chenopodiaceae) *Trigonella foenum-graecum* *Tamarindus indica* L. (Caesalpiniaceae) *Amaranthus viridis* Pollich ex. Moq. (Amaranthaceae)	Leaves	Methanol	DPPH ABTS TPh TF	DPPH (IC_{50} μg/mL) *M. arvensis* 28 μg/mL ABTS (IC_{50} μg/mL) *T. indica* 35 μg/mL TPh (mg GAE/g ext) *M. arvensis* 75 mg TF (mg QE/g Ext) *M. arvensis* 674 mg	Raghavendra et al. (2013)
Morus alba L. (Moraceae)	Stem barks (SB) Root bark (RB) Leaves (L) Fruits (F)	Methanol	DPPH Hydroxyl radical scavenging assay FRAP Lipid peroxidation inhibition assay	DPPH (IC_{50} μg/mL) SB: 37.75 μg/mL RB: 40.20 μg/mL F: 102.03 μg/mL L: 175.01 μg/mL Hydroxyl radical (IC_{50} μg/mL) SB: 58.90 μg/mL RB: 114.63 μg/mL F: 220.23 μg/mL L: 234.63 μg/mL Lipid peroxidation (IC_{50} μg/mL) SB: 145.31 μg/mL.	Khan et al. (2013)
Mangifera indica L.	Leaves of: *Gedong mangoe* (GD) *Golek mangoe* (GL) *Apel mangoe* (AP) *Arumanis mangoe* (AR)	Ethyl acetate (EA) Methanol (M)	DPPH ABTS TPh TF Carotenoid content	DPPH (% inhibition, I %) EA = GD: 98.70% M/AP:98.31%, GD: 94.95%, GL: 95.46% ABTS (% inhibition, I%) EA = AR: 70.55%, GD: 65.68%,GL: 48.95% M = GD: 66.99%, GL: 60.86% TF (g QE/100 g Ext) AR: 237.57 g TPh (g GAE/100 g Ext) GD: 230.73 g Carotenoids content (g BET/100 g Ext) GL: 16.28 g	Fidrianny et al. (2013)

Continued

Table 1.1: Recent Studies Related to Antioxidant Activity in Plant Extracts and Isolated Compounds Carried Out by Different Methods—cont'd

Plant Species	Part Used	Extract/Isolated Compound	Antioxidant Method	Results	References
Torilis leptophylla Rchb.f. (Apiaceae)	Aerial parts	n-Hexane (H) Chloroform (C) Ethyl acetate (EA) n-Butanol (B) Methanol (M) Water (W)	DPPH ABTS Lipid peroxidation Phosphomolybdate Hydroxyl radicals Superoxide radicals Hydrogen peroxide Glutathione contents in male Sprague–Dawley rat. TPh TF	In vitro assays DPPPH (IC_{50} μg/mL) B: 41.0 μg/mL ABTS B: 10.0 μg/mL Phosphomolybdate B: 10.7 μg/mL Hydroxyl radicals C: 8.0 μg/mL Superoxide radicals M: 57.0 μg/mL Hydrogen peroxide EA: 68.0 μg/mL TPh (mg GAE/g Ext) M: 121.9 mg TF (mg QE/g Ext) EA: 60.9 mg *In vivo* assays Lipid peroxidation M: 200 mg/kg Glutathione content M: 200 mg/kg Cotreatment with silymarin 50 mg/kg, effectively prevented lipid and glutathione alterations and maintained the antioxidant activity	Saeed et al. (2012)

Dioscorea alata L. (Dioscoreaceae)	Tuber	Petroleum ether (PE) Benzene (Be) Ethyl acetate (EA) Methanol (M) Ethanol (E)	DPPH Hydroxyl superoxide ABTS TPh TF	DPPH (IC_{50} µg/mL) M: 1 µg/mL ABTS (IC_{50} µg/mL) PE: 27.16 µg/mL, Be: 26.12 µg/mL, EA 30.65 µg/mL M: 25.53 µg/mL TPh (mg GAE/g Ext) M: 0.68 mg TF (mg QE/g Ext) M: 1.21 mg	Sakthidevi and Mohan (2013)
Saccocalyx satureioïdes Coss. and Durieu *Teucrium polium* Decne. ex. C.Presl *Salvia verbenaca* L. (Lamiaceae)	Aerial parts	Methanol (M)	DPPH TPh TF	DPPPH (IC_{50} µg/mL) *S. satureioïdes* 7.17 µg/mL *T. polium* 5.70 µg/mL *S. verbenaca* 9.79 µg/mL TPh (mg GAE/g Ext) *T. polium* 3.81 mg *S. satureioides* 3.41 mg *S.verbenaca* 0.85 mg TF (mg QE/g Ext) *T. polium* 2.9 mg *S. satureioides* 3.2 mg *S. verbenaca* 1.1 mg	Belmekki and Bendimera (2012)
Vismia macrophylla Kunth (Hypericaceae) *Zyzygium malaccensis* (L.) M. and P (Moraceae) *Symphonia globurifera* L. F (Clusiaceae) *Bellucia pentamera* Naudin (Melastomataceae) *Manekia sidowii* (Trel) (Piperaceae)	Leaves Bark Mature leaves and fruits	Ethanol	DPPH TPh	DPPH (IC50 µg/mL) *B. pentamera* 2.17 µg/mL *V. macrophylla* 2.54 µg/mL *Z. malaccensis* 4.93 µg/mL TPh (mg GAE/g Ext) *B. pentamera* 15.00 mg *V. macrophylla* 14.25 mg *Z. malaccensis* 11.63 mg	Gutierrez Mosquera et al. (2011)

Continued

Table 1.1: Recent Studies Related to Antioxidant Activity in Plant Extracts and Isolated Compounds Carried Out by Different Methods—cont'd

Plant Species	Part Used	Extract/Isolated Compound	Antioxidant Method	Results	References
Glycosmis mauritiana (Lam.) Tanaka (Rutaceae) *Streblus asper* Lour (Moraceae)	Leaves	Methanol (M) Water (W)	DPPH TPh TF	DPPH (% inhibition, I %) M: *G. mauritiana* 93.04% *S. asper* 89.3% W: *G. mauritiana* 61.49% *S. asper* 78.07% TF (mg QE/g Ext) M: *G. mauritiana* 0.021 mg *S. asper* 0.014 mg W: *G. mauritiana* 0.028 mg *S. asper* 0.030 mg TPh (mg GAE/g Ext) M: *G. mauritiana* 0.077 mg *S. asper* 0.056 mg W: *G. mauritiana* 0.74 mg *S. asper* 0.042 mg	Arun Kumar et al. (2011)
Cinnamomum osmophloeum Kaneh (Laureaceae)	Twigs	*n*-Hexane (H) Ethyl acetate (EA) *n*-butanol (B) Water (W)	DPPH NBT Reducing power Lipid peroxidation using (RPLP) mouse brain homogenates Metal chelating ability Photochemiluminescence TPh	DPPPH (IC_{50} μg/mL) H: 33.5 μg/mL, EA: 9.8 μg/mL, B: 6.9 μg/mL, W: 31.7 μg/mL NBT (10 μg/mL, I %) H: 10%, EA: 55.3%, B: 63.6%, W: 32.1% RPLP (50 μg/mL, I %) H: 0.3%, EA: 1.6%, B: 1.9%, W: 0.7% TPh(μmol of trolox/g Ext) H: 728.0 μmol, EA: 2278.7 μmol, B: 18889.7 μmol, W: 380.0 μmol	Meng-Thong et al. (2008)

Vismia baccifera ssp. *ferruginea* *Vismia guianensis* (Aubl.) Choisy (Hypericaceae)	Fruits	Petroleum ether (PE) Ethyl acetate (EA) Methanol (M)	DPPH ABTS TPh	DPPH (IC_{50} μg/mL) *V. baccifera* spp *ferrugiena* PE: 15.90 μg/mL, EA: 4.46 μg/mL, M: 17.63 μg/mL *V. guianensis* PE: 7.04 μg/mL, EA: 3.72 μg/mL, M: 6.52 μg/mL ABTS (IC_{50} μg/mL) *V. baccifera* spp *ferrugiena* PE: 7.00 μg/mL, EA: 4.16 μg/mL, M: 11.23 μg/mL *V. guianensis* PE: 8.73 μg/mL, EA: 5.86 g/mL, M: 10.41 μg/mL TPh (mg GA/g Ext) *V. baccifera* spp *ferrugiena* PE: 78.33 mg, EA: 350.56 mg, M: 186.67 mg *V. guianensis* PE: 205.0 mg, EA: 356.67 mg, M: 186.67 mg	Álvarez et al. (2008)

ABTS, 2,2′-azinobis(3-ethylbenzothiazoline-6-sulfonic acid) method; *BET*, *β*-carotene equivalents; *CatE*, catechin equivalents; *DPPH*, 2,2-diphenyl-1-picrylhydrazyl radical assay; *GAE*, gallic acid equivalents; *HP*, hydrogen peroxide; *I%*, inhibition percentage; *NBT*, superoxide radical scavenging assay; *QE*, quercetin equivalents; *RPLP*, reducing power lipid peroxidation; *TF*, total flavonoids; *TPh*, total phenols.

References

Álvarez, E., Jímenez, J.G., Posada, M.A., Rojano, B., Gil, J.H., García, C.M., Durango, D.L., 2008. Actividad antioxidante y contenido fenólico de los extractos provenientes de las bayas de dos especies del género *Vismia* (Guttiferae). VITAE. Revista de La Facultad de Química Farmacéutica 15 (1), 165–172.

Alves, C.Q., David, J.M., David, J.P., Bahia, M.V., Aguiar, R.M., 2010. Methods for determination of *in vitro* antioxidant activity for extracts and organic compounds. Química Nova 33 (10), 2202–2210.

Ammar, W.A., Siervo, M., Mathers, J.C., 2016. Vitamin C, Antioxidant status, and cardiovascular aging. In: Malavolta, M., Mocchegiani, E. (Eds.), Molecular Basis of Nutrition and Aging. Academic Press, New Yok, pp. 609–619.

Antolovich, M., Prenzler, P.D., Patsalides, E., McDonald, S., Robards, K., 2002. Methods for testing antioxidant activity. Analyst 127, 183–198.

Arun Kumar, G.S., Seema, K., Chandrasekhar Raju, B., Ramarao, M., 2011. Biological activity of methanolic and aqueous extract of *Glycosmis mauritiana* and *Streblus asper*. International Research Journal of Pharmacy 2 (12), 267–269.

Azzahra Lahlou, F., Hmimid, F., Loutfi, M., 2014. Antioxidant activity phenolics flavonoids and proanthocyanidins content of *Senecio anteuphorbium* and *Noureddine Bourhim*. International Journal of Biochemistry Research and Review 4 (6), 550–558.

Badarinath, A.V., Rao, K.M., Chetty, C.M.S., Ramkanth, V., Rajan, T.V.S., Gnanaprakash, K., 2010. A review on *In-vitro* antioxidant methods: comparisons, correlations and considerations. International Journal of PharmTech Research 2 (2), 1276–1285.

Barreiros, A.L.B.S., David, J.M., David, J.P., 2006. Estresse oxidativo: relação entre geração de espécies reativas e defesa do organismo. Química Nova 29 (1), 113–123.

Baruah, A.M., Barua, I., 2014. A comparative study of the *in vitro* antioxidant property of different extracts of *Acorus calamus* Linn. Journal of Natural Product and Plant Resources 4 (1), 8–18.

Belmekki, N., Bendimera, N., 2012. Antioxidant activity and phenolic content in methanol crude extracts from three Lamiaceae grown in southwestern Algeria. Journal of Natural Product and Plant Resources 2 (1), 175–181.

Benzie, I.F.F., Choi, S.W., 2014. Antioxidants in food: content, measurement, significance, action, cautions, caveats, and research needs. In: Jeyakumar, H. (Ed.). Jeyakumar, H. (Ed.), Advances in Food and Nutrition Research, vol. 71. Elsevier Science, Amsterdam, pp. 1–53.

Benzie, I.F.F., Strain, J.J., 1996. The ferric reducing ability of plasma (FRAP) as a measure of antioxidant power: the FRAP assay. Analytical Biochemistry 239, 70–76.

Benzie, I.F.F., Strain, J.J., 1999. Ferric reducing antioxidant power assay: direct measure of total antioxidant activity of biological fluids and modified version for simultaneous measurement of total antioxidant power and ascorbic acid concentration. Methods in Enzymology 299, 15–27.

Bravo, L., 1998. Polyphenols: chemistry, dietary sources, metabolism and nutritional significance. Nutrition Reviews 56 (11), 317–333.

Buitrago, A., Rojas, J., Rojas, L., Peñaloza, Y., 2016. In vitro antioxidant activity and qualitative phytochemical analysis of two *Vismia* (Hypericaceae) species collected in Los Andes, Venezuela. Revista de Biologia Tropical 6 (4), 1431–1439.

Cao, G., Alessio, H.M., Culter, R.G., 1993. Oxygen radical absorbance capacity assay for antioxidant free radicals. Biology and Medicine 14, 303–311.

Couto, N., Malys, N., Gaskell, S.J., Barber, J., 2013. Partition and turnover of glutathione reductase from *Saccharomyces cerevisiae*: a proteomic approach. Journal of Proteome Research 12 (6), 2885–2894.

Dewanto, V., Wu, X., Adom, K., Liu, R.H., 2002. Thermal processing enhances the nutritional value of tomatoes by increasing total antioxidant activity. Journal of Agricultural and Food Chemistry 50 (10), 3010–3014.

Dimitrova, M., Hristova, L., Damianova, E., Yordanova, Y., Petrova, N., Kapchina-Toteva, V., 2014. Antioxidant activity and secondary metabolites in different extracts of *Euphrasia officinalis* L. growing in Bulgaria. Science and Technologies 4 (1), 128–132.

Djeridane, A., Yousfi, M., Nadjemi, B., Boutassouna, D., Stocker, P., Vidal, N., 2006. Antioxidant activity of some Algerian medicinal plants extracts containing phenolic compounds. Food Chemistry 97 (4), 654–660.

Ellman, G.L., 1959. Tissue sulfhydryl groups. Archives of Biochemistry and Biophysics 82, 70–77.

Fagan, R.L., Palfey, B.A., 2010. Flavin-dependent enzymes. In: Lew, M., Hung-Wen (Ben), L. (Eds.), Comprehensive Natural Products II. Elsevier, Amsterdam, pp. 115–139.

Ficner, R., Huber, R., 1993. Refined crystal structure of phycoerythrin from *Porphyridium cruentum* at 0.23-nm resolution and localization of the γ subunit. European Journal of Biochemistry 218 (1), 103–106.

Fidrianny, I., Rahmiyani, I., Ruslan Wirasutisna, K., 2013. Antioxidant capacities from various leaves extracts of four varieties *Mangoes* using DPPH, ABTS assays and correlation with total phenolic, flavonoid, carotenoid. International Journal of Pharmacy and Pharmaceutical Sciences 5 (4), 189–914.

Frei, B., Stocker, R., England, L., Ames, B.N., 1990. Ascorbate the most effective antioxidant in human blood plasma. Advances in Experimental Medicine and Biology 264, 155–163.

Ghafourifar, P., Cadenas, E., 2005. Mitochondrial nitric oxide synthase. Trends in Pharmacological Sciences 26, 190–195.

Ghiselli, A., Serafini, M., Maiani, G., Azzini, E., Ferro-Luzzi, A., 1995. A fluorescence-based method for measuring total plasma antioxidant capability. Free Radical Biology and Medicine 18, 29–36.

Gupta, V.K., Verma, A.K., 2010. Comprehensive Bioactive Natural Products, vol. 4. Stadium press, Houston, USA.

Gutiérrez Mosquera, A., Pino Benitez, N., Cuesta Lemos, J.A., 2011. Actividad antioxidante y contenido de fenoles totales de varios extractos etanólicos de plantas medicinales. Investigación, Biodiversidad y Desarrollo 30 (1), 66–73.

Halliwell, B., Gutteridge, J.M.C., 2007. Free Radicals in Biology and Medicine, fourth ed. Clarendon, Oxford, UK.

Haslam, E., 1996. Natural polyphenols (vegetable tannins) as drugs: possible modes of action. Journal of Natural Products 59 (2), 205–215.

Ignat, I., Volf, I., Popa, V.I., 2011. A critical review of methods for characterization of polyphenolic compound in fruits and vegetables. Food Chemistry 126 (4), 1821–1835.

Jang, I.C., Jo, E.K., Bae, M.S., Lee, H.J., Jeon, G.I., Park, E., Yuk, H.G., Ahn, G.H., Lee, S.C., 2010. Antioxidant and antigenotoxic activities of different parts of persimmon (*Diospyros kaki* cv Fuyu) fruit. Journal of Medicinal Plants Research 4 (2), 155–160.

Jayaprakash, G.K., Singh, R.P., Sakariah, K.K., 2001. Antioxidant activity of grape seed extracts on peroxidation models *In-vitro*. Journal of Agricultural and Food Chemistry 55, 1018–1022.

Jocelyn, P.C., 1972. Biochemistry of the SH Group. Academic Press, London, p. 10.

Kakkar, P., Das, B., Viswanathan, P.N., 1984. A modified spectrophotometric assay of superoxide dismutase. Indian Journal of Biochemistry and Biophysics 21, 131–132.

Kanoun, K., Belyagoubi-Benhammou, N., Ghembaza, N., Atik Bekkara, F., 2014. Comparative studies on antioxidant activities of extracts from the leaf, stem and berry of *Myrtus communis* L IFRJ. 21 (5), 1957–1962.

Karadag, A., Ozcelik, B., Saner, S., 2009. Review of methods to determine antioxidant capacities. Food Analytical Methods 2, 41–60.

Kaur, S., Mondal, P., 2014. Study of total phenolic and flavonoid content, antioxidant activity and antimicrobial properties of medicinal plants. Journal of Microbiology 1 (1), 1–6.

Khan, M.A., Rahman, A.A., Islam, S., Khandokhar, P., Parvin, S., Badrul Islam, M.d., Hossain, M., Rashid, M., Sadik, G., Nasrin, S., Haque Mollah, M.N., Khurshid Alam, A.H.M., 2013. A comparative study on the antioxidant activity of methanolic extracts from different parts of *Morus alba* L. (Moraceae). BMC Research Notes 6 (24), 1–9.

Konan, G., Edwige, A.A., Dodehe, Y., Djaman Allico, J., Yapo Adou, F., N'guessan Jean, D., Yapi Houphouet, F., 2014. Evaluation of anti-inflammatory activities of aqueous and ethanolic extracts of *Gomphrena celosioides* Linn. International Journal of Biochemistry Research and Review 5 (2), 127–134.

Kooy, N.W., Royall, J.A., Ischiropoulos, H., Beckman, J.S., 1994. Peroxynitrite-mediated oxidation of dihydrorhodamine 123. Free Radical Biology and Medicine 16, 149–156.

Krishnaiah, D., Sarbatly, R., Nithyanandam, R., 2011. A review of the antioxidant potential of medicinal plant species. Food and Bioproducts Processing 89 (3), 217–233.

Kulisic, T., Radonic, A., Katalinic, V., Milos, M., 2004. Use of different methods for testing antioxidative activity of oregano essential oil. Food Chemistry 85 (4), 633–640.

Lu, S.C., 2013. Glutathione synthesis. Biochimica et Biophysica Acta 1830 (5), 3143–3153.

Magder, S., 2006. Reactive oxygen species: toxic molecules or spark life? Critical Care 10 (1), 1–8.

Marcocci, I., Marguire, J.J., Droy-lefaiz, M.T., Packer, L., 1994. The nitric oxide scavenging properties of Ginkgo biloba extract. Biochemical and Biophysical Research Communications 201, 748–755.

Martínez Sánchez, G., Delgado Hernández, R., Garrido Garrido, G., Guevara Garcia, M., García Rivera, D., Paéz Betancourt, E., Núñez Sellés, A.J., 2003. Mitos y Realidades de la Terapia Antioxidante. Vimang: Nuevo Producto Natural Antioxidante. Center of Pharmaceutical Chemistry, Havana, p. 88.

McCord, J., Fridovich, I., 1969. Superoxide dismutase, an enzymic function for erythrocuprin. Journal of Biological Chemistry 244, 6049–6055.

Meng-Thong, C., Yu-Tang, T., Shang-Tzen, C., 2008. Antioxidant activities of ethanolic extracts from the twigs of *Cinnamomum osmophloeum*. Bioresource Technology 99, 1918–1925.

Meyer, A.S., Isaksen, A., 1995. Application of enzymes as food antioxidants. Trends in Food Science and Technology 6, 300–304.

Montoro, P., Braca, A., Pizza, C., De Tommasi, N., 2005. Structure-antioxidant activity relationships of flavonoids isolated from different plant species. Food Chemistry 92 (2), 349–355.

Muller, F.L., Lustgarten, M.S., Jang, Y., Richardson, A., Van Remmen, H., 2007. Trends in oxidative aging theories. Free Radical Biology and Medicine 43 (4), 477–503.

Muthu Lakshmi, T., Radha, R., Jayshree, N., 2014. In vitro antioxidant activity, total phenolic and total flavonoid content in extracts from the bark of *Dalbergia sissoo* Roxb. International Journal of Pharma Sciences and Research 5 (5), 226–231.

Naczk, M., Shahidi, F., 2004. Extraction and analysis of phenolics in food. Journal of Chromatography A 1054 (1–2), 95–111.

Nur Alam, M., Jahan Bristi, N., Rafiquzzaman, M., 2013. Review on *in vivo* and *in vitro* methods evaluation of antioxidant activity. Saudi Pharmaceutical Journal 21, 143–152.

Oliveira, A.C., Valentim, I.B., Silva, C.A., Bechara, E.J.H., Barros, M.P., Mano, C.M., Goulart, M.O.F., 2009. Total phenolic content and free radical scavenging activities of methanolic extract powders of tropical fruit residues. Food Chemistry 115 (2), 469–475.

Oyaizu, M., 1986. Studies on products of browning reactions: antioxidant activities of products of browning reaction prepared from glucosamine. Journal of Nutrition 44, 307–315.

Pereira, M., Steffens, R., Jablonski, A., Hertz, P., Rios, A., Vizzotto, M., Flores, S., 2012. Characterization and antioxidant potential of Brazilian fruits from the Myrtaceae family. Journal of Agricultural and Food Chemistry 60 (12), 3061–3067.

Prior, R.L., Hoang, H., Gu, L., 2003. Assays for hydrophilic and lipophilic antioxidant capacity (oxygen radical absorbance capacity (ORAC (FL)) of plasma and other biological and food samples. Journal of Agricultural and Food Chemistry 51, 3273–3279.

Raghavendra, M., Madhusudhana Reddy, A., Raghuveer Yadav, P., Sudharshan Raju, A., Siva Kumar, L., 2013. Comparative studies on the *in vitro* antioxidant properties of methanolic leafy extracts from six edible leafy vegetables of India. Asian Journal of Pharmaceutical and Clinical Research 6 (3), 96–99.

Re, R., Pellegrini, N., Proteggente, A., Pannala, A., Yang, M., Rice-Evans, C., 1990. Antioxidant activity applying an improved ABTS radical cation decolorization assay. Free Radical Biology and Medicine 26 (9–10), 1231–1237.

Rice-Evans, C., Miller, N., Paganga, G., 1997. Antioxidant properties of phenolic compounds. Trends in Plant Science 2, 152–159.

Robak, J., Gryglewski, R.J., 1988. Flavonoids are scavengers of superoxide anions. Biochemical Pharmacology 37, 837–841.

Rondón, M., García, I., Cornejo, X., Rojas, J., Terán, W., 2015. Phytochemical screening and antioxidant activity of seven medicinal plants species from Ecuador. Pharmacologyonline 3, 19–28.

Roszer, T., 2012. The Biology of Subcellular Nitric Oxide. Springer Dordrecht Heidelberg, London, New York.

Ruch, R.J., Cheng, S.J., Klaunig, J.E., 1989. Prevention of cytotoxicity and inhibition of intercellular communication by antioxidant catechins isolated from Chinese green tea. Carcinogen 10, 1003–1008.

Saeed, N., Khan, M.R., Shabbir, M., 2012. Antioxidant activity, total phenolic and total flavonoid contents of whole plant extracts *Torilis leptophylla* L. Complementary and Alternative Medicine 12 (221), 1–12.

Sakthidevi, G., Mohan, V.R., 2013. Total phenolic, flavonoid contents and *in vitro* antioxidant activity of *Dioscorea alata* L. tuber. Journal of Pharmaceutical Sciences and Research 5 (5), 115–119.

Seeram, N.P., Henning, S.M., Lee, R., Niu, Y., Scheuller, H.S., Heber, D., 2006. Catechin and caffeine contents of green tea dietary supplements and correlation with antioxidant activity. Journal of Agricultural and Food Chemistry 54, 1599–1603.

Singh, S., Kashif Khan, M.D., Sharma, S.K., Singh, L., 2014. In vitro assessment of antimicrobial and antioxidant activity of various extracts of *Hamelia patens*. Journal of Chemical and Pharmaceutical Sciences 7 (2), 147–153.

Singleton, V.L., Rossi, J.A., 1965. Colorimetry of total phenolics with phosphomolybdic-phosphotungstic acid reagents. American Journal of Enology and Viticulture 16 (3), 144–158.

van der Weij-De Wit, C.D., Doust, A.B., van Stokkum, I.H.M., Dekker, J.P., Wilk, K.E., Curmi, P.M.G., Scholes, G.D., van Grondelle, R., 2006. How energy funnels from the phycoerythrin antenna complex to photosystem I and photosystem II in *Cryptophyte rhodomonas* CS24 cells. The Journal of Physical Chemistry B 110, 25066–25073.

Velioglu, Y.S., Mazza, G., Gao, L., Oomah, B.D., 1998. Antioxidant activity and total phenolics in selected fruits, vegetables and grain products. Journal of Agricultural and Food Chemistry 46 (10), 4113–4117.

Virginia, H., Sarah, L.E., Rachel, J.S., Nathaniel, T., Joseph, S., Adam, E., Cecilia, G., 2003. Mitochondrial nitric-oxide synthase: role in pathophysiology. IUBMB Life 55 (10 &11), 599–603.

Wannes, W.A., Mhamdi, B., Sriti, J., Jemia, M.B., Ouchikh, O., Hamdaoui, G., Kchouk, M., Wood, J.L., 1970. Biochemistry of mercapturic acid formation. In: Fishman, W.H. (Ed.), Metabolic Conjugation and Metabolic Hydrolysis. Academic Press, London and New York, pp. 261–299.

Further Reading

Marzouk, B., 2010. Antioxidant activities of the essential oil and methanol extracts from myrtle (*Myrtus communis* var. italica L.) leaf, stem and flower. Food and Chemical Toxicology 48 (5), 1362–1370.

CHAPTER 2

Phenolic Compounds: Structure, Classification, and Antioxidant Power

Milena Morandi Vuolo[1], Verena Silva Lima[1,2], Mário Roberto Maróstica Junior[3]

[1]*Department of Food and Nutrition, School of Food Engineering, University of Campinas, Campinas, Brazil;* [2]*Health and Biotechnology Institute, Federal University of Amazonas, Coari, Brazil;* [3]*Department of Food and Nutrition, Faculty of Food Engineering, University of Campinas, Campinas, Brazil*

2.1 Introduction

Phenolic compounds are defined as natural metabolites arising biogenetically from either the shikimate/phenylpropanoid pathway, which directly provides phenylpropanoids. Their structure consists of an aromatic ring, containing one or more hydroxyl substituents (Tsao, 2010). Phenolic compounds have been extensively studied for their ability to counteract oxidative stress, once chronic oxidative stress is linked to several metabolism disorders and a range of pathologies, such as obesity, diabetes, and cardiovascular diseases (Furukawa et al., 2004).

When natural antioxidant system defenses (enzymatic and nonenzymatic) are overwhelmed by an excessive generation of reactive oxygen species (ROS) or prooxidants, oxidative stress is installed, leading to damage to cellular and extracellular macromolecules (Burton et al., 2011). Phenolic compounds' antioxidant power is linked to their reducing properties as hydrogen- or electron-donating agents, which predicts their potential for action as free-radical scavengers (antioxidants), furthermore, they have the ability of metal chelation, particularly iron and copper, suppressing metal-catalyzed free radical formation. The structure of phenolic compounds is related to their radical-scavenging and metal-chelating activity. The number of hydroxyl groups and their position in relation to the carboxyl functional group influences the antioxidant activity of phenolic compounds (Balasundram et al., 2006).

The abilities of the phenolic compounds to scavenge free radicals can be assessed for a range of assays. They can be divided into categories in which a single electron transfer reaction or a hydrogen atom transfer reaction is evaluated. The most assays that have been used to evaluated antioxidant potential of phenolic compounds are the following, the copper reduction (CUPRAC) assay (Apak et al., 2005), trolox equivalent antioxidant capacity (ABTS or TEAC) assay (Miller et al., 1993), the ferric-reducing ability of plasma (FRAP) assay (Benzie and Strain, 1996), and the 2,2-diphenyl-1-picrylhydrazyl (DPPH) radical-scavenging capacity assay (Peng et al., 2000) to single electron transfer reaction and oxygen radical absorbance capacity (ORAC) assay (Prior

Bioactive Compounds. https://doi.org/10.1016/B978-0-12-814774-0.00002-5

and Cao, 1999), peroxyl scavenging capacity (PSC) (Wayner et al., 1985; Adom and Rui, 2005) and the total peroxyl radical-trapping antioxidant parameter (TRAP) assay (Wayner et al., 1985) to hydrogen atom transfer reaction. Such assays have some limitations regarding differences in the radical used or even using radicals not found in the human body, furthermore, often the antioxidant effects observed in in vitro assays do not correspond to when in vivo (Alam et al., 2013).

2.2 Phenolic Compounds: Definition and Classification

There are approximately more than 200,000 chemicals isolated and identified with diverse structures and classes from higher plants around the planet. Such chemicals are divided into two main groups: primary and secondary metabolites (Chikezie et al., 2015). The primary metabolites are essential to cell maintenance, such as fatty acids, proteins, carbohydrates, and nucleic acids. The secondary metabolites are no less important, despite not participating directly in photosynthetic or respiratory metabolism, they are known to be essential to plant survival (Chikezie et al., 2015). Their structures and chemicals are diverse when compared to primary metabolites and they are responsible for plant defense (Chikezie et al., 2015).

Secondary metabolites also act as signal compounds, attracting pollinators or animals for seed dispersion, in addition, they protect the plant from oxidants and ultraviolet radiation. The secondary metabolites are classified according to their biosynthetic routes and structure; they are divided into three major groups: (1) flavonoids, allied phenolic, and polyphenolic compounds; (2) terpenoids, and (3) nitrogen-containing alkaloids and sulfur-containing compounds. Such compounds are linked to primary metabolites by building blocks and biosynthetic enzymes (Vora, 2017) (Fig. 2.1).

Phenolic compounds (flavonoids, allied phenolic, and polyphenolic compounds) are one of the secondary metabolites more widely distributed in plants. They are derived from pentose phosphate, shikimate, and phenylpropanoid pathways in plants (Fig. 2.2) (Harborne, 1980). Such compounds perform an important role in the growth and reproduction of plants, giving protection against pathogens and predators. Furthermore, they contribute to color and sensory characteristics of vegetables and fruits (Chikezie et al., 2015).

Their structure comprises an aromatic ring, containing one or more hydroxyl substituents. They can range from simple phenolic molecules to highly polymerized compounds. The most phenolic compounds occur naturally as conjugates with mono- and polysaccharides, associated with one or more phenolic groups. In addition, they also can be linked to esters and methyl esters. Due to their structure diversity, there is a wide range of phenolic compounds that occur in nature. Currently, more than 8000 phenolic compound structures are known. They can be categorized into several classes as shown in Table 2.1 (Del Rio et al., 2013).

The most important phenolic compound classes found in the human diet are phenolic acids, flavonoids, and tannins. Chemically, phenolic acids have at least one aromatic ring where at least one hydrogen is substituted by a hydroxyl group (Heleno et al., 2015). They consist of

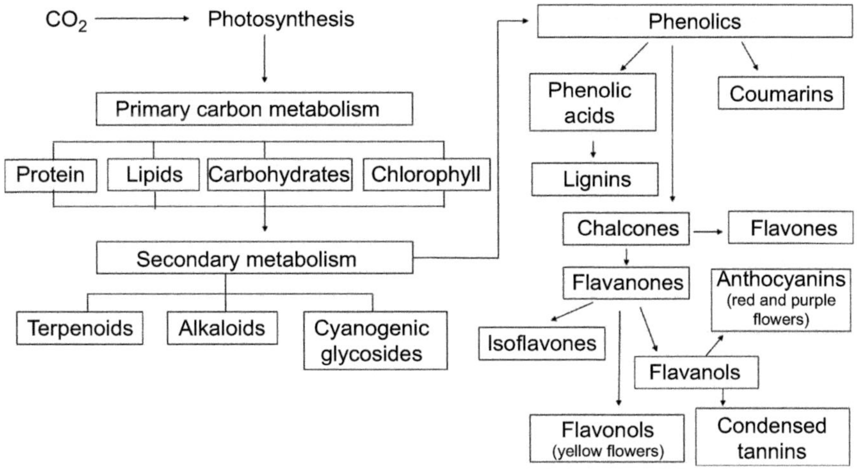

Figure 2.1
Primary and secondary metabolism linkage in plants.

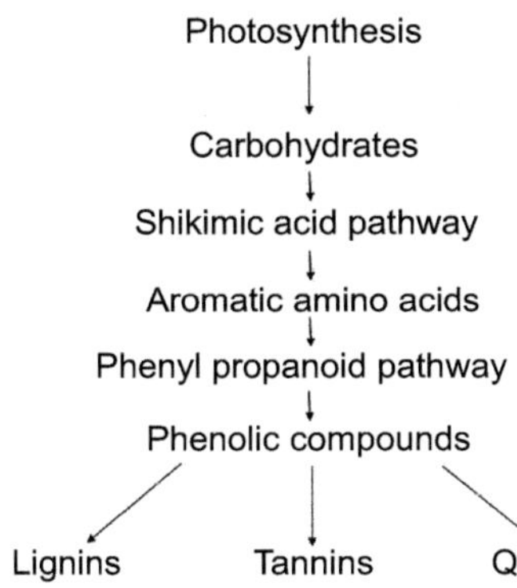

Figure 2.2
Biosynthetic pathways leading to the synthesis of phenolic compounds.

Table 2.1: Phenolic Compounds in Plants: Categories

Class	Structure
Simple phenolics, benzoquinones	C6
Hydroxybenzoic acids	C6–C1
Hydroxycinnamic acids, phenylpropanoids	C6–C3
Acetophenones, phenylacetic acids	C6–C2
Xanthones	C6–C1–C6
Stilbenes, anthraquinones	C6–C2–C6
Flavonoids, isoflavonoids	C6–C3–C6
Lignans, neolignans	$(C6–C3)_2$
Lignins	(C6–C3)n
Condensed tannins (proanthocyanidins or flavolans)	(C6–C3–C6)n

two groups: hydroxybenzoic acids (HBAs) and hydroxycinnamic acids (HCAs), which are derived from nonphenolic molecules of benzoic and cinnamic acid, respectively. They are synthesized through the shikimate pathway, in which L-phenylalanine or L-tyrosine is the precursory substance (Williamson, 2017).

The main reactions responsible for phenolic acid synthesis from L-phenylalanine or L-tyrosine are deamination, hydroxylation, and methylation. Firstly, phenylalanine and/or tyrosine cause deamination, resulting in cinnamic and/or p-coumaric acids, respectively. Aromatic rings of cinnamic and/or p-coumaric acids are then hydroxylated and methylated to form ferulic and caffeic acid. Cinnamic acid side chain degradation results in benzoic acid, which can undergo hydroxylation and methylation, producing protocatechuic and p-hydroxybenzoic acids (Rice-Evans et al., 1996).

HBAs have the general structure C_6–C_1, with some variations in their basic structure such as hydroxylations and methoxylations of the aromatic rings. HBAs are salicylic acid, 4-hydroxybenzoic acid, protocatechuic acid, gentisic acid, vanillic acid, syringic acid, gallic acid, ellagic acid, and hexahydroxydiphenic acid (ellagic acid dilactone). They are present in plant food as conjugates, whereas they can be free in some fruits (e.g., gallic acid in persimmons) or can be released in fruit or vegetable processing (Tomás-Barberán and Clifford, 2000). An example of conjugated HBAs is gallic acid, which can be a dimer, trimer, or tetramer (ellagic acid, tergallic acid, and gallagic acid, respectively) (Clifford and Scalbert, 2000). The main food sources of such compounds are fruits, vegetables, teas, and cereals. Ellagic acid is found in blackberries and strawberries; salicylic acid can be found in apricot, blueberry, and black tea; gentisic acid is present in citrus and grapes, and also some Solanaceae fruits as tomato, egg-plant, pepper, and Cucurbitaceae, such as melon and cucumber; gallic acid is present in a range of teas, including Japanese and Chinese green and black teas and semifermented teas (Clifford and Scalbert, 2000).

HCAs have the basic structure as C_6–C_3, with a double bond in the side chain that may have a *cis* or a *trans* configuration, in food they usually occur as monomers, dimers, or polymers; they can be found as condensates with alcohols, hydroxy acids, or mono/disaccharides producing esters, or can be amides by condensation with amines. They are rarely found in free form, but processed foods can have free forms from sterilization, freezing, or fermentation processes (Rommel and Wrolstad, 1993). The main dietary sources of HCAs are fruits, such as apples, cherries, various berries, peaches, plums, and some citrus fruits. Caffeic acid and p-coumaric acid represent between 75% and 100% of the total HCA content in fruits, being the most abundant HCAs found in fruits (D'Archivio et al., 2007). Caffeic acid is found in coffee, blueberry, grape, pear, cranberry, apple, and orange (Naczk and Shahidi, 2006) (which is estimated to account for 92% of the caffeic acid intake); for coumaric acid the main fruit sources are grapes, cherries, and berries (strawberry mainly) (Mattila et al., 2005).

The pericarp of wheat grain contains 98% of the total ferulic acid present in seeds (Manach et al., 2004). Berries such as strawberries are abundant in coumaric acid, and peanuts are also

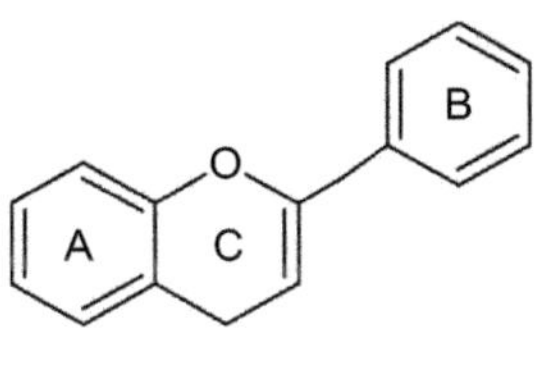

Figure 2.3
Structure of a flavonoid molecule.

rich in this compound. Sinapic acid is more abundant in cereals than in other food plants, Brassica vegetables are the most abundant source of this compound in vegetables (Mattila et al., 2005). The estimation of hydroxycinnamic acid consumption in an average person is 211 mg per day, in which coffee is the main dietary source (Lafay and Gil-Izquierdo, 2008).

Flavonoids constitute the largest group of phenolic compounds. They comprise more than 6000 compounds within more than 8000 phenolic compounds found in plant food. They are low molecular weight, characterized by a 15-carbon skeleton, arranged as C6–C3–C6, with different substitutions, unsaturation degree, and arrangement of the basic skeleton, resulting in different subclasses (Lafay and Gil-Izquierdo, 2008). Flavonoid structures are basically constituted of two aromatic rings, A and B, joined by a three-carbon bridge, frequently in the form of a heterocyclic ring, C (Fig. 2.3). The origins of flavonoid molecules are from the acetate/malonate pathway, in which ring A is derivated and the shikimate pathway in which ring B is derived from phenylalanine (Birt and Jeffery, 2013). In addition, C rings are mainly responsible for the varieties of flavonoid classes; variations in their substitution patterns provide the major flavonoid classes, such as flavonols, flavones, flavanones, flavanols (catechins), isoflavones, flavanonols, and anthocyanidins. Variations in the A and B rings give rise to the different compounds within the class of flavonoids. Such variations are due to the substitutions which can be oxygenation, alkylation, glycosylation, sulfation, and acylation (Birt and Jeffery, 2013).

They are often stored in the plant bound to sugar(s), called glycosides; the bonded form of flavonoids are more stable than the free version, however, they have relatively poor bioavailability when ingested. Bond flavonoids are conjugated with *O*- or *C*-glycosides, which can be glucose, galactose, rhamnose, apiose, etc., for *O*-glycosides, the sugar residues are usually linked to 3-, 7-, or 4-hydroxyl groups, while for *C*-glycosides, the sugar residues are linked directly to C-6 or C-8 (Cook and Samman, 1996).

A large part of flavonoids have a yellow to red color, this is because of conjugated chromophores present in the molecules, which are responsible for the color range of flowers, seeds, and fruits (Erlund, 2004). An example is the anthocyanidins, such as Cyanidin, which demonstrates colors from red to magenta (Navarre et al., 2013); Perlargodin for orange to red; and Delphinidin for magenta to purple. Flavones and flavonols generally present with a red color, while flavanones tend to be colorless, white, or brownish. They have low solubility in water, in which their contribution to food color is reduced when compared to anthocyanidins, flavanols, and proanthocyanidins (Birt and Jeffery, 2013; Metabolism and Compounds, 2009).

Flavonoid biosynthesis is through the phenylpropanoid pathway, the first step produces 4-coumaroyl-CoA from phenylalanine, which by the action of chalcone synthase enzyme produces chalcone scaffolds, a flavonoid backbone for generation of all flavonoids, being a central pathway to flavonoid biosynthesis. The different flavonoid subclasses are synthesized from a basic flavonoid skeleton by the action of an enzyme group, such as reductases, isomerases, hydroxylases, and several Fe^{2+}/2-oxoglutarate-dependent dioxygenases (Martens et al., 2010). Such enzymes modify the basic flavonoid structure by adding sugars (D-glucose, lignin, arabinose, glucorhamnose, L-rhamnose, galactose), methyl groups, and/or acyl moieties; these modifications can alter their interaction, solubility, and reactivity with cells, leading to diverse physiological activities (Ferrer et al., 2008).

The number of flavonoids present in fruits and vegetables may vary depending on species variety, edaphoclimatic conditions, part of the plant, cultivation, and degree of ripeness. Flavonoids in foods are not only responsible for the color, they also participate in taste, protection of lipid peroxidation, enzymes, and vitaminic compounds. The cooking process can modify the content of flavonoids, furthermore, storage conditions of food can also change the flavonoid amount (Ferrer et al., 2008).

The main dietary sources of flavonoids are the following foods; **flavones** (apigenin, luteonin) in spices and herbs, fats and oils fruits, vegetables and vegetable products, cereal grains and pasta; **flavonols** (kaempherol, myricetin, quercetin, isorhemetin, and rutin) in dairy products, spices and herbs, vegetables and vegetable products, nuts and seeds, beverages; **flavanones** (eriodictyol, hesperidin, naringenin) in spices and herbs, fruits, vegetables and vegetable products, nuts and seeds, beverages, cereal grains and pasta; **flavan-3-ols** ((−)-cathecin, (+)-gallocatechin, (−)- epicatechin, (−)-epigallocatechin, (−)-epigallocatechingallate, epigallocatechin gallate, theaflavine, theaflavine gallate) in dairy products, fruits, vegetables and vegetable products, nuts and seeds, beverages, legumes and legume products; **anthocyanidins** (cyanidin, delphinidin, malvidin, pelargonidin, peonidin, petunidin) in fruits, vegetables and vegetable products, nuts and seeds, beverages, legumes and legume products, and cereal grains and pasta (Faggio et al., 2017).

Tannins, commonly referred to as tannic acid, have value in interactions between plants and their ecosystems, as an example, they can act against herbivores or have a exert role as antimicrobial agents. They are water-soluble and their molecular weight is from 500 to 3000 Da, furthermore, they can make an insoluble complex with water and proteins, gelatin, and alkaloids. These compounds contain a large number of hydroxyl or other functional groups and therefore are found in the form of esters or heterosis (Ferrer et al., 2008).

Such compounds are very chemically reactive and form intra- or intermolecular hydrogen bridges. Then, they have the ability of precipitate proteins; they can precipitate salivary glycoproteins, leading to loss of lubricant capability, which is responsible for the astringency of many fruits and plant products (Ferrer et al., 2008). The bond between proteins and tannins takes place through hydrogen bridges between site-specific proteins and the phenolic group from tannins, the stability of this interaction depends on the molecular weight of the tannins (with low molecular weight the phenolic compound is not able to establish interactions to give stability to this combination, in addition high molecular weight prevents interleaving between fiber spaces in proteins) (Ferrer et al., 2008). Tannins are easily oxidized by metals, ferric chloride, or even specific vegetable enzymes, leading to darkening of solutions (Chung et al., 1998).

They can be classified chemically into two groups: hydrolyzable and nonhydrolyzable or condensed tannins. Hydrolyzable tannins are formed from shikimate producing gallic acid esters (gallotannins), partially or wholly esterified by gallic acid and glycosylated ellagic acid (ellagitannins). After hydrolysis by certain enzymes, acids, bases, and gallic acid esters yield glucose and gallic acids. The ellagitannins undergo lactonization to produce ellagic acid (Ferrer et al., 2008; Ozawa et al., 1987).

Condensed tannins are structurally more complex and uniform than hydrolyzable tannins. They are oligomers and polymers of flavan-3-old flavan and/or flavan-3,4-diols, the structure of condensed tannins in food plants differs in stereochemistry, type of flavan linkages intramolecularly, the degree of polymerization, and hydroxylation (De Bruyne et al., 1999). Procyanidins and prodelphinidins are the most characterized tannins, both are linked via flavan-3-ol monomers (C4–C8), but differ in the flavonoid B-ring by the degree of hydroxylation. The degree of polymerization can vary from dimers to polymers by up to 30 or more subunits. Condensed tannins are widely distributed in food plants such as fruits, vegetables, red wine, cocoa, and some grains (sorghum and finger millets) and legumes (Tsao, 2010).

Flavan-3-oils are also known as catechins, they present four isomers due to asymmetrical carbon atoms at the C-2 and C-3 positions. They are (+) and (−)-catechins: (−)-epicatechin, (+)-gallocatechin, and (+)-catechin. Flavan-3,4-diols present eight isomers due to asymmetric carbon atoms at C-2, C-3, and C-4. Some of these compounds include leucocyanidin, leucopelargonidin, leucodelphinidin, guibourtacacidin, (+)-leucorobinetinidin, (−)-melacacidin, and (−)-teracacidin (Tsao, 2010).

2.3 Phenolic Compounds: Correlation Between Structure and Antioxidant Power

Phenols are potent antioxidants, they are compounds that prevent biomolecules (proteins, nucleic acids, polyunsaturated lipids, and sugars) from undergoing oxidative damage through free radical-mediated reactions (Heleno et al., 2015) and their beneficial effects include antiinflammatory, antidiabetic, cardioprotective, neuroprotective, antitumor, and antiaging properties (Zhang et al., 2017).

There are two main mechanisms by which antioxidants perform these properties: free radical inactivation and electron transfer (Valko et al., 2007). In the first mechanism the free radical (R *) can remove a hydrogen atom from the antioxidant (ArOH), which becomes radical. The lower the bond dissociation energy (BDE) of the O–H bonds, the easier the reaction of the inactivation of the free radical and therefore the greater the antioxidant action.

$$R^{\cdot} + ArOH \rightarrow RH + ARO^{\cdot}$$

In the second mechanism, the antioxidant can donate an electron to the free radical, which becomes a cation radical. In this mechanism, the lower the potential ionization (IP), the easier the electron abstraction, meaning greater antioxidant activity.

$$R^{\cdot} + ArOH \rightarrow RH^{-} + ARO^{\cdot +}$$

The antioxidant action of phenolics is established by the structure–activity relationship (SAR) (Bendary et al., 2013), including the number and positions of the hydroxyl group (-OH), presence of double bond (C2=C3), glycosylation, and the presence of substituents in the rings (Wang et al., 2018).

As previously demonstrated, hydrogens and electrons are donated by hydroxyl groups, forming stable radicals. Thus, the position and the hydroxylation number correlate with the antioxidation of flavonoid. The presence of two hydroxyl groups in the ring suggests an improved antioxidant effect, while the presence of 3-OH clearly contributes to the suppression of antioxidant activity. Hydrophilicity is increased by increasing the number of hydroxyls, keeping the nucleus of the flavonoid in the hydrophobic cavity, which may constitute a connection with the active site of enzymatic relevance (Wang et al., 2018).

The existence of a C2=C3 double bond in conjunction with a 4-carbonyl group also significantly increases antioxidant activity, as it provides planarity, electron expansion, and displacement between adjacent rings. The 4-carbonyl group can further induce electron shift by resonance effects, influencing the dissociation constant of the hydroxyl groups and the stability of radicals (Wang et al., 2018).

As for phenolic glycosylation, different forms have different capacities. The C-glycoside form has demonstrated greater antioxidant ability when compared to O-glycoside in chemical assays (Wang et al., 2018). Glycosylation also interferes with planarity, methylation, and electron displacement (Heim et al., 2002).

It is also known that the antioxidant activity of the glycosides is weaker than the corresponding aglycones, however, the bioavailability of the former is higher, thus increasing the antioxidant activity (Heim et al., 2002).

Some rings are more susceptible to O-methyl substitution, which in turn decreases antioxidant activity by causing steric hindrance. Methylation affects planarity, electron donation, and molecular hydrophobicity. This hydrophobicity may still be related to absorption through the bio-membrane and may further influence the interactions between the compounds, in which also facilitates this decrease. Some studies indicate that in alternation with the hydroxyl group O-methyl can reduce the antioxidant capacity of the phenolic substance (Silva et al., 2000).

There is still a discrepancy in SAR for phenolic antioxidant activity, requiring further investigations for later application and better utilization. It should not be forgotten that besides these factors, the functional portion may also influence the antioxidant activity of the compound, as well as the synergistic and antagonistic interactions (Williamson, 2017).

2.4 Methods to Evaluate Antioxidant Activity of Phenolic Compounds

The study of the antioxidant potential of samples (phenolic compounds, plant extracts, and commercial antioxidants) is one of the most popular topics of study among the scientific community (Kasote et al., 2015). The most common in vitro methods used to evaluate antioxidant activity of phenolic compounds and plant extracts are based on a single electron transfer reaction or a hydrogen atom transfer reaction. The first are reactions of redox type that present variations of coloring correlated with antioxidant concentration species in the sample (Moharram and Youssef, 2015). Such methods include the copper reduction (CUPRAC) assay (Apak et al., 2005), Trolox equivalent antioxidant capacity (ABTS or TEAC) assay (Miller et al., 1993), the ferric-reducing ability of plasma (FRAP) assay (Benzie and Strain, 1996), and the 2,2-diphenyl-1-picrylhydrazyl (DPPH) radical scavenging capacity assay (Peng et al., 2000). The second are reactions based on thermodynamics competition between substrates and antioxidants by peroxyl radical and, consequently, fluorescent probe formation, later monitored and interpreted by kinetic curves (Moharram and Youssef, 2015), including oxygen radical absorbance capacity (ORAC) assay (Prior and Cao, 1999), peroxyl scavenging capacity (PSC) (Wayner et al., 1985; Adom and Rui, 2005), and the total peroxyl radical-trapping antioxidant parameter (TRAP) assay (Wayner et al., 1985).

These methods have limitations and could be useful if well understood. Such in vitro studies do not account for biochemical, metabolic, and other physiological parameters (Moharram and Youssef, 2015). Phenolic compounds pass through numerous biochemical reactions from ingestion, digestion, and absorption in the organisms (Tufarelli et al., 2018). This fact could explain why phenolic compounds and plant extracts present high antioxidant activity in vitro studies while this is not found in vivo studies. Furthermore, in vitro antioxidant activity assay measures only the antioxidant activity against one radical, not to all reactive oxygen species (Moharram and Youssef, 2015).

In vivo studies has been mainly conducted in eukaryotic cells, rabbits, guinea-pigs, mice, and fish (Alam et al., 2013). The in vivo antioxidant activity evaluation of phenolic compounds or extracts is assessed by the effect on a range of biochemical parameters associated with oxidant/antioxidant equilibrium in organisms, such as catalase (CAT), superoxide dismutase (SOD), glutathione peroxidase (GSH-Px), which are involved in the direct elimination of

ROS, furthermore, gluthatione (GSH), glutathione peroxidase (GSH-Px), malondialdehyde (MDA), glutathione reductase (GR), glutathione-S-transferase (GST), xanthine oxidase (OX), peroxidase (Px), FRAP, ABTS, and ORAC, among others (Kasote et al., 2015).

2.4.1 In Vitro Evaluation of Antioxidant Activity of Phenolic Compounds: Single Electron Transfer Reaction

2.4.1.1 Copper Reduction (CUPRAC) Assay

This method is based on the application of copper(II)-neocuproine (2,9-dimethyl-1,10-phenanthroline) reagent as the chromogenic oxidizing agent. In the presence of antioxidants copper(II)-neocuproine complex is reduced to copper(I)-neocuproine complex and detected at 450 nm (Apak et al., 2005). In phenolic compounds, copper(II)-neocuproine complex reacts with n-electron reductant leading to the corresponding quinone formation. This assay is carried in conditions similar to those found in vivo at pH 7 and temperature 37°C, as opposed to the unrealistic acidic conditions (pH 3.6), as an example in FRAP assays (Fig. 2.4). The results are presented as Trolox (6-hydroxy-2,5,7,8-tetramethylchroman2-carboxylic acid) equivalent, a potent analogous water-soluble vitamin E. CUPRAC assay is able to oxidize thiol-type antioxidants, others methods such as FRAP assay do not measure thiol-type antioxidants such as glutathione; CUPRAC reagent does not oxidize sugars and citric acid. Additionally, this method can simultaneously measure hydrophilic as well as lipophilic antioxidants (Apak et al., 2005).

2.4.1.2 Radical ABTS Assay (ABTS or TEAC)

The first ABTS assay was described in 1997 by Miller et al.; the method was developed based on absorbance of the ABTS+ radical cation for the evaluation of the total antioxidant capacity

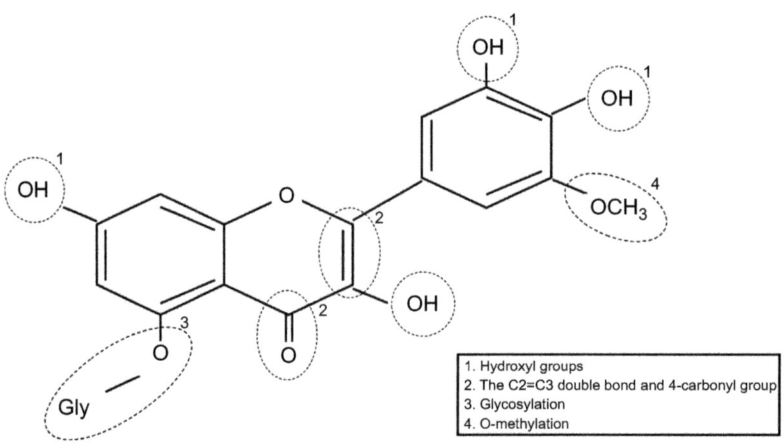

Figure 2.4
Summary of SAR for phenolic-induced antioxidant.

of body fluids and drug solutions. This method was first based on the production of ABTS radical cation by activation of metmyoglobin with hydrogen peroxide in the presence of ABTS [2,2′-azinobis-(3-ethylbenzothiazoline-6-sulfonic acid)], with or without antioxidants. Then, antioxidants arrest $ABTS^{\cdot+}$ radical [2,2-azino-bis(ethylbenzene-thiazoline-6-sulfonic acid)], leading to a decrease in absorbance, which is detected by the antioxidant combination of antioxidant with radicals at different times (Miller et al., 1993)

The radical $ABTS^{\cdot+}$ is generated through 2,2-azino-bis(3-etilbenzotiazolin)-6-sulfonic acid. This radical is a chemically stable chromophore compound with a wide pH range, which is water soluble and exhibits strong absorption in the range of 600–750 nm. $ABTS^{\cdot+}$ radical is soluble in aqueous and organic media, therefore antioxidant activity can be determined in water-soluble and fat-soluble samples. However, antioxidants may also react with ABTS giving the cation radical, intervening in the results of the assay (Miller et al., 1993).

Other ways to obtain ABTS radicals have been proposed. ABTS can also be generated from the reaction between ABTS and potassium persulfate, with direct production of the blue/green $ABTS^{\cdot+}$ chromophore, with absorption maxima at wavelengths of 645, 734, and 815 nm, or more commonly a maximum at 415 nm (Fig. 2.4). The radical is then generated directly in a stable form prior to reaction with antioxidants. The decolorization of ABTS extent is determined by the percentage inhibition of the $ABTS^{\cdot+}$ radical cation as a function of concentration and time after the reactivity relative of Trolox as a standard is measured (Nenadis et al., 2004).

2.4.1.3 The Ferric-Reducing Ability of Plasma (FRAP) Assay

The ferric-reducing ability of plasma was developed to evaluate the antioxidant effect of nonenzymatic defense in biological fluids, in which the response could provide a measure of antioxidant ability (Benzie and Strain, 1996). The assay consists of the reduction of ferric to ferrous ions at low pH (3.6), providing a colored complex. Ferric tripyridyltriazine (TPTZ 2,4,6-Tris(2-pyridyl)-*s*-triazine) has a yellow color and is reduced to ferrous (Fe II) form by antioxidants and a violet-blue color is formed with an absorption maximum at 593 nm (Antolovich et al., 2002) (Fig. 2.4). The pH values are very important in the reduction of antioxidant capacity since protonation of antioxidant compounds happens in acid conditions, leading to a decrease in the antioxidant activity, in basic conditions the opposite happens: protons are dissociated from phenolic compounds and an increase in antioxidant activity can be observed (Benzie and Strain, 1996). Further, the method was extended to evaluate the antioxidant power of plant foods (Pulido et al., 2000). The change in absorbance is proportional to the total ferric-reducing/antioxidant power of the antioxidants in the sample (Benzie and Strain, 1996). FRAP assay does not have a radical in the system, creating a problem in comparing the antioxidant power towards different kinds of radicals (Roginsky and Lissi, 2005).

2.4.1.4 2,2-Diphenyl-1-Picrylhydrazyl (DPPH) Radical Scavenging Capacity Assay

The DPPH assay was first suggested in the 1950s to find electron donors in natural products (Blois, 1958). Later it was used to determine the antioxidant activity of phenolic compounds and plant food (Boylan et al., 2015).

The DPPH (2,2-diphenyl-1-picrylhydrazyl) is a radical of organic nitrogen, which has an unpaired valence electron at one atom of nitrogen bridge, and is a stable, soluble organic medium (especially alcoholic) and is an insoluble aqueous medium, with a purple color and absorption in the range 515–520 nm. The radical is available commercially, enabling its use and decreasing possible interferences in the reactions through radical production (Pulido et al., 2000). The solution with radical in the presence of an electron donor loses its intensity and becomes yellow, according to the number of electrons detainees, in which color changes happen when the nitrogen atom of DPPH receives electrons from antioxidant compounds (Fig. 2.4). The antioxidant activity is expressed as Trolox equivalent as used in other antioxidant assays (Bondet et al., 1997). Also, the EC_{50} technique is carried out, based on the amount of antioxidant necessary to decrease by 50% the initial concentration of DPPH radical, thus determining the efficiency of antiradical antioxidants. The assay has some limitations, such as DPPH• interacts with other radicals and the curve response to reach a stationary stage is not linear (relations between antioxidant and DPPH are variable); the antioxidant activity is expressed as Trolox equivalent (Chen et al., 2013)

2.4.2 In Vitro Evaluation of Antioxidant Activity of Phenolic Compounds: Single Hydrogen Atom Transfer Reaction

2.4.2.1 Oxygen Radical Absorbance Capacity (ORAC) Assay

The ORAC assay was developed by Cao and Prior (1998). The method is based on the utilization of a protein, the B or R phycoerythrin, as radicals target molecules, which are highly fluorescent. This protein is found in red algae, has a molecular weight of 240,000 Da, a red photoreceptor pigment, and is from species of purple algae and cyanobacteria. These proteins have a high absorption coefficient, and are readily detected by fluorescence spectroscopy at concentrations as low as 10^{-12} M. Radical peroxyl (ROO•) is generated through thermal decomposition of 2,2′-azobis(2-amidinopropane) dihydrochloride (AAPH) at 37°C (Cao and Prior, 1998). Then, the exposure of phycoerythrin to peroxyl free radicals at a constant rate leads to a linear decrease in fluorescence emission; the measurement of the decline in fluorescence protein is the way to evaluate the antioxidant activity in the assay (Prior and Cao, 1999). Trolox is used as a standard, in which one molecule of Trolox traps two peroxyl radicals, a known amount of Trolox provides the rate of peroxyl radical generation in the reaction mixture, from these data the complete loss of fluorescence results from the reaction of peroxyl radical and molecule of phycoerythrin can be estimated (Wu et al., 2004).

The loss of phycoerythrin fluorescence is not linear with time, but is exponential with time, therefore, for quantification of fluorescence the area under the curve (AUC) is used. This method developed a reaction between reactive species and substrate until the end, furthermore, the AUC calculation provides the inhibition percentage in combination with the total length of time inhibition of reactive species by antioxidants (Cao and Prior, 1998).

Peroxyl is a reactive oxygen species (ROS) that is biologically the most important in the human body, which is the most abundant and is responsible for oxidative damage. ORAC using peroxyl radical is a more realistic method than others that use radicals not found in the human body (Cao and Prior, 1998).

The major limitation of the ORAC assay is the use of phycoerythrin as a probe. Phycoerythrin is not photostable, with a variable reactivity to peroxyl radicals and interactions with polyphenols due to nonspecific protein binding (Ou et al., 2001). Thus, fluorescein (FL) (3′,6′-dihydroxyspiro[isobenzofuran-1[3H],9′[9H]-xanthen]-3-one) was tested as a fluorescent probe to replace phycoerythrin. The basis of the reaction is the same as phycoerythrin, with a hydrogen atom transfer reaction (Fig. 2.4). The fluorescein is photostable and thermostable, in which homogeneity is achieved after exposition of excitation light and it does not interact with antioxidants from the sample, hence reducing experiment cost. ORAC using fluorescein can measure the antioxidant activity of hydrophilic and lipophilic compounds of a sample, by lipophilic measuring, methylated β-cyclodextrin can be used to augment the water solubility of such compounds due to be an amphipathic molecule, furthermore, it is chemically inert without any interfering side reactions (Ou et al., 2001).

ORAC uses fluorescence instead of absorbance, as others methods that measure antioxidant activity cause less interference to colored compounds available in the samples. This is essential to analyses in which the antioxidant activity of colored foods (wine, fruits, and vegetables) or supplements of natural products is to be assessed (Ou et al., 2001).

2.4.2.2 Peroxyl Scavenging Capacity (PSC) Assay

Peroxyl scavenging capacity (PSC) assay is based on the degree of inhibition of dichlorofluorescein oxidation by antioxidants that scavenge peroxyl radicals, produced by thermal degradation of 2,2′-azobis(amidinopropane) (ABAP). The reaction mechanism is thermal degradation of ABAP at 37°C, which produces peroxyl radicals, leading to oxidation of nonfluorescent dichloro-dihydro-fluorescein (DCFH), producing fluorescent dichlorofluorescein (DCF). Antioxidants that scavenge peroxyl radicals are the basis for calculating the antioxidant activity as they inhibit DCFH oxidation and can be used to calculate the degree of inhibition (Fig. 2.4). Dichloro-dihydro-fluorescein diacetate (DCFH-DA) is stable to oxidation and needs to be treated with KOH in which the diacetate portion of the molecule is broken, forming DCFH which is very slowly oxidized at ambient conditions without ABAP (Adom and Rui, 2005). Fluorescence in PSC assay is monitored at 485 nm excitation and 538 nm emission. The area under the curve fluorescence–reaction time is used as the basis for calculating antioxidant activity, in which the PSC unit is given by subtracting 1 from the AUC/AUC ratio from control reaction with the only buffer. PSC can be expressed as micromoles of vitamin C equivalents or by $EC_{50.}$ Vitamin C in this assay is demonstrated to react rapidly with peroxyl radicals until the ascorbic acid present is exhausted and then other compounds are analyzed (catechin, ferulic acid, Trolox, chlorogenic acid, caffeic acid, EGCG, quercetin, and gallic acid). This assay is very sensitive, being able to run in the low micromolar range (Adom and Rui, 2005).

PSC assay analyzes the antioxidant activity of both hydrophilic and lipophilic compounds, to lipophilic evaluation, methylated β-cyclodextrin was also used in an ORAC assay to augment the water solubility of such compounds and extracts (Adom and Rui, 2005).

2.4.2.3 Peroxyl Radical-Trapping Antioxidant Parameter (TRAP): Phycoerythrin-Based Assay

The peroxyl radical-trapping antioxidant parameter (TRAP) assay is based on the capacity of antioxidant to capture ROS generated by a substrate, as the thermal decomposition of 2,2′-azobis(amidinopropane) dihydrochloride (AAPH) at 37°C leads to peroxyl radical production (DeLange and Glazer, 1989). The loss maintenance of fluorescence of R-phycoerythrin is related to the antioxidant potential when the sample is added to the system. Antioxidants inhibit the decomposition of R-phycoerythrin and thus delay the decrease in fluorescence (Fig. 2.4). When the fluorescence reaches a regularity in decay the measurement is stopped. Kinetics of phycoerythrin quenching is not linear in the presence of peroxyl or hydroxyl radicals. Then, for quantification of fluorescence the area under the curve (AUC) is used, which is inversely proportional to antioxidant capacity (Ghiselli et al., 1995).

One of the great criticisms of indirect methods is that the capacity of reduction obtained does not necessarily reflect the antioxidant activity of samples. Such methods do not consider a real oxidative degradation (Amorati and Valgimigli, 2015). Furthermore, antioxidant activity validation in vivo studies are scarce or even are not found, in other words, the results from in vitro assays are not substantiated by in vivo assays. The scientific literature shows a rise of in vitro studies, for example, from 2000 to 2014 there was a pronounced increase in in vitro studies in the literature, which was not followed by the same augmentation of in vivo studies as validation of the antioxidant power of phenolic compounds or extracts in vivo (Alam et al., 2013; Prior and Cao, 1999). Thus, in vivo experiments and clinical trials should also be carried out as validation of the antioxidant effect from phenolic compounds and food plant extracts evaluated by in vitro methods (Alam et al., 2013).

2.5 Conclusion and Future Perspectives

Phenolic compounds are widely found in plant food, they are classified in a range of groups according to their structure. Such variations give them diverse characteristics, one of which is antioxidant activity that is linked to the molecular structure of phenolic compounds. The extent of conjugation and degree of hydroxylation seem to be the main factors that give the reducing power of dietary polyphenols. The antioxidant activity evaluation of phenolic compounds has been studied for years and several assays have been proposed, however, a gap exists between the results observed in vitro studies and those observed in vivo studies. In vitro methods have numerous divergences and, for this reason, a significant variation is observed for the same sample when tested by different assays. There is another important point: in vitro methods do not consider the contribution of human metabolism to the antioxidant potential.

Thus, it is essential to show the effective antioxidant power of phenolic compounds and extracts, considering their bioavailability and bioefficacy. Therefore, in vivo studies must be carried out for the evaluation of the antioxidant activities of phenolic compounds and extracts. Furthermore, biological and physiological variables analyzed in vivo studies, including differences in metabolism (ingestion, digestion, and absorption), as well as the bioavailability of phenolic compounds, which could influence the antioxidant power response, are necessary to improve the health and well-being of people in the future.

References

Adom, K.K., Rui, H.L., 2005. Rapid peroxyl radical scavenging capacity (PSC) assay for assessing both hydrophilic and lipophilic antioxidants. Journal of Agricultural and Food Chemistry. 53, 6572–6580. https://doi.org/10.1021/jf048318o.

Alam, M.N., Bristi, N.J., Rafiquzzaman, M., 2013. Review on in vivo and in vitro methods evaluation of antioxidant activity. Saudi Pharmaceutical Journal. 21, 143–152. https://doi.org/10.1016/j.jsps.2012.05.002.

Amorati, R., Valgimigli, L., 2015. Advantages and limitations of common testing methods for antioxidants 49, 633–649. https://doi.org/10.3109/10715762.2014.996146.

Antolovich, M., Prenzler, P.D., Patsalides, E., McDonald, S., Robards, K., 2002. Methods for testing antioxidant activity. Analyst. 127, 183–198. https://doi.org/10.1039/b009171p.

Apak, R., Güçlü, K., Özyürek, M., Esîn Karademîr, S., Altun, M., 2005. Total antioxidant capacity assay of human serum using copper(II)-neocuproine as chromogenic oxidant: the CUPRAC method. Free Radical Research. 39, 949–961. https://doi.org/10.1080/10715760500210145.

Balasundram, N., Sundram, K., Samman, S., 2006. Phenolic compounds in plants and agri-industrial by-products: antioxidant activity, occurrence, and potential uses. Food Chemistry. 99, 191–203. https://doi.org/10.1016/j.foodchem.2005.07.042.

Bendary, E., Francis, R.R., Ali, H.M.G., Sarwat, M.I., Hady, S.E., 2013. Antioxidant and structure – activity relationships (SARs) of some phenolic and anilines compounds. Annals of Agricultural Science. 58, 173–181. https://doi.org/10.1016/j.aoas.2013.07.002.

Benzie, I.F.F., Strain, J.J., 1996. The ferric reducing ability of plasma (FRAP) as a measure of "antioxidant power": the FRAP assay. Analytical Biochemistry. 239, 70–76. https://doi.org/10.1006/abio.1996.0292.

Birt, D.F., Jeffery, E., 2013. Flavonoids. Advances in Nutrition. 4, 576–577. https://doi.org/10.3945/an.113.004465.Diet.

Blois, M., 1958. Antioxidant determinations by the use of a stable free radical. Nature. 181, 1199–1200. https://doi.org/10.1038/1811199a0.

Bondet, V., Brand-Williams, W., Berset, C., 1997. Kinetics and mechanisms of antioxidant activity using the DPPH•free radical method. Lebensmittel-Wissenschaft und -Technologie- Food Science and Technology. 30, 609–615. https://doi.org/10.1006/fstl.1997.0240.

Boylan, F., Menezes, S., Leita, G.G., 2015. Screening of Brazilian plant extracts for antioxidant activity by the use of DPPH free radical method. Phytotherapy Research. 130, 127–130. https://doi.org/10.1002/ptr.687.

Burton, G.J., Jauniaux, E., Medicine, F., 2011. Best practice & research clinical obstetrics and gynaecology oxidative stress. Best Practice and Research Clinical Obstetrics and Gynaecology. 25, 287–299. https://doi.org/10.1016/j.bpobgyn.2010.10.016.

Cao, G., Prior, R.L., 1998. Measurement of oxygen radical absorbance capacity in biological samples. Methods in Enzymology. 299, 50–62. https://doi.org/10.1016/S0076-6879(99)99008-0.

Chen, Z., Bertin, R., Froldi, G., 2013. EC50estimation of antioxidant activity in DPPH* assay using several statistical programs. Food Chemistry. 138, 414–420. https://doi.org/10.1016/j.foodchem.2012.11.001.

Chikezie, P.C., Ibegbulem, C.O., Mbagwu, F.N., 2015. Bioactive principles from medicinal plants. Research Journal of Phytochemistry. 9, 88–115. https://doi.org/10.3923/rjphyto.2015.88.115.

Chung, K.T., Wong, T.Y., Wei, C.I., Huang, Y.W., Lin, Y., 1998. Tannins and human health: a review. Critical Reviews in Food Science and Nutrition. 38, 421–464. https://doi.org/10.1080/10408699891274273.

Clifford, M.N., Scalbert, A., 2000. Review Ellagitannins – nature, occurrence and dietary burden. Journal of the Science of Food and Agriculture. 80, 1118–1125. https://doi.org/10.1002/(SICI)1097-0010(20000515).

Cook, N.C., Samman, S., 1996. Flavonoids—chemistry, metabolism, cardioprotective effects, and dietary sources. The Journal of Nutritional Biochemistry. 7, 66–76. https://doi.org/10.1016/S0955-2863(95)00168-9.

De Bruyne, T., Pieters, L., Deelstra, H., Vlietinck, A., 1999. Condensed vegetable tannins: biodiversity in structure and biological activities. Biochemical Systematics and Ecology. 27, 445–459. https://doi.org/10.1016/S0305-1978(98)00101-X.

Del Rio, D., Rodriguez-Mateos, A., Spencer, J.P.E., Tognolini, M., Borges, G., Crozier, A., 2013. Dietary (Poly) phenolics in human health: structures, bioavailability, and evidence of protective effects against chronic diseases. Antioxidants and Redox Signaling. 18, 1818–1892. https://doi.org/10.1089/ars.2012.4581.

DeLange, R.J., Glazer, A.N., 1989. Phycoerythrin fluorescence-based assay for peroxy radicals: a screen for biologically relevant protective agents. Analytical Biochemistry. 177, 300–306. https://doi.org/10.1016/0003-2697(89)90056-0.

D'Archivio, M., Filesi, C., Benedetto, R., Di, Gargiulo, R., Giovannini, C., Masella, R., 2007. Polyphenols, dietary sources and bioavailability. Ann Ist Super Sanita Polyphenols, dietary sources and bioavailability. Annali dell'Istituto Superiore di Sanita 43, 348–361 https://doi.org/ 43 (4), 348–361.

Erlund, I., 2004. Review of the flavonoids quercetin, hesperetin, and naringenin. Dietary sources, bioactivities, bioavailability, and epidemiology. Nutrition Research. 24, 851–874. https://doi.org/10.1016/j.nutres.2004.07.005.

Faggio, C., Sureda, A., Morabito, S., Sanches-Silva, A., Mocan, A., Nabavi, S.F., Nabavi, S.M., 2017. Flavonoids and platelet aggregation: a brief review. European Journal of Pharmacology. 807, 91–101. https://doi.org/10.1016/j.ejphar.2017.04.009.

Ferrer, J.L., Austin, M.B., Stewart, C., Noel, J.P., 2008. Structure and function of enzymes involved in the biosynthesis of phenylpropanoids. Plant Physiology and Biochemistry. 46, 356–370. https://doi.org/10.1016/j.plaphy.2007.12.009.

Furukawa, S., Fujita, T., Shimabukuro, M., Iwaki, M., Yamada, Y., Nakajima, Y., Nakayama, O., Makishima, M., 2004. Increased oxidative stress in obesity and its impact on metabolic syndrome. 114. https://doi.org/10.1172/JCI200421625.1752.

Ghiselli, A., Serafini, M., Maiani, G., Azzini, E., Ferro-Luzzi, A., 1995. A fluorescence-based method for measuring total plasma antioxidant capability. Free Radical Biology and Medicine. 18, 29–36. https://doi.org/10.1016/0891-5849(94)00102-P.

Harborne, J.B., 1980. Secondary plant products: encyclopedia of plant physiology. New Series Volume 8 Phytochemistry. 19, 2803–2804. https://doi.org/10.1016/S0031-9422(00)83983-7.

Heim, K.E., Tagliaferro, A.R., Bobilya, D.J., 2002. Flavonoid antioxidants: chemistry, metabolism and structure-activity relationships. 13, 572–584.

Heleno, S.A., Martins, A., Queiroz, M.J.R.P., Ferreira, I.C.F.R., 2015. Bioactivity of phenolic acids: metabolites versus parent compounds: a review. Food Chemistry. 173, 501–513. https://doi.org/10.1016/j.foodchem.2014.10.057.

Kasote, D.M., Katyare, S.S., Hegde, M.V., Bae, H., 2015. Significance of antioxidant potential of plants and its relevance to therapeutic applications. International Journal of Biological Sciences. 11, 982–991. https://doi.org/10.7150/ijbs.12096.

Lafay, S., Gil-Izquierdo, A., 2008. Bioavailability of phenolic acids. Phytochemistry Reviews. 7, 301–311. https://doi.org/10.1007/s11101-007-9077-x.

Manach, C., Scalbert, Augustin, Morand, Christine, Rémésy, Christian, Jimenez, Liliana, 2004. Polyphenols - food sources and Bioavailability.pdf. American Journal of Clinical Nutrition 79, 727–747.

Martens, S., Preuß, A., Matern, U., 2010. Multifunctional flavonoid dioxygenases: flavonol and anthocyanin biosynthesis in *Arabidopsis thaliana* L. Phytochemistry. 71, 1040–1049. https://doi.org/10.1016/j.phytochem.2010.04.016.

Mattila, P., Pihlava, J.M., Hellström, J, 2005. Contents of phenolic acids, alkyl- and alkenylresorcinols, and avenanthramides in commercial grain products. Journal of Agricultural and Food Chemistry. 53, 8290–8295. https://doi.org/10.1021/jf051437z.

Metabolism, T.S., Compounds, P., 2009. Phenolic Compounds, pp. 1–6.

Miller, N.J., Rice-Evans, C., Davies, M.J., Gopinathan, V., Milner, A., 1993. A novel method for measuring antioxidant capacity and its application to monitoring the antioxidant status in premature neonates. Clinical Science. 84, 407–412. https://doi.org/10.1042/cs0840407.

Moharram, H.A., Youssef, M.M., 2015. Methods for Determining the Antioxidant Activity: A Review.

Naczk, M., Shahidi, F., 2006. Phenolics in cereals, fruits and vegetables: occurrence, extraction and analysis. Journal of Pharmaceutical and Biomedical Analysis. https://doi.org/10.1016/j.jpba.2006.04.002.

Navarre, D.A., Payyavula, R.S., Shakya, R., Knowles, N.R., Pillai, S.S., 2013. Changes in potato phenylpropanoid metabolism during tuber development. Plant Physiology and Biochemistry. 65, 89–101. https://doi.org/10.1016/j.plaphy.2013.01.007.

Nenadis, N., Wang, L.-F., Tsimidou, M., Zhang, H.-Y., 2004. Estimation of scavenging activity of phenolic compounds using the $ABTS^{\bullet+}$ assay. Journal of Agricultural and Food Chemistry. 52, 4669–4674. https://doi.org/10.1021/jf0400056.

Ou, B., Hampsch-Woodill, M., Prior, R.L., 2001. Development and validation of an improved oxygen radical absorbance capacity assay using fluorescein as the fluorescent probe. Journal of Agricultural and Food Chemistry. 49, 4619–4626. https://doi.org/10.1021/jf010586o.

Ozawa, T., Lilley, T.H., Haslam, E., 1987. Polyphenol interactions: astringency and the loss of astringency in ripening fruit. Phytochemistry. 26, 2937–2942. https://doi.org/10.1016/S0031-9422(00)84566-5.

Peng, C.-L., Chen, S.-W., Lin, Z.-F., Lin, G.-Z., 2000. Detection of antioxidative capacity in plants by scavenging organic free radical DPPH. Progress in Biochemistry and Biophysics 27.

Prior, R.L., Cao, G., 1999. In vivo total antioxidant capacity: comparison of different analytical methods. Free Radical Biology and Medicine. 27, 1173–1181. https://doi.org/10.1016/S0891-5849(99)00203-8.

Pulido, R., Bravo, L., Saura-Calixto, F., 2000. Antioxidant activity of dietary polyphenols as determined by a modified ferric reducing/antioxidant power assay. Journal of Agricultural and Food Chemistry. 48, 3396–3402. https://doi.org/10.1021/jf9913458.

Rice-Evans, C.A., Miller, N.J., Paganga, G., 1996. Structure-antioxidant activity relationships of flavonoids and phenolic acids. Free Radical Biology and Medicine. 20, 933–956. https://doi.org/10.1016/0891-5849(95)02227-9.

Roginsky, V., Lissi, E.A., 2005. Review of methods to determine chain-breaking antioxidant activity in food. Food Chemistry. 92, 235–254. https://doi.org/10.1016/j.foodchem.2004.08.004.

Rommel, A., Wrolstad, R.E., 1993. Composition of flavonols in red raspberry juice as influenced by cultivar, processing, and environmental factors. Journal of Agricultural and Food Chemistry. 41, 1941–1950. https://doi.org/10.1021/jf00035a025.

Silva, F.A.M., Borges, F., Guimara, C., 2000. Phenolic Acids and Derivatives: Studies on the Relationship Among Structure, Radical Scavenging Activity, and Physicochemical Parameters, pp. 2122–2126. https://doi.org/10.1021/jf9913110.

Tomás-Barberán, F.A., Clifford, M.N., 2000. Dietary hydroxybenzoic acid derivatives - nature, occurrence and dietary burden. Journal of the Science of Food and Agriculture https://doi.org/10.1002/(SICI)1097–0010(20000515)80:7<1024::AID-JSFA567>3.0.CO;2-S.

Tsao, R., 2010. Chemistry and biochemistry of dietary polyphenols. Nutrients. 2, 1231–1246. https://doi.org/10.3390/nu2121231.

Tufarelli, V., Casalino, E., D'Alessandro, A.G., Laudadio, V., 2018. Dietary phenolic compounds: biochemistry, metabolism and significance in animal and human health. Current Drug Metabolism. 18, 905–913. https://doi.org/10.2174/1389200218666170925124004.

Valko, M., Leibfritz, D., Moncol, J., Cronin, M.T.D., Mazur, M., Telser, J., 2007. Free radicals and antioxidants in normal physiological functions and human disease 39, 44–84. https://doi.org/10.1016/j.biocel.2006.07.001.

Vora, D.J.D., 2017. "Insight into the biochemical link between biodiversity and nutraceuticals. IOSR Journal of Environmental Science, Toxicology and Food Technology. 11, 22–25. https://doi.org/10.9790/2402-1103032225.

Wang, T., Li, Q., Bi, K., 2018. Bioactive flavonoids in medicinal plants : structure, activity and biological fate. Asian Journal of Pharmaceutical Sciences. 13, 12–23. https://doi.org/10.1016/j.ajps.2017.08.004.

Wayner, D., Burton, G., Ingold, K., Locke, S., 1985. Quantitative Measurment of the total, peroxil radical trapping antioxidant capability of human blood plasma by controlled peroxidation. FEBS Letters 187, 33–37.

Williamson, G., 2017. The role of polyphenols in modern nutrition. Nutrition Bulletin. 42, 226–235. https://doi.org/10.1111/nbu.12278.

Wu, X., Beecher, G.R., Holden, J.M., Haytowitz, D.B., Gebhardt, S.E., Prior, R.L., 2004. Lipophilic and hydrophilic antioxidant capacities of common foods in the United States. Journal of Agricultural and Food Chemistry. 52, 4026–4037. https://doi.org/10.1021/jf049696w.

Zhang, Y., Wu, S., Qin, Y., Liu, J., Liu, J., Wang, Q., Ren, F., Zhang, H., 2017. Interaction of phenolic acids and their derivatives with human serum albumin: structure affinity relationships and effects on antioxidant activity. Food Chemistry. https://doi.org/10.1016/j.foodchem.2017.07.100.

PART 2

Antidiabetic Activity of Bioactive Compounds

CHAPTER 3

Antihyperglycemic, Hypoglycemic, and Lipid-Lowering Effect of Peptide Fractions of M. pruriens L. in an Obese Rat Model

Armando M. Martín Ortega[1], Edwin Eenrique Martínez Leo[1], Juan José Acevedo Fernández[2], Maira Rubi Segura Campos[1]

[1]*Facultad de Ingeniería Química, Universidad Autónoma de Yucatán, Mérida, Mexico;* [2]*Facultad de Medicina, Universidad Autónoma del Estado de Morelos, Cuernavaca, Mexico*

3.1 Introduction

Obesity is the great epidemiological pandemic of the 21st century, in the last three decades the number of people with this disease has tripled, creating a global public health crisis. In 2014 alone, more than 2500 million adults aged 18 or over were obese, representing 13% of the world's adult population (11% men and 15% women). Currently, the statistical data of the Global Nutrition Report of 2017 report 2 billion people with obesity (Global Nutrition Reports, 2017).

Obesity is classified as a complex, chronic disease of multifactorial origin. The foregoing implies an important relationship between an environmental component and a multigene genetic one. The most recent evidence points out that, shared with other common pathologies and a complex therapeutic approach, the existence of a state of chronic low-grade inflammation that perpetuates the disease and is associated with multiple complications (Lee et al., 2016).

Malnutrition and genetic factors are the cornerstones of the development of obesity. Hypertrophy of adipose tissue, excessive consumption of saturated and/or *trans* fats and excessive consumption of simple carbohydrates, generates a state of cellular and systemic glycolipotoxicity, responsible for the production of reactive oxygen species (ROS) and inflammation mediators. The above is defined by a saturation of the oxidative pathways in mitochondria, due to a high intake of simple carbohydrates in a chronic manner, leading to an increase in free oxygen, responsible for the nonenzymatic production of ROS and decreasing ATP production. Given the oxidative alterations generated in the mitochondria, this state of toxicity is termed mitochondrial dysfunction. The collapse of the mitochondria, due to the establishment of oxidative stress, causes disturbances at the level of subcellular organelles, cellular lesions that end with the rupture of the plasma membrane and cell necrosis (Hernández et al., 2013).

Bioactive Compounds. https://doi.org/10.1016/B978-0-12-814774-0.00003-7

Necrosis is an irreversible state of the cell, where the integrity of the plasma membrane can not be maintained and there is an escape of cytoplasmic elements, denaturation of the proteins by autolysis or coming from lytic enzymes of neighboring leukocytes, since necrosis attracts the components of inflammation. The severity of the necrosis is exacerbated when the damage is in cells that do not have regenerative capacity, such as neurons, nephrons, and retinocytes, associated with the microvascular complications of diabetes mellitus (DM).

The excessive circulation of free fatty acids, mainly saturated ones, as part of lipotoxicity, contributes to the metabolic activation of cells of the immune system mediated by toll-like receptors which induces the activation of nuclear factor kappa beta (NF-κB) and with it the synthesis of proinflammatory cytokines, such as tumor necrosis factor alpha (TNF-α), interleukin-1 (IL-1), interleukin-6 (IL-6), and the chemoattractant protein of macrophages-1 (MCP-1) as well as the reduction or inhibition of the expression of antiinflammatory cytokines (Hirai et al., 2010); generating a low-grade proinflammatory state. The chronic inflammatory state, generated by obesity and glycolipotoxicity, is the main trigger of insulin resistance (Nieto-Vázquez et al., 2008). The increased circulation of free fatty acids increases the synthesis rate of triacylglycerides in the liver, thus increasing the concentration of plasma lipoproteins and contributing to the characteristic dyslipidemia in the context of the development of obesity.

Likewise, the increase in plasma concentrations of proinflammatory cytokines activates the vascular endothelium (VE), which will express adhesion molecules (ICAM/VACM) and the alteration of its metabolism in its apical membranes. The activation of the VE induces the adhesion of leukocytes and with it the generation of ROS, this act favors endothelial dysfunction, atherosclerosis, and an increase in arterial pressure. In the VE, cell signaling mediated by IRS-1 is necessary for the production of nitric oxide synthase (NOS) stimulated by insulin. Therefore, the phosphorylation of IRS-1 brings about the inhibition of synthesis of NOS, the main enzyme producing nitric oxide (NO), the regulator of vasodilation. In turn, other vasoconstrictor molecules, such as endothelin-1 (ET-1) and angiotensin-II, maintain the vascular endothelium stimulation, promoting alterations in the regulation of vascular diameter and thus blood pressure (Arce-Esquivel et al., 2013).

The above described, comprises a set of symptoms that encompass a set of metabolic disorders called metabolic syndrome (MS). This process is characterized by a disorder that starts with obesity and is constituted by an increase in visceral adipose tissue, a decrease in high-density lipoprotein concentrations (c-HDL), an elevation in triglyceride and low-density lipoprotein concentration (c-LDL), endothelial dysfunction with a sustained increase in blood pressure, and chronic hyperglycemia (Rani et al., 2016).

3.2 Beyond Conventional Feeding

At the beginning of the new millennium, a new era in the area of food science and nutrition was present with greater intensity, the important interaction between food and the physiological functions of the organism was recognized, emerging in the sideboards of supermarkets, called "functional foods" (Pelcastre et al., 2017).

The concept of functional foods was developed in Japan during the 1980s, as a necessity to reduce the high cost of health insurance that was increased by the need to provide coverage to an increasingly elderly population. Over time, components with biological activity in foods have been identified, supported by an increase in scientific evidence supporting physiological effects or health benefits. The National Academy of Science of the United States has defined functional foods as "any food or modified food ingredient" that can provide a health benefit superior to the traditional nutrients contained (Herrera et al., 2014).

Currently, functional foods can be seen in various supermarkets under a wide range of compounds that are attributed to retard or prevent disease. Companies such as Nestle, Kellogs, Nutrigem, USANA, among others, offer consumers various products for diseases such as diabetes, obesity, hypertension, and cancer (Martínez et al., 2016).

A wide variety of vegetables are highly valued for their therapeutic potential attributed to the content of components known as bioactive phytochemicals. Many of these chemical substances with biological activity are products of the secondary metabolism of the plant and are considered as non-nutritious compounds, because they do not have any energy contribution (Herrera et al., 2014). It is considered a bioactive compound to all the chemical substances of a food, which provides a health benefit beyond those considered as basic nutrition. These compounds are generally found in products of vegetable origin and in foods rich in polyunsaturated fatty acids. These components are found in foods in small quantities and contribute to their functional value, which is why world food trends focus on consuming foods that, in addition to nutritional value, contribute to the health of the human body (Díaz et al., 2008).

Considering the therapeutic importance of the bioactive components derived from vegetables and included in the daily diet, among the most studied and with proven evidence on a positive effect on health, we can highlight polyphenols, phytoestrogens, polyunsaturated fatty acids, isoprenoids, and peptides with biological activity (Drago et al., 2006)

3.3 Bioactivity of Peptides Derived From Food Proteins

Currently, proteins and peptides with biological activity constitute one of the most important categories within the functional food sector. These peptides can be generated during food processing, by hydrolysis in vitro, or during gastrointestinal digestion. However, these mechanisms are often insufficient to generate a functional effect, which is why enzymatic hydrolysis is used (Segura et al., 2013).

Protein hydrolysis or cleavage of the peptide bond can be produced by biological methods using enzymes. This process is widely used in obtaining biopeptides because it is a system that mimics the digestive process, making it an ideal way to metabolize proteins, easy to operate, process control and preserving the quality of the protein as well as its biological value (Segura et al., 2013).

Obtaining protein hydrolysates can be by chemical and enzymatic methods. Because chemical methods lead to adverse effects such as the oxidation and destruction of some amino acids and generation of toxic compounds, enzymatic methods are preferred, which minimize the formation of these compounds (Segura et al., 2009). The generation of protein hydrolysates enzymatically involves the use of enzymes that break peptide bonds, generating smaller peptides or even free amino acids (Gallegos et al., 2013).

Any source of food protein is capable of providing functional peptides and structural and physiological aspects have been isolated and characterized from those obtained from very diverse proteins such as milk, corn, soybeans, and chickpeas, by in vitro digestion, with proteolytic enzymes of vegetable, animal, and microbial origin. This is exemplified by the whey protein hydrolysate in which 18 peptides of different molecular size have been found that have antihypertensive properties. One of the first products to be marketed was the Evolus in Finland in 2000, currently produced are LH (Iceland), Kaiku Vita (Spain), and Emmi Evolus (Portugal), which are basically fermented milks by different species of lactobacilli on casein. In Japan, Calpis is produced, which also results from the hydrolysis of casein by *Aspergillus niger* proteases with hypotensive action associated with biopeptides (IPP VPP); as well as the CholesteBlock beverage marketed in Japan by KyowaHakko with hypocholesterolemic action and associated with biopeptides isolated from soy (Hartmann et al., 2007).

3.4 Peptides Derived From M. pruriens L. (Velvet Bean), as a Potential Functional Ingredient

Mexico has a wide biodiversity of plants and foods including leguminous grains, rich in proteins and amino acids, that have been used for years in human and animal feed, so they may be feasible to use as raw material for the generation of biopeptides, enhancing agricultural production and giving new opportunities for farmers (Herrera et al., 2013).

Velvet bean belongs to the genus *Mucuna*, which includes approximately 100 species of vines and shrubs that are found throughout the tropical regions of the world. *M. pruriens* L. is native to India and Southeast Asia, but is now widely distributed in the tropics. There are four botanical varieties of *M. pruriens*; *M. pruriens* var. *utilis* is the cultivated variety that does not itch, *M. pruriens* var. *pruriens* ("pica pica") has urticating hairs that contain the irritating compound mucanain, *M. pruriens* var. *hirsuta*, from India, and *M. pruriens* var. *sericophylla*, from the Philippines (Brunner et al., 2011).

M. pruriens is an annual cycle shrub or vine that can measure up to 1 m in height and 15 m in length. Its leaves are large, trifoliate, oviform or rhomboid, with purple or white flowers.

Figure 3.1
Variety of colors and patterns of *Mucuna* seeds (Brunner et al., 2011).

Their cylindrical, long, or linear pods are covered with dense orange velvety hairs and each contains seeds that may be black, white, reddish, brown, or mottled, and with a raised thread (Fig. 3.1) (Brunner et al., 2011).

Velvet bean seeds have a very homogeneous chemical composition. Its most abundant components are starch (44%–59%), protein (23%–38%), and some lipid varieties (2%–14%). Of the total content of proteins present, 70% are globulins, 10% albumins, and 20% small amounts of glutelins and prolamins (Balasubiramanian and Veerabahu, 2010).

M. pruriens L., planted in the Mexican southeast, has proven to be a good source of protein (23%–38%) and therefore of peptides with biological activity through enzymatic hydrolysis, resulting in an alternative in the treatment and/or prevention of various metabolic diseases associated with overweight and obesity (Segura et al., 2013).

Studies carried out with protein hydrolysates of *M. pruriens* L. using enzymes such as pepsin, pancreatin, alcalase, and flavourzyme, have shown several important biological activities, among which are hypotensive, lipid-lowering, antioxidant, and antithrombotic effects (Cruz et al., 2010; Ganem et al., 2011; Herrera et al., 2014).

Studies conducted by Galicia et al. (2013) and Herrera et al. (2014) demonstrate that the protein hydrolysate and peptide fractions generated from *M. pruriens* L. are pharmacologically active, with the ability to lower blood pressure. These results propose the incorporation of these bioactive compounds to certain foods as an alternative to the treatment of hypertension.

Herrera et al. (2014) also reported an inhibitory effect of platelet aggregation of 33.33% and 31.72% of peptide fractions of *M. pruriens* L. >10 kDa of hydrolysates of Alcalase-Flavourzyme and Pepsin-Pancreatin, respectively, which they correspond to a moderate antithrombotic activity and similar to that presented by drugs such as aspirin and indometacin (Herrera et al., 2014).

In vivo studies also report an important hypotensive activity. Evaluating the hypotensive effect in normotensive rats, it was found that the highest percentage of blood pressure reduction is obtained from hydrolysates of *M. pruriens* L. with a Pepsin-Pancreatin system at a dose of 5 mg/kg administered intraperitoneally (Herrera et al., 2014).

Likewise, the lipid-lowering effect has also been evaluated and it was found that the peptide fractions <1 kDa in concentrations of 10 mg/kg peritoneally, had greater hypocholesterolemic and hypertriglyceridemic effects, even comparable with those released by pravastatin in the same doses, in diabetic rats induced with alloxan (Herrera et al., 2014).

The various studies carried out on the hydrolysates and peptide fractions of *M. pruriens*, show their biological potential on health and are the scientific basis for the development of more information about their role in biological media, with the aim of developing functional foods, incorporating these peptides in specific products, and thus offering to the public products that provide an additional benefit to their health.

The aim of this chapter is to evaluate the in vivo antihyperglycemic, hypoglycemic, and lipid-lowering potential effect of peptide fractions (>10 kDa and 1–3 kDa) of *M. pruriens* L.

3.5 Materials and Methods

3.5.1 Grains

The grains of *M. pruriens* L. (velvet bean) were obtained from the February 2014 harvest, from Xmaben, Hopelchén, Campeche (latitude 19.24°N, longitude 89.38°W).

3.5.2 Reactives

The experimental analysis was developed with reagents of registered trademarks, such as Sigma and J.T. Baker. The hydrolysis system was carried out with enzymes acquired from Sigma laboratories.

3.5.3 Obtaining Flour From M. pruriens L.

The *Mucuna* grains were cleaned manually; the best ones were selected and the largest quantity of impurities present was eliminated. Subsequently, they were crushed in a disk mill and the shell removed manually. Finally, a finer milling was carried out in a Cyclotec 1093 mill (TecatorSweden) to obtain flour with a smaller particle size (80 mesh).

3.5.4 Obtaining the Protein Concentrate of M. pruriens L.

The method reported by Herrera et al. (2014) was used, which consisted of suspending 1 kg of *Mucuna* flour in 3% sodium bisulfite in a ratio of 1:10 p/v, the resulting suspension was adjusted to pH 11 with 1 N NaOH and kept under stirring for 1 h. Subsequently, the suspension was filtered through an 80 mesh, separating the fiber from the liquid part with protein and starch. The solid residue was washed five times with 200 mL of the bisulfite solution, which was filtered again through an 80 mesh and the supernatant was bound with the previous one. The fiber residue was discarded and the supernatant was passed through a 100 mesh to remove the finer fiber. The suspension was allowed to stand until complete sedimentation of the starch (30 min) and the solubilized protein present in the supernatant was separated. The protein-rich supernatant was adjusted to pH 4.6 with 1 N HCl and centrifuged at 700 × g for 20 min. Finally, the precipitate was dried at −47°C in a lyophilizer, thus obtaining the protein fraction of *M. pruriens* L.

3.5.5 Enzymatic Hydrolysis

Protein hydrolysis was carried out using the method reported by Herrera et al. (2014). An enzymatic system Pepsin -Pancreatin was used sequentially for 90 min, as this has been shown to be an effective amount of time for obtaining peptides with biological activity. The hydrolysis parameters of the enzymatic system were: substrate concentration 4%, substrate enzyme ratio 1/10 p/v, temperature 37°C, pH 2 for Pepsin and pH 7.5 for Pancreatin. The reaction time was 90 min. During the first 45 min, the hydrolysis was carried out with the Pepsin suspension at the aforementioned hydrolysis conditions; followed by hydrolysis with Pancreatina for the second 45 min.

The hydrolysis was stopped by placing the samples in a water bath at 80°C for 20 min and finally centrifuged at 11,200 × g for 20 min in order to obtain the soluble portion. Since the process of enzymatic hydrolysis has been shown to reduce the amount of antinutrient compounds, it is not necessary to previously cook *M. pruriens* L, since temperatures and times similar to cooking are handled.

3.5.6 Fractionation by Ultrafiltration

The soluble fraction of the enzymatic hydrolysate was subjected to fractionation by ultrafiltration (UF) according to the methodology proposed by Cho et al. (2004) using three membranes with different molecular weight cuts: 1, 3, and 10 kDa. The soluble fractions of the hydrolysates were passed through each of the membranes, beginning with the largest pore (10 kDa), collecting separately the retentate and permeate thereof. The permeate of the 10 kDa membrane was again filtered using the 3 kDa membrane and, in the same way, the permeate of the 3 kDa membrane was passed through the 1 kDa membrane. The result of the ultrafiltration was three peptide fractions with different molecular weights: (1) F > 10 kDa, (2) F 3–10 kDa, and (3) F 1–3 kDa. For the purpose of the study, only two peptide fractions of the three obtained were used, the one with the highest molecular weight (F > 10 kDa) and the one with the lowest molecular weight (F 1–3 kDa), taking into consideration that the peptides of lower molecular weight usually exhibit higher biological activity and those of higher molecular weight are more susceptible to gastrointestinal digestion, presenting less activity.

Once the peptide fractions were obtained, these and the hydrolysate were lyophilized, to later suspend in aqueous solution and obtain a 50 mg/mL stock concentration, from which the working dilutions for the determination of biological activities were derived.

3.5.7 Animals

Obese Wistar rats aged 7 months old (600–800 g) were used in this study. The induction to obesity was carried out using a 20% solution of sucrose, instead of water, from 2 months of age. For the study, five groups n = 5 rats/group were formed. Two groups for peptide fractions: FP > 10 kDa and FP 1–3 kDa. There were also a basal group of healthy nonobese rats, weighing between 200 and 300 g, a negative control (C−) (obese), and a positive control group (C+) (obese). The positive control group was used exclusively during the glucose tolerance test. The model was performed for 4 weeks with daily oral administration of 10 mg/kg of each treatment to the corresponding groups. The basal and negative control groups did not receive treatment with any fraction, but they were administered the treatment vehicle.

The distribution of the animals to the groups was carried out randomly and obtained from the bioterium of the Faculty of Medicine of the Autonomous University of the State of Morelos.

3.5.8 Preparation and Administration of Treatments

The preparation of the treatments was carried out daily, using distilled water as a solvent. For the calculation of the doses, weights were taken weekly. The corresponding dose for each rat was dissolved in 1 mL of solvent. The administration of the treatments was carried out with a cannula for the feeding of rats of the Orchid brand, model 22G. The treatments were administered in single doses in the morning.

The experimental protocol was carried out in accordance with the Official Mexican Standard (NOM-062-ZOO-1999), "Technical Specifications for the Care and Use of Laboratory Animals," as well as all federal and institutional regulations.

3.5.9 Antihyperglycemic, Hypoglycemic, and Lipid-Lowering Activity

The antihyperglycemic effect was evaluated by the glucose tolerance curve (GTC), before starting the treatment for 4 weeks. The rats were subjected to 8 h of fasting, prior to the performance of the test. One g/kg of starch plus 5 mg/kg of the respective treatment of each group were administered. For this test a positive control group was used which, in addition to the starch, was administered with 0.5 mg/kg of acarbose (inhibitor of intestinal alpha glucosidase). The glycemia was evaluated at 0, 15, 30, 45, 60, and 120 minutes.

Fasting blood glucose (FBG) and triacylglycerides (TAG) were determined in blood taken from the tip of the tail with the portable glucose analyzer Optimum Neo by Freestyle, and lipid analysis with the cholesterol analyzer from Mission. The biochemical parameters were taken weekly.

3.5.10 Statistic Analysis

The analysis of the data was made by descriptive statistics by the calculation of mean and standard deviation. The statistical difference was analyzed by means of a parametric Student's *t* test between the study and negative control groups with $P<.05$ in the GraphPad Prism 7.0 program.

3.6 Results and Discussion

3.6.1 Evaluation of Antihyperglycemic Activity

To carry out the GTC, a positive control group was added, which, in addition to the starch, received acarbose, an alpha glucosidase inhibitor drug. Acarbose is a drug used for the treatment of diabetes mellitus as it reduces the digestion of dietary carbohydrates, thus reducing the amount and speed of absorption of this nutrient, maintaining controlled levels of glycemia (DiNicolantonio et al., 2015).

The results of the GTC are shown in Fig. 3.2. The results of the glycemias of the groups are presented below with an order of initial blood glucose and final glycemia, respectively: negative control 84.6±3.21–94.2±7.98 mg/dL; positive control 95.8±9.73–92.2±6.94 mg/dL; F>10 kDa 98.4±11.30–115.8±24.23 mg/dL; F 1–3 kDa 95.8±15.9–114±22.84 mg/dL. The C+ had a reduction of 2.12% of the final glycemia compared to the C−. The groups F>10 kDa and

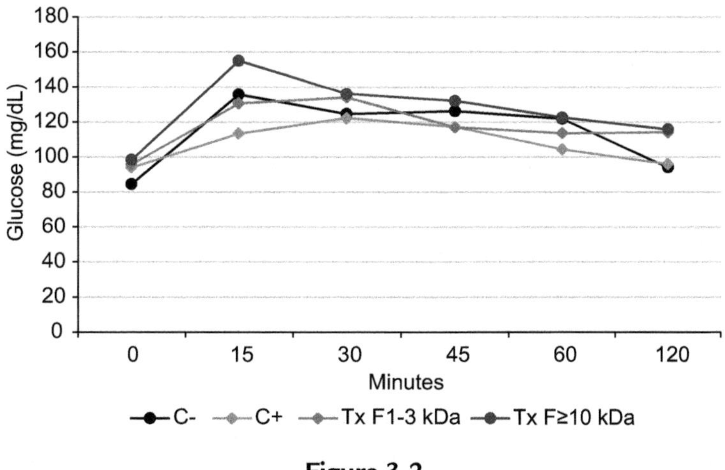

Figure 3.2
Glucose tolerance curve in Wistar rats, after treatment with peptide fractions of *M. pruriens* L.

F1-3 kDa show a final glycemia higher by 22.92% and 21.01%, respectively, compared to C–. According to the previous results, the groups F > 10 kDa and F 1–3 kDa did not show an antihyperglycemic effect at the end of the study, with respect to the negative control.

In comparison with the C+, the treatment groups did not show a reduction of the final glycemia, on the contrary, they showed a significant elevation of the final glycemia. The elevation of the final glycemia in the treatment groups is likely to be related to its protein composition, being consistent with the results of human studies where the consumption of carbohydrates, together with protein sources, raises the postprandial glycemia in greater proportion compared to carbohydrate-only consumption (Smart et al., 2013; Paterson et al., 2015).

3.6.2 Evaluation of Hypoglycemic Activity

The fasting blood glucose (Fig. 3.3) of the basal groups, C–, Tx F 1–3 kDa and Tx F > 10 kDa exhibited levels of 82 ± 4.55, 110.4 ± 16.99, 88.6 ± 5.18, and 103.2 ± 14.41 mg/dL, respectively. After the 4-week treatment, the basal group, Tx F1-3 kDa and Tx F > 10 kDa showed an increase in FBG by 10.67%, 13.09%, and 10.27%, respectively, being statistically nonsignificant changes. On the other hand, the C– group showed a nonsignificant reduction in FBG by 5.43%.

To analyze the previous results, it must be taken into account that any experimental manipulation of animals exerts a certain level of stress on them. According to Rostamkhani et al. (2012) experimental manipulation of murine models explains the elevation of FBG in rats of the basal group (nonobese), due to an increase in cortisol levels, which increases the mobilization of fatty acids and induces the expression of gluconeogenic enzymes, thus generating an increase in the levels of TAG and glycemia. However, the obesity present in the other

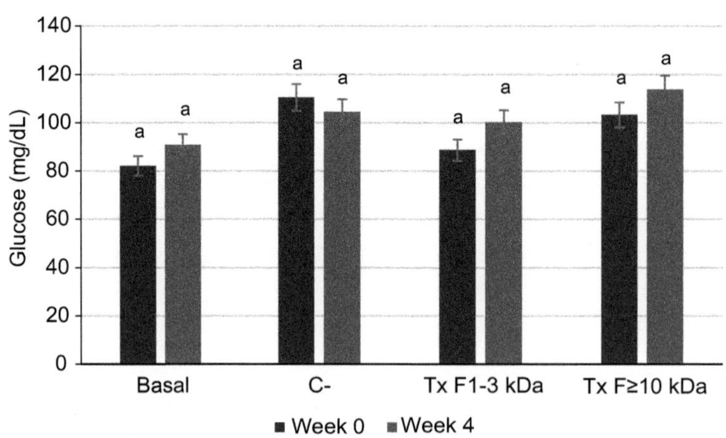

Figure 3.3
Fasting blood glucose levels in Wistar rats, after treatment with peptide fractions of *M. pruriens* L. Results expressed as mean ± SD. Different letters indicate statistical difference between week 0 and week 4 of treatment ($P < .05$).

experimental groups conditions them to have previous neuroendocrine alterations (increase in basal levels of cortisol and reduction in the expression of corticoid receptors) (Mattsson et al., 2003) that modify their response to chronic stress by manipulation. Thus, according to Aslani et al. (2015), the metabolic state of rats modifies their response to stress, being higher in nonobese rats (normal feeding) and lower in obese rats (hyperenergic feeding), which explains the elevation of the FBG in the basal group and that the FBG in C– are not altered and even showed a reduction.

Moreover, the FBG increase in the treatment group, as in the GTC, is related to the protein composition of the treatments. The ingestion of high-protein meals (in this case the treatments) has been shown to increase circulating plasma glucagon (pancreatic hormone), which stimulates the release of glucose by the liver (Paterson et al., 2015), thereby altering the FBG in the treatments group, but not in the C–. These results suggest that there is gastrointestinal digestion of the peptide fractions.

The hypoglycemic and antihyperglycemic properties attributed to the peptides are mainly due to their ability to inhibit the pancreatic enzyme α-amylase and the intestinal enzyme α-glucosidase, responsible for the digestion of carbohydrates. However, according to Estrada-Salas et al. (2014), some peptides such as those derived from the seeds of *Phalaris canriensis* L. have been shown to have inhibitory effects of the enzyme dipeptidyl peptidase IV (DPP-IV), an enzyme whose function is to hydrolyze the intestinal hormone peptide similar to glucagon type-1 (GLP-1), incretin with an important function in glucose-mediated insulin secretion (Tasyurek et al., 2014). The inhibition of DPP-IV reduces the degradation of GLP-1, maintaining its antidiabetic effects for longer.

Likewise, the peptides derived from the fungus *Aspergillus awamori* and from the seeds of *Capsicum annuum*, in in vitro studies, have been shown to have an inhibitory capacity of α-amylase, in 81% and 100%, respectively (Singh and Kaur, 2016; Vieira Bard et al., 2015).

The lack of antihyperglycemic and hypoglycemic effects of the peptide fractions, is likely to be related to its rapid degradation. In the other hand, a study by Uenishi et al. (2012), evaluated the antihyperglycemic effect in rats, of peptides derived from the casein of gouda cheese, finding a significant effect from 30 min ($P < .01$) until the end of the glucose tolerance test, in comparison of the control group. The octapeptide LPQNIPPL evaluated by Uenishi et al. (2012) presented the inhibitory capacity of DPP-IV in vitro, and according to the results, it is likely that this peptide is resistant to gastrointestinal digestion, which is why it is necessary to evaluate the capacity of *Mucuna pruriens* L. peptide fractions to inhibit the DPP-IV enzyme in the future, then it would give the guideline that the use of the peptides by the parenteral route could demonstrate pharmacological effects.

It's likely that the amino acid composition of these peptide fractions make them more susceptible to the gastrointestinal digestion of the rat. However, little is known about the bioavailability of the peptides. In this regard, Polanco-Lugo et al. (2014) evaluated the bioactivity and functional properties of peptides derived from the protein of *Phaseolus lunatus*, finding that protein isolates with a higher degree of hydrolysis and of lower molecular weight had greater biological activity (antioxidant and inhibitor of the angiotensin-converting enzyme), however, the gastrointestinal digestion resistance and absorption are not completely elucidated.

3.6.3 Evaluation of Lipid-Lowering Activity

The baseline group at the beginning of the study presented TAG values of 67 ± 15.01 mg/dL, while in the C− and treatments F 1–3 kDa and F ≥ 10 kDa, they showed values of 147.6 ± 66.75, 199 ± 109.94, and 161 ± 41.03 mg/dL, respectively, at week 0. After 4 weeks of treatment, a significant decrease in TAG was observed for F ≥ 10 kDa (17.08%), but not for F 1–3 kDa, which even increased by 35% (Fig. 3.4).

As mentioned above, the experimental manipulation of murine models explains the elevation of TAG in rats of the basal group by increasing the cortisol blood level (Rostamkhani et al., 2012).

The lipid-lowering effects exhibited by F ≥ 10 kDa are comparable with other peptides of vegetable origin, such as soybean (Ruiz et al., 2014) and velvet bean (Herrera et al., 2016). For their part, these authors point out as possible mechanisms of lipid-lowering action by which the peptides could exert biological effect are: (1) reduction of the expression of genes that regulate lipogenesis; (2) increase in the activity of TAGs and lipoprotein lipases; and (3) inhibition of pancreatic lipase. Thus, the peptide fractions ≥10 kDa of *M. pruriens* L. present a lipid-lowering potential.

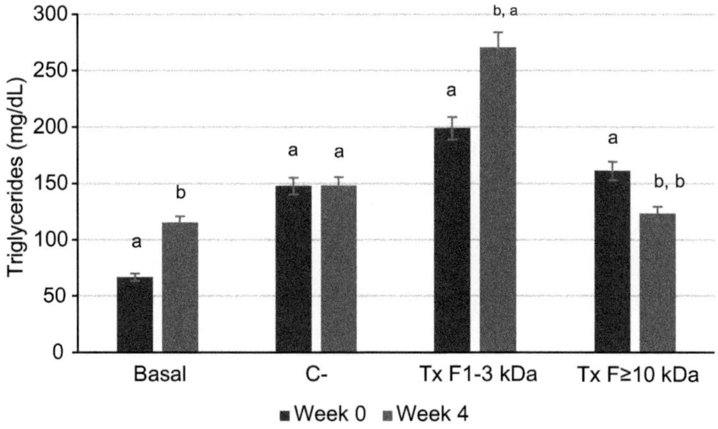

Figure 3.4
Triglyceride levels in Wistar rats, after treatment with peptide fractions of *M. pruriens* L. Results expressed as mean ± SD. Different letters indicate statistical difference between week 0 and week 4 of treatment ($P < .05$).

The increase in TAG in the F 1–3 kDa group is likely to be related to the elevation of their FBG, being consistent with studies conducted in humans where a correlation is found between the alterations in FBG levels and an alteration of the blood lipids (Zhang et al., 2009).

3.7 Conclusion

The peptide fraction >10 kDa showed a significant lipid-lowering activity (reduction of 17.08%) after 4 weeks of treatment, but did not present antihyperglycemic or hypoglycemic activities and even showed an increase in FBG (10.27%). It's likely that the lipid-lowering activity of this peptide fraction is related to the inhibition of pancreatic lipase, since its hyperglycemic effect shows that it is digested. Thus, it is possible that the gastrointestinal digestion of this peptide fraction of higher molecular weight results in bioactive peptide inhibitors of pancreatic lipase.

The peptide fraction 1–3 kDa did not present lipid-lowering, antihyperglycemic, or hypoglycaemic activities. However, the group that was treated with this fraction, after 4 weeks of treatment, had an elevation of 35% and 13.09% in their TAG and FBG levels, respectively. Probably, the elevation of the TAG is related to the elevation of FBG, which indicates a greater digestion and consequent absorption of this peptide fraction.

The biological activities presented by the peptide fractions derived from *M. pruriens* L. demonstrate that the peptides of dietary origin retain part of their pharmacological properties after gastrointestinal digestion. Thus, further pharmacological studies are needed to elucidate the mechanisms of action and bioavailability of the peptide fractions, in order to know their potential effectiveness in the treatment of obesity and its comorbidities.

Acknowledgments

This research was part of the project "Scientific research aimed at the development of protein derivatives of *Mucuna pruriens* with potential biological activity for the prevention and/or treatment of chronic diseases associated with overweight and obesity," funded by the National Council of Science and Technology (CONACYT) with number 154307.

References

Arce-Esquivel, A., Bunker, A., Mikus, C., Laughlin, M., 2013. In: Kazuko Masuo (Ed.), Insulin Resistance and Endothelial Dysfunction: Macro and Microangiopathy, Type 2 Diabetes. InTech. https://doi.org/10.5772/56475. Available from: https://www.intechopen.com/books/type-2-diabetes/insulin-resistance-and-endothelial-dysfunction-macro-and-microangiopathy.

Aslani, S., Vieira, N., Marques, F., Costa, P.S., Sousa, N., Palha, J.A., 2015. The effect of high-fat diet on rat's mood, feeding behavior and response to stress. Translational Psychiatry 5 (11), e684. http://doi.org/10.1038/tp.2015.178.

Balasubiramanian, K., Veerabahu, R., 2010. Nutritional and anti-nutritional potential of three accessions of itching bean (*Mucuna pruriens* (L.) DC var. pruriens): an under-utilized tribal pulse. International Journal of Food Sciences and Nutrition 61 (5), 497–511.

Brunner, B., Beaver, J., Flores, L., 2011. Mucuna. Proyecto de agricultura orgánica. Departamento de cultivos y ciencias agroambientales.

Cho, M.J., Unklesbay, N., Hsieh, F., Clarke, A.D., 2004. Hydrophobicity of bitter peptides from soy protein hydrolysates. Journal of Agricultural and Food Chemistry 52 (19), 5895–5901.

Cruz, G., Castellano, A., Rosado, G., Chel, L., 2010. Capacidad antioxidante de hidrolizados enzimáticos de la proteína de frijol terciopelo *Mucuna pruriens*. Revista de la Facultad de Ingeniería Química 50, 17–25.

Development Initiatives, 2017. Global Nutrition Report 2017: Nourishing the SDGs. Development Initiatives, Bristol, UK.

DiNicolantonio, J.J., Bhutani, J., O'Keefe, J.H., 2015. Acarbose: safe and effective for lowering postprandial hyperglycaemia and improving cardiovascular outcomes. Open Heart 2 (1), e000327. http://doi.org/10.1136/openhrt-2015-000327.

Drago, M., López, M., Saínz, T., 2006. Componentes bioactivos de alimentos funcionales de origen vegetal. Revistas científicas de América Latina 37, 58–68.

Estrada-Salas, P.A., Montero-Morán, G.M., Martínez-Cuevas, P.P., González, C., Barba De La Rosa, A.P., 2014. Characterization of antidiabetic and antihypertensive properties of canary seed (*Phalaris canariensis* L.) peptides. Journal of Agricultural and Food Chemistry 62 (2), 427–433. https://doi.org/10.1021/jf404539y.

Galicia Martínez, S., Torruco Uco, J., Negrete León, E., Cadena Pino, M.L., Acevedo Fernández, J.J., Angeles Chimal, J.S., Santa-Olalla Tapia, J., Petricevich López, V.L., 2013. Actividad de los hidrolizados proteínicos de *Mucuna pruriens* en modelos *in vivo* que revierten enfermedades incluidas dentro del síndrome metabólico. In: Maria Segura-Campos, Luis Chel Guerrero, David Betancur Ancona (Eds.), Bioactividad de péptidos derivados de proteínas alimentarias. OmniaScience, Barcelona, pp. 155–173.

Gallegos, S., Chel, L., Corzo, L., Martínez, L., 2013. Péptidos con actividad antioxidante de proteínas vegetales. In: En Segura, M., Chel, L., Betancur, D. (Eds.), Bioactividad de péptidos derivados de proteínas alimentarias. OmniaScience, Barcelona, pp. 111–122.

Ganem, J., Pérez, E., Lima, J., Mary, L., Cándido, B., 2011. In vitro evaluation of *Mucuna pruriens* (L.) DC. antioxidant activity. Brazilian Journal of Pharmaceutical Sciences 47, 535–544.

Hartmann, R., Wal, J.M., Bernard, H., Pentzien, A.K., 2007. Cytotoxic and allergenic potential of bioactive proteins and peptides. Current Pharmaceutical Desing 13, 897–920.

Hernández, A., Rull, A., Rodríguez, E., Riera, M., Luciano, F., Camps, J., et al., 2013. Mitochondrial dysfunction: a basic mechanism in inflammation-related non-communicable diseases and therapeutic opportunities. Mediators of Inflammation 1–13. https://doi.org/10.1155/2013/135698.

Herrera, F., Betancur, D., Segura, M., 2013. Compuestos bioactivos de la dieta con potential en la prevención de patologías relacionadas con sobrepeso y obesidad: péptidos biológicamente activos. Nutrición Hospitalaria 29, 10–20.

Herrera, F., Ruiz, J., Acevedo, J., Betancur, Segura, M., 2014. ACE inhibitory, hypotensive and antioxidant peptide fractions from Mucuna pruriens proteins. Process Biochemistry 49, 1691–1698.

Herrera, F., Ruiz, J., Santaolalla, J., Acevedo, J., Betancur, D., Segura, M., 2016. The hypolipidemic effect and antithrombotic activity of *Mucuna pruriens* protein hydrolysates. Food Function 7, 434–444.

Hirai, S., Takahashi, N., Goto, T., Lin, S., Uemura, T., Yu, R., Kawada, T., 2010. Functional food targeting the regulation of obesity-induced inflammatory responses and pathologies. Mediators of Inflammation 367838, 8.

Lee, J.J., Pedley, A., Hoffmann, U., Massaro, J.M., Fox, C.S., 2016. Associations of changes in abdominal fat quantity and quality with incident cardiovascular disease risk factors. Journal of the American College of Cardiology 68, 1509–1521.

Martínez, E., Acevedo, J., Segura, M., 2016. Biopeptides with antioxidant and anti-inflammatory potential in the prevention and treatment of diabesity disease. Biomedicine and Pharmacotherapy 83, 816–826.

Mattsson, C., Lai, M., Noble, J., McKinney, E., Yau, J.L., Seckl, J.R., Walker, B.R., 2003. Obese zucker rats have reduced mineralocorticoid receptor and 11β-hydroxysteroid dehydrogenase type 1 expression in Hippocampus—implications for dysregulation of the hypothalamic-pituitary-adrenal Axis in obesity. Endocrinology 144 (7), 2997–3003. https://doi.org/10.1210/en.2002-221015.

Nieto-Vazquez, I., Fernández-Veledo, S., Krämer, D.K., Vila-Bedmar, R., Garcia-Guerra, L., Lorenzo, M., 2008. Insulin resistance associated to obesity: the link TNF-alpha. Archives of Physiology and Biochemistry 114 (3), 183–194. https://doi.org/10.1080/13813450802181047.

Paterson, M., Bell, K.J., O'Connell, S.M., Smart, C.E., Shafat, A., King, B., 2015. The role of dietary protein and fat in glycaemic control in type 1 diabetes: implications for intensive diabetes management. Current Diabetes Reports 15 (9), 61. http://doi.org/10.1007/s11892-015-0630-5.

Pelcastre, D., Martínez, E., Segura, M., 2017. Functional and Biological Potential of Bioactive Compounds in Foods for the Dietary Treatment of Type 2 Diabetes Mellitus. InTech. ISBN: 978-953-51-3440-4.

Polanco-Lugo, E., Dávila-Ortiz, G., Betancur-Ancona, D.A., Chel-Guerrero, L.A., 2014. Effects of sequential enzymatic hydrolysis on structural, bioactive and functional properties of Phaseolus lunatus protein isolate. Food Science and Technology 34 (3), 441–448. Epub September 09, 2014 https://dx.doi.org/10.1590/1678-457x.6349.

Rani, V., Deep, G., Singh, R., Palle, K., Yadav, U., 2016. Oxidative stress and metabolic disorders: pathogenesis and therapeutic strategies. Life Sciences 148 (1), 183. https://doi.org/10.1016/j.lfs.2016.02.002.

Rostamkhani, F., Zardooz, H., Zahediasl, S., Farrokhi, B., 2012. Comparison of the effects of acute and chronic psychological stress on metabolic features in rats. Journal of Zhejiang University–Science B 13 (11), 904–912. https://doi.org/10.1631/jzus.B1100383.

Ruiz Ruiz, J.C., Betancur Ancona, D.A., Segura Campos, M.R., 2014. Bioactive vegetable proteins and peptides in lipid-lowering; nutraceutical potential. Nutrici6n Hospitalaria 29 (4), 776–784. https://doi.org/10.3305/nh.2014.29.4.7208.

Segura, M., Chel, L., Betancur, D., 2009. Péptidos bioactivos como promotores de la salud. Industria Alimentaria 31, 10–15.

Segura, M., Chel, L., Betancur, D., 2013. Bioactividad de péptidos derivados de proteínas alimentarias. OmniaScience, Barcelona.

Singh, B., Kaur, A., 2016. Antidiabetic potential of a peptide isolated from an endophytic *Aspergillus awamori*. Journal of Applied Microbiology 120 (2), 301–311. https://doi.org/10.1111/jam.12998.

Smart, C.E.M., Evans, M., O'Connell, S.M., McElduff, P., Lopez, P.E., Jones, T.W., et al., 2013. Both dietary protein and fat increase postprandial glucose excursions in children with type 1 diabetes, and the effect is additive. Diabetes Care 36 (12), 3897–3902. http://doi.org/10.2337/dc13-1195.

Tasyurek, H.M., Altunbas, H.A., Balci, M.K., Sanlioglu, S., 2014. Incretins: their physiology and application in the treatment of diabetes mellitus. Diabetes Metabolism Research and Reviews 32 (30), 13–23. https://doi.org/10.1002/dmrr.

Uenishi, H., Kabuki, T., Seto, Y., Serizawa, A., Nakajima, H., 2012. Isolation and identification of casein-derived dipeptidyl-peptidase 4 (DPP-4)-inhibitory peptide LPQNIPPL from gouda-type cheese and its effect on plasma glucose in rats. International Dairy Journal 22 (1), 24–30. https://doi.org/10.1016/j.idairyj.2011.08.002.

Vieira Bard, G.C., Nascimento, V.V., Ribeiro, S.F.F., Rodrigues, R., Perales, J., Teixeira-Ferreira, A., Gomes, V.M., 2015. Characterization of peptides from *Capsicum annuum* hybrid seeds with inhibitory activity against α-amylase, serine proteinases and fungi. The Protein Journal 34 (2), 122–129. https://doi.org/10.1007/s10930-015-9604-3.

Zhang, L., Qiao, Q., Tuomilehto, J., Hammar, N., Janus, E.D., Söderberg, S., et al., 2009. Blood lipid levels in relation to glucose status in seven populations of Asian origin without a prior history of diabetes: the DECODA study. Diabetes/Metabolism Research and Reviews 25, 549–557. https://doi.org/10.1002/dmrr.

PART 3

Anti-inflammatory Effect of Medicinal Plants

CHAPTER 4

Protein Derivatives From Commercial Grains and Their Antiinflammatory Activity

Ivan Chan Zapata[1], Víctor Ermilo Arana Argáez[2], Maira Rubi Segura Campos[1]
[1]*Facultad de Ingeniería Química, Universidad Autónoma de Yucatán, Mérida, Mexico;*
[2]*Facultad de Química, Universidad Autónoma de Yucatán, Mérida, Mexico*

4.1 Introduction

Inflammation is a protective response to ensure removal of detrimental stimuli and a healing process for repairing damaged body tissues (Osamu and Shizuo, 2010). Many years ago, the essential signs and symptoms of inflammation were described as redness, heat, swelling, and pain. Nowadays, it has been described that most of the signs and symptoms of inflammation are caused by changes in the local vasculature of the affected tissue, leading to vasodilation, erythema, and heat (Rock et al., 2010).

The function of inflammation may be either acute or chronic. Acute inflammation constitutes a primary defense against infections and a great stimulus in the healing process (Laveti et al., 2013). Also, acute inflammation encompasses the immediate and early immune responses to an antigen and is quickly resolved. On the other hand, chronic inflammation results when the injurious agent persists and is also associated with a variety of cardiovascular, metabolic, and neurodegenerative diseases (Ashley et al., 2012).

4.2 Mechanism of the Inflammatory Process

The inflammatory response is a key participant in the great majority of human diseases due to proinflammatory mediators produced either from endogenous leukocytes (macrophages, dendritic cells, lymphocytes, and others) or from tissue cells (Janssen and Henson, 2012). The inflammatory process starts with the recognition of the infectious or damaging agent. This is achieved by the detection of pathogen-associated molecular patterns (PAMPs), which are molecules expressed by pathogens that are essential for survival of those, as well as damage-associated molecular patterns (DAMPs), which are endogenous molecules that signal tissue damage or necrosis (Ashley et al., 2012). These structures are recognized through a limited number of germ line-encoded pattern recognition receptors, such as transmembrane Toll-like receptors (TLRs) and cytosolic nucleotide-binding oligomerization domain (NOD)-like receptors (NLRs) (Mogensen, 2009).

Bioactive Compounds. https://doi.org/10.1016/B978-0-12-814774-0.00004-9

Once recognition of ligands occurs, TLRs activate a MyD88-dependent signal transduction pathway that culminates in the activation of the nuclear factor κB (NF-κB). Upon transduction of the signal, NF-κB is released from inhibitory κB protein (IκB) and translocates to the nucleus, where transcription and translation of genes induce expression of proinflammatory cytokines, such as interleukin 1β (IL-1β), interleukin 6 (IL-6), tumor necrosis factor α (TNF-α), and others. Also, NLRs signal the inflammasome, which activates caspase-1 and converts some cytokines (mainly IL-1β) into active forms, eliciting inflammation after being released from the cell (Ashley et al., 2012). Termination of the transcriptional activity of NF-κB is achieved by inhibitors of IκB family, mainly IκBα. In acute inflammation, newly synthesized IκBα enters the nucleus, removes NF-κB from the deoxyribonucleic acid (DNA), and relocates it to the cytosol, ending the NF-κB activity. Nevertheless, in chronic inflammation, the persistent presence of NF-κB-activating stimuli overcomes the inhibitory feedback circuits, leading to elevated activity of NF-κB (Hoesel and Schmid, 2013).

Upon antigen recognition, a variety of cytokines and chemokines are secreted by tissue cells. These factors immediately trigger a local increase of blood flow, capillary permeability, and recruit additional circulating leukocytes via extravasation, such as monocytes that differentiate into macrophages at the site of inflammation (Lawrence and Fong, 2010).

Once in the site of inflammation, leukocytes exert their inflammatory mechanisms. Phagocytosis is an important process in the immune response and requires two components: particle internalization and phagosomal maturation. Professional phagocytes can engulf large particles (≥0.5 μm) in a vacuole or phagosome, which maturates until the formation of phagolysosome (Flannagan et al., 2009). Macrophages produce reactive oxygen species (ROS) within phagolysosomes, contributing to the progression of inflammation. The major ROS include hydrogen peroxide (H_2O_2), superoxide anions ($O_2^{\cdot -}$), hydroxyl anions (OH^-), hydroxyl radicals ($OH^{\cdot}$), and hypochlorous acid (HOCl), metabolites that possess strong oxidizing capabilities (Mittal et al., 2014). Those cells can synthesize reactive nitrogen species (RNS), mainly nitric oxide (NO) and other molecules, such as peroxynitrite ($ONOO^-$), peroxynitrous acid (ONOOH), and nitrogen dioxide radicals ($NO_2^{\cdot -}$) (Lugrin et al., 2014).

Differentiated macrophages also secrete a variety of proinflammatory cytokines, which activate defense mechanisms and help to drive the inflammatory response (Wynn et al., 2013). IL-1β stimulates the production of acute-phase proteins and increases the expression of cell adhesion molecules on leukocytes and endothelial cells. IL-6 promotes the recruitment of monocytes to the inflammation site, differentiation of B lymphocytes into plasma cells, activation of cytotoxic T lymphocytes, and also leads to the production of acute-phase proteins. Finally, TNF-α triggers the expression of chemokines and augments the expression of cell adhesion molecules, as well as inducing vasodilation and loss of vascular permeability that help the recruitment of leukocytes to the inflammation site (Duque and Descoteaux, 2014). The mechanism in the inflammatory process is summarized in Fig. 4.1.

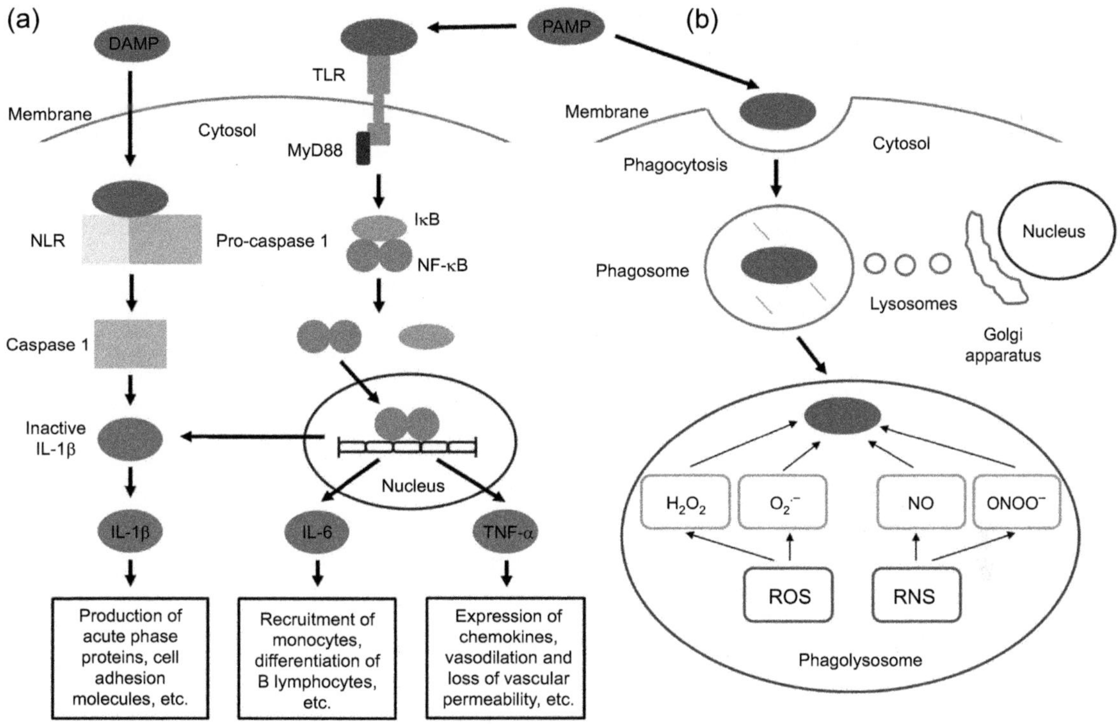

Figure 4.1
Inflammatory response. (a) The recognition of antigen leads the production of proinflammatory cytokines. (b) The synthesis of ROS and RNS collaborate in the progression of inflammation.

4.3 Antiinflammatory Treatments

The detection of transcription factors (NF-κB and others) and inflammatory markers (IL-1β, IL-6, and TNF-α, for example) has laid important molecular bases on the role of inflammation in chronic diseases. This has marked a possible way for the development of new therapeutic treatments using synthetic and natural compounds that might eventually decrease the prevalence of these diseases (Laveti et al., 2013).

Glucocorticoids have been used widely to treat most autoimmune diseases, as they enter cells, bind to cytoplasmic receptors, and translocate to the nucleus, being recognized by specific DNA sequences. As a result, the suppression of NF-κB and other transcription factors is carried out, avoiding the expression of genes encoding proinflammatory cytokines (Dinarello, 2010). Moreover, multiple strategies have been developed to prevent and ameliorate the harmful side effects of the proinflammatory cytokines. For example, anticytokine antibodies that can inhibit the binding of the cytokines to their receptors and prevent them from binding to the corresponding receptors on the cell surface (Venkatesha et al., 2015). Yet another treatment may be therapeutic administration of lipid mediators implicated in the inflammation resolution process (for example, resolvin E1 and neuroprotectin D1), particularly in view of evidence that some chronic inflammatory diseases lower the levels or actions of these molecules (Tabas and Glass, 2013).

Nevertheless, the clinical use of antiinflammatory therapies suffers from the disadvantage of side effects and high cost of treatment. Recent researches have isolated many compounds with great structural diversity that is not commonly seen in synthetic compounds and have evaluated the antiinflammatory activity of these natural products on particular targets like NO, NF-κB, cytokines, chemokines, and others. Because of this, natural products have become an alternative to such therapies (Gautam and Jachak, 2009). In this sense, functional foods have provided an opportunity to improve health, since they contain numerous biologically active compounds that may contribute to the health-promoting properties of these foods (El Sohaimy, 2012). Several functional components derived from food have demonstrated the capacity to modulate some metabolic processes or pathways in the body, reducing the risk of inflammation and resulting in health benefits (Abuajah et al., 2015). Specifically, peptide-rich protein hydrolysates and bioactive peptides have represented a new direction in functional foods due to their potential as antiinflammatory agents (Chakrabarti et al., 2014).

4.4 Protein Hydrolysates and Bioactive Peptides

By definition, bioactive peptides derived from food are peptides that possess beneficial pharmacological properties beyond normal and adequate nutrition (Udenigwe and Aluko, 2012). During food digestion, proteins are hydrolyzed into a large variety of peptides, which share structural characteristics with endogenous peptides that act as hormones, neurotransmitters, or regulators in the human body. Therefore, these exogenous peptides can interact with the same receptors in the organism and exert an agonistic or antagonistic activity (Hernández-Ledesma et al., 2014).

Studies that investigate the biological effects of peptides apply them in two different forms: hydrolysates of precursor proteins or bioactive peptides (Sarmadi and Ismail, 2010). The production of a protein hydrolysate involves the release of peptide fragments through hydrolysis of peptide bonds, usually by the proteolytic action of enzymes sourced endogenously (autolysis) or exogenously (commercial enzyme preparations), as well as via fermentation. Finally, the resulting protein hydrolysate may undergo fractionation processes (Li-Chan, 2015). Ultrafiltration processes using membranes with a low-molecular-weight cut-off have been found to be useful for separating out small peptides from high-molecular-mass residues and remaining enzymes, yielding fractions rich in small bioactive peptides (Korhonen and Pihlanto, 2006). Since the bioactivity and functionality of peptides depend on their sequence, molecular mass, and amino acid composition, peptides with varying effects might be derived from a single hydrolysate. Hence, additional stages of purification are required to obtain the peptide with desired effect. Protein hydrolysates and peptides can be used as functional foods and nutraceuticals because of their bioactivity, as well as technological components due to their functional properties. Then, these products may contain the whole hydrolysate or isolated peptides (Hajfathalian et al., 2017).

Excessive animal protein intake and associated unhealthy levels of saturated fats exposure are increasingly common in developed countries. In response, research interest has grown in the search for new sources of proteins (Segura Campos et al., 2013). For this reason, plants have been employed as a source of low-cost protein with many health benefits due to their nutritional value and functional properties (Garba and Sawinder, 2014).

Recently, numerous studies have shown that cereals, pseudocereals, and legumes are important sources of proteins and bioactive peptides. Therefore, these proteins and peptides have been evaluated in different models and have been demonstrated to exert important biological effects in the prevention of various diseases (Malaguti et al., 2014).

4.5 Generalities About Commercial Grains

The cereals are annual common grain members of the Gramineae family, which are grown in greater quantities and provide more food energy worldwide than any other type of crop (Sarwar et al., 2013). On the other hand, pseudocereals are similar in function and composition to the true cereals, but they are dicotyledonous plants as opposed to cereals, which are monocotyledonous (Alvarez-Jubete et al., 2010). Finally, legumes belong to the Leguminosae or Fabaceae families and their seeds have been a staple food for part of the world population (Sánchez-Chino et al., 2015).

In their natural form (as whole grain), cereals are a rich source of essential vitamins, minerals, fats, carbohydrates, and proteins (Sarwar et al., 2013). Likewise, the seeds of legumes also provide valuable amounts of fiber, minerals, carbohydrates, lipids, and proteins (Sánchez-Chino et al., 2015). Thus, cereals and legumes are the main components of some diets and significantly contribute to the daily protein requirement (Malaguti et al., 2014).

Recent evidence suggests that the consumption of whole grains, partial grains, or their bioactive components may protect against numerous diseases and disorders. Many research groups have also demonstrated the defensive roles in inflammation of some of those grains (Lee et al., 2015). These studies have shown the antiinflammatory effect of hydrolysates or peptides derived from cereals in different inflammatory models (Chakrabarti et al., 2014). Also, other reports suggest that peptides from legumes can also modulate inflammatory processes (Malaguti et al., 2014).

4.6 Antiinflammatory Activity of Protein Derivatives From Cereals

A recent study carried out by Wen et al. (2016) evaluated the inhibitory effect of protein hydrolysates produced from rice (*Oryza sativa* L.) on the inflammatory response of RAW 264.7 macrophages. They demonstrated that protein hydrolysates attenuated the inflammatory markers and highlighted the effects of a fraction called RPHs-C-7-3, which suppressed the release of NO and TNF-α by the cells, as well as the phagocytic ability of the activated macrophages. Also, the same fraction decreased the transcription of inducible nitric oxide synthase (iNOS), IL-1β, IL-6, and TNF-α in a concentration-dependent manner. In other research, Saisavoey et al. (2016) showed the antiinflammatory effects of a peptide fraction (<3 kDa) obtained of a protein hydrolysate from defatted rice bran. The fraction inhibited NO production by RAW 264.7 cells and suppressed iNOS, IL-6, and TNF-α expression, suggesting that interference is mediated at a transcriptional level. In addition, other researches have evaluated the effect of rice in animal models. Boonloh et al. (2015) determined the antiinflammatory activity of protein hydrolysates from rice bran in rats with metabolic syndrome induced by a high-carbohydrate/high-fat diet. The results showed that protein hydrolysate significantly decreased the expression of genes of monocyte chemoattractant protein-1 (MCP-1), IL-6, and TNF-α in cells isolated from adipose tissues, while an increase in the expression of interleukin 10 (IL-10) was observed.

The antiinflammatory effect of peptides from other cereals has also been determined. The research of Udenigwe et al. (2009) evaluated the peptide fractions of protein hydrolysates from flaxseed (*Linum usitatissimum*) and showed that a low molecular peptide fraction (<1 kDa) inhibited the NO production by RAW 264.7 macrophages in a concentration-dependent manner. Finally, Hirai et al. (2014) demonstrated that pyroglutamyl-leucine, a peptide usually contained in the gluten of wheat (*Triticum aestivum*), significantly decreased the secretion of NO, IL-6, and TNF-α by RAW 264.7 macrophages activated with lipopolysaccharides (LPS). Also, this peptide inhibited the phosphorylation of c-Jun N-terminal kinase (JNK) and extracellular signal-regulated kinase (ERK) in a concentration-dependent manner, which are markers related to mitogen-activated protein kinases (MAPKs)-signaling pathway. The antiinflammatory effects of proteins and peptides obtained from cereals are summarized in Table 4.1.

Table 4.1: Antiinflammatory Effect of Proteins and Peptides Obtained From Cereals

Cereal	Protein or Peptide	Biological Model	Antiinflammatory Effect	References
Rice	Peptide fraction	RAW 264.7 macrophages	Phagocytic activity NO and TNF-α production IL-1β, IL-6, TNF-α, and iNOS expression	Wen et al. (2016)
Rice	Peptide fraction (<3 kDa)	RAW 264.7 macrophages	NO production IL-6, TNF-α, and iNOS expression	Saisavoey et al. (2016)
Rice	Protein hydrolysates	Rats with metabolic syndrome	MCP-1, IL-6, and TNF-α expression Increase of IL-10 expression	Boonloh et al. (2015)
Flaxseed	Peptide fraction (<1 kDa)	RAW 264.7 macrophages	NO production	Udenigwe et al. (2009)
Wheat	Pyroglutamyl-leucine	RAW 264.7 macrophages	NO, IL-6, and TNF-α production Phosphorylation of JNK and ERK	Hirai et al. (2014)

4.7 Antiinflammatory Activity of Protein Derivatives From Pseudocereals

Evaluation of the antiinflammatory effect from some pseudocereals has been done, mainly with amaranth (*Amaranthus hypochondriacus*). For example, Montoya-Rodríguez et al. (2014) evaluated the effect of protein hydrolysates produced from amaranth on THP-1 and RAW 264.7 macrophages, demonstrating a reduction in the secretion of NO, TNF-α, and prostaglandin E2 (PGE_2). The amaranth hydrolysates also inhibited the expression of proinflammatory proteins by the cells, such as iNOS and cyclooxygenase 2 (COX-2). In the same way, another study carried out by Montoya-Rodríguez and González De Mejía (2015) showed that peptides from amaranth proteins suppressed some inflammatory markers on THP-1 cells. The HGSEPFGPR, RDGPFPWPWYSH, and RPRYPWRYT peptides reduced the expression of intracellular adhesion molecule 1 (ICAM-1) and matrix metalloproteinase 9 (MMP-9). RPRYPWRYT reduced the expression of MCP-1, TNF-α, IL-6, interleukin 1α (IL-1α), and interferon γ (IFN-γ), while HGSEPFGPR showed significant inhibition in the expression of IL-1α, IL-6, and MCP-1. In the same way, the antiinflammatory effects of proteins and peptides obtained from pseudocereals are summarized in Table 4.2.

4.8 Antiinflammatory Activity of Protein Derivatives From Legumes

Legumes are another type of grain with antiinflammatory properties. Ndiaye et al. (2012) investigated the effects of a protein hydrolysate with peptides (<3 kDa) from yellow field pea (*Pisum sativum* L.) in RAW 264.7 cells stimulated with LPS/IFN-γ, observing a decrement in the concentrations of NO, IL-6, and TNF-α produced by the activated cells.

On the other hand, López-Barrios et al. (2016) determined that protein hydrolysates from common dry bean (*Phaseolus vulgaris* L.) significantly inhibited the NO production by RAW 264.7 cells induced with LPS. Similarly, the research of Garcia-Mora et al. (2015) evaluated the antiinflammatory activity of protein hydrolysates from bean and the results showed that protein hydrolysates decreased the secretion of IL-6 by myofibroblasts activated with IL-1β.

The study of Martinez-Villaluenga et al. (2009) demonstrated that protein hydrolysates obtained from soybean (*Glycine* max L.) exerted antiinflammatory effects, since the protein hydrolysates

Table 4.2: Antiinflammatory Effect of Proteins and Peptides Obtained From Pseudocereals

Pseudocereal	Protein or Peptide	Biological Model	Antiinflammatory Effect	References
Amaranth	Protein hydrolysates	THP-1 macrophages RAW 264.7 macrophages	NO, TNF-α, and PGE_2 production COX-2 and iNOS expression	Montoya-Rodríguez et al. (2014)
Amaranth	HGSEPFGPR, RDGPFPWPWYSH, and RPRYPWRYT peptides	THP-1 macrophages	ICAM-1 and MMP-9 expression IFN-γ, IL-1α, IL-6, TNF-α, and MCP-1 expression	Montoya-Rodríguez and González De Mejía (2015)

Table 4.3: Antiinflammatory Effect of Proteins and Peptides Obtained From Legumes

Legume	Protein or Peptide	Biological Model	Antiinflammatory Effect	References
Yellow field pea	Protein hydrolysate with peptides (<3 kDa)	RAW 264.7 macrophages	NO, IL-6, and TNF-α production	Ndiaye et al. (2012)
Bean	Protein hydrolysates	RAW 264.7 macrophages	NO production	López-Barrios et al. (2016)
Bean	Protein hydrolysates	Myofibroblasts	IL-6 production	Garcia-Mora et al. (2015)
Soybean	Protein hydrolysates	RAW 264.7 macrophages	NO and PGE_2 production COX-2 and iNOS expression	Martinez-Villaluenga et al. (2009)
Soybean	Protein	Rats with arthritis	IL-6 and TNF-α serum levels	Shahi et al. (2012)
Soybean	VPY peptide	Caco-2 intestinal cells THP-1 macrophages Mice model	IL-8 and TNF-α production IFN-γ, IL-1β, IL-6, IL-17, and TNF-α expression	Kovacs-Nolan et al. (2012)
Soybean	Lunasin	RAW 264.7 macrophages	NO and PGE_2 production COX-2 and iNOS expression	Dia et al. (2009)
Soybean	Lunasin	RAW 264.7 macrophages	ROS, IL-6, and TNF-α production	Hernández-Ledesma et al. (2009)

reduced the production of NO and PGE_2 by RAW 264.7 macrophages, as well as the gene expression of iNOS and COX-2 by the same cells. Shahi et al. (2012) exhibited the activity of a soybean protein on collagen-induced arthritic rats, whose values showed that the protein significantly suppressed the progression of arthritis and decreased the serum levels of IL-6 and TNF-α.

Likewise, other research groups have evaluated the effect of peptides isolated from soybean. Kovacs-Nolan et al. (2012) evaluated the antiinflammatory properties of VPY peptide, obtaining that the peptide inhibited the secretion of interleukin 8 (IL-8) and TNF-α by Caco-2 intestinal epithelial cells and THP-1 macrophages, respectively. In addition, the VPY peptide decreased the gene expression of TNF-α, IFN-γ, IL-1β, IL-6, and interleukin 17 (IL-17) in mice. Dia et al. (2009) isolated, purified, and characterized lunasin from defatted soybean flour and found that the peptide decreased the production of NO and PGE_2 by RAW 264.7 cells. The treatments with lunasin also inhibited the expression of iNOS and COX-2. In another study, Hernández-Ledesma et al. (2009) demonstrated that lunasin reduced the production of intracellular ROS by RAW 264.7 macrophages activated with LPS, in a concentration-dependent manner. Equally, the peptide inhibited the release of IL-6 and TNF-α. Finally, the antiinflammatory effect of proteins and peptides obtained from legumes are summarized in Table 4.3.

4.9 Conclusions

Protein hydrolysates and peptides isolated from cereals, pseudocereals, and legumes have shown important biological effects, including over the immune system. Specifically, the antiinflammatory effects of those protein derivatives have been determined on various cell and animal models, highlighting the properties of low-molecular-weight proteins and peptides. These studies have laid a basis for the possible antiinflammatory mechanism that these peptides could follow, although more researches are needed before contemplating the possibility of their use in humans. However, the obtained results in the previously described studies have shown the potential of proteins and peptides isolated from commercial grains, allowing the evaluation for their future use in the treatment of inflammatory diseases.

References

Abuajah, C.I., Ogbonna, A.C., Osuji, C.M., 2015. Functional components and medicinal properties of food: a review. Journal of Food Science and Technology 52 (5), 2522–2529.

Alvarez-Jubete, L., Arendt, E.K., Gallagher, E., 2010. Nutritive value of pseudocereals and their increasing use as functional gluten-free ingredients. Trends in Food Science and Technology 21 (2), 106–113.

Ashley, N.T., Weil, Z.M., Nelson, R.J., 2012. Inflammation: mechanisms, costs, and natural variation. Annual Review of Ecology, Evolution, and Systematics 43, 385–406.

Boonloh, K., Kukongviriyapan, V., Kongyingyoes, B., Kukongviriyapan, U., Thawornchinsombut, S., Pannangpetch, P., 2015. Rice bran protein hydrolysates improve insulin resistance and decrease pro-inflammatory cytokine gene expression in rats fed a high carbohydrate-high fat diet. Nutrients 7 (8), 6313–6329.

Chakrabarti, S., Jahandideh, F., Wu, J., 2014. Food-derived bioactive peptides on inflammation and oxidative stress. BioMed Research International 2014, 1–11.

Dia, V.P., Wang, W., Oh, V.L., De Lumen, B.O., Gonzalez De Mejia, E., 2009. Isolation, purification and characterisation of lunasin from defatted soybean flour and *in vitro* evaluation of its anti-inflammatory activity. Food Chemistry 114 (1), 108–115.

Dinarello, C.A., 2010. Anti-inflammatory agents: present and future. Cell 140 (6), 935–950.

Duque, G.A., Descoteaux, A., 2014. Macrophage cytokines: involvement in immunity and infectious diseases. Frontiers in Immunology 5 (491), 1–12.

El Sohaimy, S.A., 2012. Functional foods and nutraceuticals-Modern approach to food science. World Applied Sciences Journal 20 (5), 691–708.

Flannagan, R.S., Cosío, G., Grinstein, S., 2009. Antimicrobial mechanisms of phagocytes and bacterial evasion strategies. Nature Reviews Microbiology 7, 355–366.

Garba, U., Sawinder, K., 2014. Protein isolates: production, functional properties and application. International Research Journal of Chemistry 4, 22–36.

Garcia-Mora, P., Frias, J., Peñas, E., Zielinski, H., Giménez-Bastida, J.A., Wiczkowski, W., Zielinska, D., Martínez-Villaluenga, C., 2015. Simultaneous release of peptides and phenolics with antioxidant, ACE-inhibitory and anti-inflammatory activities from pinto bean (*Phaseolus vulgaris* L. var. pinto) proteins by subtilisins. Journal of Functional Foods 18, 319–332.

Gautam, R., Jachak, S.M., 2009. Recent developments in anti-inflammatory natural products. Medicinal Research Reviews 29 (5), 767–820.

Hajfathalian, M., Ghelichi, S., García-Moreno, P.J., Moltke Sørensen, A.D., Jacobsen, C., 2017. Peptides: production, bioactivity, functionality, and applications. Critical Reviews in Food Science and Nutrition 1–33.

Hernández-Ledesma, B., Hsieh, C.C., De Lumen, B.O., 2009. Antioxidant and anti-inflammatory properties of cancer preventive peptide lunasin in RAW 264.7 macrophages. Biochemical and Biophysical Research Communications 390 (3), 803–808.

Hernández-Ledesma, B., García-Nebot, M.J., Fernández-Tomé, S., Amigo, L., Recio, I., 2014. Dairy protein hydrolysates: peptides for health benefits. International Dairy Journal 38 (2), 82–100.

Hirai, S., Horii, S., Matsuzaki, Y., Ono, S., Shimmura, Y., Sato, K., Egashira, Y., 2014. Anti-inflammatory effect of pyroglutamyl-leucine on lipopolysaccharide-stimulated RAW 264.7 macrophages. Life Sciences 117 (1), 1–6.

Hoesel, B., Schmid, J.A., 2013. The complexity of NF-κB signaling in inflammation. Molecular Cancer 12 (86), 1–15.

Janssen, W.J., Henson, P.M., 2012. Cellular regulation of the inflammatory response. Toxicologic Pathology 40 (2), 166–173.

Korhonen, H., Pihlanto, A., 2006. Bioactive peptides: production and functionality. International Dairy Journal 16 (9), 945–960.

Kovacs-Nolan, J., Zhang, H., Ibuki, M., Nakamori, T., Yoshiura, K., Turner, P.V., Matsui, T., Mine, Y., 2012. The PepT1-transportable soy tripeptide VPY reduces intestinal inflammation. Biochimica et Biophysica Acta 1820 (11), 1753–1763.

Laveti, D., Kumar, M., Hemalatha, R., Sistla, R., Naidu, V.G.M., Talla, V., Verma, V., Kaur, N., Nagpal, R., 2013. Anti-inflammatory treatments for chronic diseases: a review. Inflammation and Allergy - Drug Targets 12 (5), 349–361.

Lawrence, T., Fong, C., 2010. The resolution of inflammation: anti-inflammatory roles for NF-κB. The International Journal of Biochemistry and Cell Biology 42 (4), 519–523.

Lee, Y.M., Han, S.I., Song, B.C., Yeum, K.J., 2015. Bioactives in commonly consumed cereal grains: implications for oxidative stress and inflammation. Journal of Medicinal Food 18 (11), 1–8.

Li-Chan, E.C.Y., 2015. Bioactive peptides and protein hydrolysates: research trends and challenges for application as nutraceuticals and functional food ingredients. Current Opinion in Food Science 1, 28–37.

López-Barrios, L., Antunes-Ricardo, M., Gutiérrez-Uribe, J.A., 2016. Changes in antioxidant and antiinflammatory activity of black bean (*Phaseolus vulgaris* L.) protein isolates due to germination and enzymatic digestion. Food Chemistry 203, 417–424.

Lugrin, J., Rosenblatt-Velin, N., Parapanov, R., Liaudet, L., 2014. The role of oxidative stress during inflammatory processes. Biological Chemistry 395 (2), 203–230.

Malaguti, M., Dinelli, G., Leoncini, E., Bregola, V., Bosi, S., Cicero, A.F.G., Hrelia, S., 2014. Bioactive peptides in cereals and legumes: agronomical, biochemical and clinical aspects. International Journal of Molecular Sciences 15 (11), 21120–21135.

Martinez-Villaluenga, C., Dia, V.P., Berhow, M., Bringe, N.A., Gonzalez De Mejia, E., 2009. Protein hydrolysates from β-conglycinin enriched soybean genotypes inhibit lipid accumulation and inflammation *in vitro*. Molecular Nutrition and Food Research 53 (8), 1007–1018.

Mittal, M., Siddiqui, M.R., Tran, K., Reddy, S.P., Malik, A.B., 2014. Reactive oxygen species in inflammation and tissue injury. Antioxidants and Redox Signaling 20 (7), 1126–1167.

Mogensen, T.H., 2009. Pathogen recognition and inflammatory signaling in innate immune defenses. Clinical Microbiology Reviews 22 (2), 240–273.

Montoya-Rodríguez, A., González De Mejía, E., Dia, V.P., Reyes-Moreno, C., Milán-Carrillo, J., 2014. Extrusion improved the anti-inflammatory effect of amaranth (*Amaranthus hypochondriacus*) hydrolysates in LPS-induced human THP-1 macrophage-like and mouse RAW 264.7 macrophages by preventing activation of NF-κB signaling. Molecular Nutrition and Food Research 58 (5), 1028–1041.

Montoya-Rodríguez, A., González De Mejía, E., 2015. Pure peptides from amaranth (*Amaranthus hypochondriacus*) proteins inhibit LOX-1 receptor and cellular markers associated with atherosclerosis development *in vitro*. Food Research International 77 (2), 204–214.

Ndiaye, F., Vuong, T., Duarte, J., Aluko, R.E., Matar, C., 2012. Anti-oxidant, anti-inflammatory and immunomodulating properties of an enzymatic protein hydrolysate from yellow field pea seeds. European Journal of Nutrition 51 (1), 29–37.

Osamu, T., Shizuo, A., 2010. Pattern recognition receptors and inflammation. Cell 140 (6), 805–820.

Rock, K.L., Latz, E., Ontiveros, F., Kono, H., 2010. The sterile inflammatory response. Annual Review of Immunology 28, 321–342.

Saisavoey, T., Sangtanoo, P., Reamtong, O., Karnchanatat, A., 2016. Antioxidant and anti-inflammatory effects of defatted rice bran (*Oryza sativa* L.) protein hydrolysates on RAW 264.7 macrophage cells. Journal of Food Biochemistry 40 (6), 731–740.

Sánchez-Chino, X., Jiménez-Martínez, C., Dávila-Ortiz, G., Álvarez-González, I., Madrigal-Bujaidar, E., 2015. Nutrient and nonnutrient components of legumes, and its chemopreventive activity: a review. Nutrition and Cancer 67 (3), 1–10.

Sarmadi, B.H., Ismail, A., 2010. Antioxidative peptides from food proteins: a review. Peptides 31 (10), 1949–1956.

Sarwar, M.H., Sarwar, M.F., Sarwar, M., Qadri, N.A., Moghal, S., 2013. The importance of cereals (Poaceae: Gramineae) nutrition in human health: a review. Journal of Cereals and Oilseeds 4 (3), 32–35.

Segura Campos, M.R., Peralta González, F., Chel Guerrero, L., Betancur Ancona, D., 2013. Angiotensin I-converting enzyme inhibitory peptides of chia (*Salvia hispanica*) produced by enzymatic hydrolysis. International Journal of Food Science 2013, 1–8.

Shahi, M.M., Rashidi, M.R., Mahboob, S., Haidari, F., Rashidi, B., Hanaee, J., 2012. Protective effect of soy protein on collagen-induced arthritis in rat. Rheumatology International 32 (8), 2407–2414.

Tabas, I., Glass, C.K., 2013. Anti-inflammatory therapy in chronic disease: challenges and opportunities. Science 339 (6116), 166–172.

Udenigwe, C.C., Lu, Y.L., Han, C.H., Hou, W.C., Aluko, R.E., 2009. Flaxseed protein-derived peptide fractions: antioxidant properties and inhibition of lipopolysaccharide-induced nitric oxide production in murine macrophages. Food Chemistry 116 (1), 277–284.

Udenigwe, C.C., Aluko, R.E., 2012. Food protein-derived bioactive peptides production, processing, and potential health benefits. Journal of Food Science 71 (1), 11–24.

Venkatesha, S.H., Dudics, S., Acharya, B., Moudgil, K.D., 2015. Cytokine-modulating strategies and newer cytokine targets for arthritis therapy. International Journal of Molecular Sciences 16 (1), 887–906.

Wen, L., Chen, Y., Zhang, L., Yu, H., Xu, Z., You, H., Cheng, Y., 2016. Rice protein hydrolysates (RPHs) inhibit the LPS-stimulated inflammatory response and phagocytosis in RAW264.7 macrophages by regulating the NF-κB signaling pathway. RSC Advances 6, 71295–71304.

Wynn, T.A., Chawla, A., Pollard, J.W., 2013. Macrophage biology in development, homeostasis and disease. Nature 496, 445–455.

PART 4

Anticancer Activity of Plants Metabolites

CHAPTER 5

Medicinal Plants and Their Bioactive Metabolites in Cancer Prevention and Treatment

Armando M. Martín Ortega, Maira Rubi Segura Campos
Facultad de Ingeniería Química, Universidad Autónoma de Yucatán, Mérida, Mexico

5.1 Introduction

The term cancer refers to a set of diseases in which the cells of an organ or tissue divide uncontrollably and acquire the ability to invade other tissues. This group of diseases represents a global health problem, leading to it being the top cause of death in all countries. According to the latest global cancer report of the World Health Organization 2014, it is estimated that during 2012 there were 14.1 million new cases of cancer and 8.2 million deaths due to cancer worldwide (Greenwell and Rahman, 2015).

Despite the fact that cancer is among the leading causes of death worldwide, the most frequent cancers in each country vary. According to GLOBOCAN 2012, prostate cancer is the most commonly diagnosed type of cancer among men in 87 countries, especially in the Americas, northern, southern and western Europe, and Oceania. In Africa and Asia, there is considerable variation in the main types of cancer that occur in men. In Africa, the most common types of cancer among men are prostate, lung, colorectal, liver, esophagus, Kaposi's sarcoma, leukemia, and others. In Asia, cancers of the lung, lips, and oral cavity are among the main ones, followed by cancer of the liver, stomach, and prostate. As for women, there is greater homogeneity between countries, with breast cancer and cervical cancer taking first place in North America, Latin America, Africa, and most of Asia. However, in China and North Korea, lung cancer ranks first (Ferlay et al., 2015; Torre et al., 2016).

The medical treatment aimed at cancer focuses on three main aspects: surgery, radiotherapy, and chemotherapy. This type of treatment is usually accompanied by a large number of side effects and deleterious effects on patient health, such as nausea, loss of appetite and weight, anemia, spinal cord injury, kidney damage, mucositis, etc. These side effects affect in an important way the quality of life of people who undergo these treatments. In addition, it is important to take into account that less than 25% of treatments reach a complete response, so it is usually necessary for several cycles of these treatments (Baba and Câtoi, 2007).

Bioactive Compounds. https://doi.org/10.1016/B978-0-12-814774-0.00005-0

Thus, modern medicine offers an increase in the life expectancy of cancer patients, but with a high cost in their quality of life; therefore, a large number of people with this condition resort to the use of medicinal plants, to treat the cancer itself and/or to reduce the side effects of modern medical treatment (Foster et al., 2017; Jaradat et al., 2016).

Medicinal plants have been known and used for traditional millenary medicines, such as Chinese medicine and medicine from India. Currently, the potential of these plants for the treatment of various diseases and symptoms is recognized. This potential lies in the bioactive metabolites contained in medicinal plants (Sen and Samanta, 2014). Thus, between 25% and 28% of the drugs currently used by modern medicine are directly or indirectly derived from plant metabolites (e.g., opium-derived morphine, *Papaver somniferum* L.) (Fridlender et al., 2015). Also, the development of medicines for the treatment of cancer has also been influenced and derived from medicinal plants; the best examples are the drugs paclitaxel (Taxol) and docetaxel (Taxotere), derived from the plants *Taxus brevifolia* and *Taxus baccata*, respectively (Fridlender et al., 2015; Francis et al., 1995).

A great variety of plants/medicinal foods are found daily in the kitchen and even in city surroundings, the knowledge of their metabolites and their respective biological activities in cancer plays an important role in the recommendation or indication for consumption as adjuvant therapy in cancer. Therefore, this document aims to summarize the anticancer activities in vitro and in vivo of different metabolites derived from medicinal plants and their roles in the prevention and/or treatment of cancer. In the same way, a part of the molecular alterations present in the malignant cells is presented.

5.2 Carcinogenesis

Carcinogenesis is the process by which a normal cell is transformed into a malignant cell and its progression to a clinically observable tumor with high probability of metastasis. Currently, it is considered a process with three steps: initiation, promotion, and progression (Weston and Harris, 2003). This process can vary depending on the etiology, since viruses induce some types of cancer. The review and comprehension of this process are important for understanding of the preventive function of the bioactive metabolites of medicinal plants (Tanaka et al., 2013).

It is important to mention that each step is characterized by morphological and biochemical changes at the cellular level and these will depend on those affected genes (Weston and Harris, 2003; Tanaka et al., 2013). Currently, other mechanisms besides mutations are considered part of the process, such as epigenetic changes (DNA methylation and histone acetylation) (Harrington, 2016).

5.2.1 Initiation

Initiation is the process by which a healthy cell undergoes irreparable DNA damage by producing a mutated or initiated cell (Weston and Harris, 2003). This process occurs through the interaction of cells with mutagenic components (e.g., polycyclic aromatic hydrocarbons), such as chemical agents that can form adducts with DNA, altering their structure and causing mutations during the process of repair (Weston and Harris, 2003; Tanaka et al., 2013). Some physical agents, such as ultraviolet light radiation, can generate mutations through direct damage to DNA or through the generation of reactive oxygen species (Weston and Harris, 2003).

Likewise, epigenetic changes, such as DNA methylation, in promoter areas of tumor suppressor genes produce their silencing, causing a preneoplastic cell in the same way. Meanwhile, structural changes in histones can facilitate the exposure and transcription of oncogenic genes (Sharma et al., 2010).

An initiated cell is one that has a mutated gene or a genomic alteration that causes the activation of a proto-oncogene or the inactivation of a tumor suppressor gene.

5.2.2 Promotion

Promotion is the step that encompasses the clonal expansion of the cells initiated. Because the generation and accumulation of mutations is proportional to the rate of cell division, clonal expansion increases the susceptibility to acquire genetic changes that catalyze the transformation to malignant cells (Weston and Harris, 2003; Tanaka et al., 2013). This process is promoted by molecules that induce or increase cell division, including hormones and growth factors. However, some exogenous components also have this effect (Tanaka et al., 2013).

Similarly, it has been found that the activation of proinflammatory intracellular signaling pathways is an essential part of the promotion process (Tanaka et al., 2013; Okada, 2014). Obesity, glycolipotoxicity, lipopolysaccharides, and alpha lipoteichoic acid are some factors that contribute to the inflammation process.

Tumor promotion depends on the concentration and exposure time with the promoter agent, so the promoter effects are reversible. Likewise, it is important to consider that some promoter agents have an effect on specific organs, while others have a generalized effect (Tanaka et al., 2013).

5.2.3 Progression

The cells or tumors found in the previous steps are considered benign (Weston and Harris, 2003). Progression is the step in which a preneoplastic cell acquires a malignant phenotype (Weston and Harris, 2003; Tanaka et al., 2013). This last step is sometimes called conversion or malignant transformation. The increase in the number of mutations and genomic instability generates the malignant conversion, being those with the capacity to metastasize and with a proliferative capacity independent of the environmental factors, likewise, its apoptotic mechanisms are absent or diminished (Weston and Harris, 2003; Tanaka et al., 2013).

This last step is accompanied by several aspects, in which we must emphasize the epithelial–mesenchymal transition (EMT) and neoangiogenesis (NEO) (Tanaka et al., 2013). EMT is the biological process by which an epithelial cell undergoes alterations in the expression of genes and in the degree of differentiation, reaching a mesenchymal cell (MC) phenotype (Kalluri and Weinberg, 2009). The MC has migratory capacity, invasive, high resistance to apoptosis, and production of extracellular matrix. EMT is mediated by the activation of different transcription factors and the expression of specific peripheral proteins of the cell membrane, which gives it stem cell qualities (Lamouille et al., 2014). On the other hand, NEO is the process by which new blood and lymphatic vessels are generated (Nishida et al., 2006).

Tumor growth depends on the supply of nutrients and oxygen, so NEO is an essential process for tumor progression. Thus, NEO also participates in the process of metastasis, allowing tumor cells to have contact with the vascular system for subsequent intravasation and extravasation (Lamouille et al., 2014; Nishida et al., 2006).

5.3 Angiogenesis

Angiogenesis is the term that refers to the process in which new capillary vessels are generated from existing blood vessels (Nishida et al., 2006). It is a complex process composed of several activities at the molecular and cellular levels, among them: proliferation of endothelial cells, degradation of the cellular basal membranes of the endothelium and extracellular matrix that is around, in addition, the migration of endothelial cells is necessary (Nishida et al., 2006; Hoff and Kalil, 2012; Yadav et al., 2015). At the molecular level an increase in the synthesis of those mitogenic factors that stimulate the division of endothelial cells is required, as well as the production of metalloproteinases (MMPs) that allow the degradation of the extracellular matrix (Yadav et al., 2015).

It has been shown that tumors can exist and remain in situ, for months or even years, without a blood supply, surviving fundamentally from interstitial fluid (Hoff and Kalil, 2012). However, the systemic conditions of the organism may favor or disfavor tumor angiogenesis. Among the main conditions are: chronic inflammation, hypoxia and ischemia, and a change in the phenotype of tumor cells. The latter has been fundamentally recognized with the mutation and inactivation of the *p53* gene, which fulfills an inhibitory function of the synthesis of proangiogenic factors (Hoff and Kalil, 2012). The change in phenotype has been called “angiogenic switch,” and is characterized mainly by the increase in the cellular secretion of proangiogenic factors (growth factors of the vascular endothelium) and/or a reduction of the antiangiogenic factors (Yadav et al., 2015).

Chronic inflammation is a factor that favors the synthesis of vascular endothelial growth factors (VEGFs), through the regulation of gene expression by proinflammatory interleukins (IL), including: IL-1, IL-6, tumor necrosis factor alpha (TNF-α), and transforming growth factor beta (TGF-β). However, by itself, chronic inflammation is a pathological condition that is a precursor to tumor development since it also favors a tumorigenic environment (Hoff and Kalil, 2012). An antiinflammatory environment, catalyzed by antiinflammatory cytokines such as IL-10, IL-12, retinoids, or by interferon-like cytokines, reduces the expression of proangiogenic factors (Hoff and Kalil, 2012).

5.4 Molecular Alterations of Cancer Cells

The molecular changes in the cells of a malignant tumor depend to a large extent on the mutations and epigenetic changes that present, so that each cancer is different, although they share a large part of its clinical characteristics. There are even differences between cells from different areas of the same tumor. The molecular changes in each type of cancer make each tumor respond differently to treatments. In addition, important molecular changes also occur in the malignant cells before the morphological changes characteristic of a tumor lesion can be observed (Shang and Wang, 2013). Thus, a general knowledge of the main molecular alterations is necessary to understand the mechanism of action of the metabolites derived from medicinal plants and their potential in the prevention and treatment of cancer.

In addition, it is important to take into account that cancer cells suffer from a greater number of molecular alterations as they advance from the clinical stage, which makes them more aggressive and resistant to conventional treatment and probably to the bioactive metabolites of medicinal plants.

Among the main molecular alterations present in the malignant cells of a tumor are modifications at the genomic level (including chromosomes and nucleic acids), epigenetic and in their proteins. In addition, the alteration in the expression of enzymatic proteins and/or a change in their functionality can lead to metabolic changes (Shang and Wang, 2013).

5.4.1 Molecular Alterations at the Chromosomal Level

Aneuploidy, an alteration in the number of chromosomes of a cell, is one of the main findings in most solid tumors and those derived from blood cells. Aneuploidy originates in the alteration of the processes corresponding to mitosis, among them the chromosomal instability defined as a persistent loss or gain in the number of chromosomes. Even more than the change in the number of chromosomes, alterations in its structure are also found as multiple sites of rupture and/or translocations of different sites. This class of molecular alterations is found in most of the more aggressive and advanced cancers, with an unfavorable prognosis, metastasis, and resistance to chemotherapeutic treatment (Thompson and Compton, 2011).

The increase in the number of chromosomes can increase the expression of various genes that promote tumorigenesis. In addition, the translocation of genes in the chromosome usually fuses two genes, generating an oncogene, such as the *BCR-ABL* gene of chronic myeloid leukemia, in addition, in some cases it can also inactivate tumor suppressor genes (Vogelstein et al., 2013).

5.4.2 Molecular Alterations at DNA Level and Gene Expression

Alterations in the DNA of the tumor cells are mainly mutations, where approximately 95% is due to the substitution of a single base (as G>C) and the remaining corresponds to the deletion or insertion of one or more bases (as CAT>CA). In most solid tumors, such as those derived from the colon or pancreas, on average there are between 33 and 66 mutated genes in which their protein product would be expected to be altered. However, today it has been identified that approximately 140 genes within the genome, its mutation, could promote the process of tumorigenesis. Thus, on average it is estimated that tumors harbor between two and eight of these mutated promoter genes, while the rest of the mutated genes do not confer an advantage on growth and survival issues (Vogelstein et al., 2013).

The mutation in the promoter genes or "drivers" gives an advantage of tumor growth. These genes correspond mainly to various regulatory proteins of intracellular growth pathways (oncogene) such as PIK3CA, SMAD4, APC, and KRAS; and proteins that are responsible for promoting apoptotic processes (suppressor gene) or repairing DNA (Vogelstein et al., 2013; Shang and Wang, 2013).

On the other hand, certain DNA mutations can increase the expression of enzymes such as extracellular matrix metalloproteinases (MMPs) that promote a modification in the cellular periphery, increasing the risk of metastasis. Likewise, other mutations can increase the expression of enzymes such as cyclooxygenase-2 (COX-2) by increasing the synthesis of proinflammatory prostaglandins, which in turn increase the expression of certain IL and VEGF, promoting tumorigenesis (Shang and Wang, 2013; Hoff and Kalil, 2012).

5.4.3 Epigenetic Alterations

The main epigenetic changes include posttranslational modifications of histones and DNA methylation, which alter or reduce the accessibility of the transcription machinery to DNA. These changes can increase or reduce the expression of various genes, regulating processes of cell differentiation and growth (Shang and Wang, 2013; Miozzo et al., 2015; Jones et al., 2016).

DNA methylation is considered to be the main epigenetic mechanism since it orchestrates the remodeling of chromatin. These methylations occur mainly in CpG sites, which are usually located in the promoter areas of the genes. Methylation of CpG inhibits the transcription of the gene, while demethylation promotes its transcription. Methylation in non-CpG zones plays an important role in the maintenance of genomic stability (Miozzo et al., 2015).

The posttranslational modifications of histones modify their interaction with DNA and various nuclear proteins. These modifications mainly involve covalent changes in several amino acid residues, through their methylation, acetylation, phosphorylation, ubiquitination, and sumoylation. In general, the transcriptionally active zones of chromatin are characterized by having trimethylations or acetylations in lysine residues in their histones (Miozzo et al., 2015).

Noncoding RNAs (ncRNAs) are another epigenetic mechanism immersed in the regulation of gene expression. Among this class of RNAs are microRNAs (miRNAs), long ncRNAs, short interfering RNAs (siRNAs), and PIWI-interacting RNAs (piRNAs). However, this kind of epigenetic regulation is still not fully understood and its influence is still under investigation (Miozzo et al., 2015).

Cancer cells usually have several areas of hyper/hypomethylated DNA, which modify the expression of their genes participating in an important way in carcinogenesis. Next-generation sequencing has revealed that approximately 50% of cancers in humans have mutations in enzymes that regulate epigenetic changes, including DNA methyltransferases (DNMT), histone deacetylases (HDAC), and histone acyltransferases (HAT). Therefore, malignant cells are not only generated by genetic mutations, but also by epigenetic alterations (Miozzo et al., 2015; Jones et al., 2016).

5.5 Medicinal Plants in Cancer Prevention and Treatment

5.5.1 Brief History

Since ancient times, humans have sought to cure their diseases in nature, using plants, roots, stems, or fruits. At first the use of medicinal plants was instinctive, as in the case of animals. Subsequently, based on observation and experience, the use of specific plants to treat certain diseases took place (Petrovska, 2012).

The oldest written evidence of the use of medicinal plants to treat diseases comes from the Sumer civilization of Nagpur, India, approximately 5000 years ago. This letter contained 12 recipes for the preparation of various medicines using more than 250 varieties of plants (Petrovska, 2012).

Subsequently, in the year 2500 BC, the Chinese emperor Shen Nung wrote the Chinese book "Pen T'Sao" with leaves and stems of plants, which contained the recipes of 365 different medicines. This book includes some currently used products such as cinnamon bark, ephedra, and ginseng (Petrovska, 2012).

Around 1550 BC, in Egypt, El Papiro de Ebers was written, which contains a collection of 800 prescriptions that refer to 700 species of plants and drugs used for therapy such as pomegranate, castor oil, aloe, senna, garlic, onion, fig, willow, cilantro, etc. (Petrovska, 2012).

In Athens, Greece, Theophrastus (371–287 BC) founded botanical science with his books "De Causis Plantarium" and "De Historia Plantarium." Both books contained a classification of more than 500 medicinal plants known at that time. Among others, these plants included cinnamon, iris rhizome, false hellebore, mint, pomegranate, cardamom, fragrant hellebore, aconite, etc. (Petrovska, 2012).

Followed by the aforementioned, is the most famous medicinal plant document "De Materia Medica," which was written by the considered "father of pharmacognosy" the Greek Dioscorides in 77 BC. This book has a description of 944 medications, of which 657 are derived from medicinal plants. In addition, it contains the place and manner of collection, the preparation, and the therapeutic effect. This book was used as a basis for medical treatment in different cultures for more than 500 years (Petrovska, 2012).

Nowadays, the use of isolated chemical compounds is the medicine of choice for most people/countries. However, conventional medicines are equally divided between those derived from medicinal plants, and those synthesized or modified in laboratories. In certain diseases, such as cancer, when the use of conventional medicine does not offer the results expected by patients, they may resort to the use of medicinal plants or other types of "alternative" therapies (Fridlender et al., 2015; Petrovska, 2012). However, for the sake of adequate and successfully applied therapy, knowledge of the precise diagnosis of the disease and its pathophysiological bases is essential, in addition to knowing the biological effects of medicinal plants and their metabolites. Thus, this section is intended to summarize the main biological activities of various bioactive metabolites derived from medicinal plants, emphasizing their anticancer properties.

5.5.1.1 Curcuma longa

Curcuma longa is a yellow rhizome commonly known as turmeric. The medicinal use of this plant is part of traditional Chinese medicine and traditional medicine of India (Ayurveda), used to treat different conditions of inflammatory origin such as joint pain and skin inflammation, and is also used for diarrheal conditions and fever (Henrotin et al., 2013; Basnet and Skalko-Basnet, 2011).

Curcumin is the yellow pigment that is found in the rhizomes of *Curcuma longa*. This pigment is the main phytochemical with anticancer properties found in turmeric and belongs to the family of polyphenols (Henrotin et al., 2013; Basnet and Skalko-Basnet, 2011). In the alcoholic extraction of turmeric, there are three main types of curcuminoids, these polyphenols are called curcumin I (diferuloylmethane), curcumin II (desmetoxicurcumin), and curcumin III (bisdemetoxicurcumin). That said, it is important to take into account that the commercial extracts of "curcumin" are actually a mixture of the three different curcuminoids, with curcumin I having the highest concentration (approximately 77%) (Basnet and Skalko-Basnet, 2011).

At present, the anti-cancer potential of curcumin has been intensively investigated, compared to other phytochemicals, which has allowed much of its intracellular mechanisms to be defined. The anticancer potential of curcumin is mainly due to its ability to inhibit and/or activate various intracellular transcription factors, thus regulating the expression of various proteins that participate in tumor growth and development. The main mechanisms of action and their effect are summarized below.

5.5.2 Inhibition of the Nuclear Factor κβ (NF-κβ)

Nuclear factor kappa-light-chain-enhancer of activated B cells, also known as NF-κβ, is composed of a family of five transcription factors, which can form different protein complexes and bind promoter regions in DNA, regulating the expression of various genes to regulate cellular processes. This family of transcription factors regulates the synthesis of proinflammatory cytokines, increases the expression of COX-2, promotes cell proliferation, inhibits apoptotic processes, and can induce epigenetic changes in the cells that regulate EMT, which facilitates metastasis. They also regulate the expression of MMP-9, IL-8, and basic fibroblast growth factor (bFGF), important molecules involved in the process of angiogenesis (Panahi et al., 2016; Xia et al., 2014).

The activation of this family of transcription factors is regulated by the phosphorylation of its inhibitor protein, IκBα, by the serine-threonine kinase IKK. The activation of IKK is regulated by the cellular response to various stimuli, including proinflammatory cytokines (TNF-α and IL-1β), activation of toll-like receptors (TLRs), and T-cell receptors (Panahi et al., 2016).

Many of the cancers are constitutively activated to NF-κβ due to a proinflammatory and oxidative tumor microenvironment (Shrihari, 2017). Its chronic activation plays a fundamental role in all stages of carcinogenesis, so it not only participates in the development of cancer, but also in its progression (Panahi et al., 2016).

Curcumin, in several studies, has been shown to have the ability to inhibit the activation of NF-κβ, preventing the phosphorylation of IkBα by IKK. Inhibition of NF-κβ reduces the expression of proinflammatory cytokines, COX-2, nitric oxide synthase, MMP-1, MMP-3, MMP-9, and cyclin D1, thus reducing proliferation, growth, and tumor angiogenesis (Shanmugam et al., 2015). In addition, curcumin has an inhibitory capacity directly on COX-2 and 5-lipoxygenase, so it has a direct antiinflammatory activity, actively participating in the reduction of inflammation in the tumor microenvironment (Panahi et al., 2016).

Several in vitro and in vivo studies, in lines of melanoma and pancreatic cancer, have shown that the inhibition of NF-κβ activation is one of the main mechanisms by which curcumin exerts a potent antiproliferative and proapoptotic effect in cancer (Siwak et al., 2005; Li et al., 2004, 2005; Marin et al., 2007; Carbone and Melisi, 2012).

This mechanism of action of curcumin has also been shown to improve the anticancer response of the chemotherapeutic agent paclitaxel and to inhibit metastasis in human breast cancer in nude mice (Aggarwal et al., 2005; Kang et al., 2009). Likewise, curcumin sensitizes human colorectal cancer xenografts in nude mice to radiation, by the same mechanism (Kunnumakkara et al., 2008).

5.5.3 Inhibition of Activating Protein-1 Transcription Factor (AP-1)

AP-1 is a dimeric transcription factor that includes members of the JUN family of proteins, FOS, ATF (activating transcription factor), and MAF (musculoaponeurotic fibrosarcoma). Its regulatory activity depends on its dimeric composition. These transcription factors have a double role in cancer, promoting tumorigenesis or inhibiting it (Eferl and Wagner, 2003).

The dimers composed with c-JUN, a member of the JUN family, have proliferative activities increasing the expression of cyclin D1, thus promoting the progression of the cell cycle. Dimers composed of c-FOS, a member of the FOS family of proteins, promote tumor angiogenesis by increasing the expression of *VEGFD*, a gene that codes for VEGF. The AP-1 transcription factors composed of c-JUN or c-FOS are considered tumorigenic (Eferl and Wagner, 2003).

For its part, curcumin has the activity of inhibiting these protein complexes (c-JUN and c-FOS), reducing the expression of cyclin D1 and the *VEGFD* gene. This activity by curcumin reduces the growth rate of cancer cells and the process of angiogenesis, playing an important role in the treatment of cancer (Shanmugam et al., 2015; Balasubramanian and Eckert, 2007). The inhibition of the transcription factor of AP-1 has also been shown to be an important regulator in the apoptosis of epidermal keratinocytes, so this mechanism also has a role in the prevention of cancer (Balasubramanian and Eckert, 2007).

5.5.4 Activation of the Peroxisome Proliferator-Activated Receptor Gamma (PPAR)

PPAR are a class of transcription factors that are activated by different ligands and their function is the regulation of different genes involved in the metabolism of lipids, carbohydrates, inflammation, and differentiation of adipocytes. There are three main isoforms of PPARs in mammals, known as PPARα, PPARγ, and PPARα/γ, each regulating the expression of different genes through their DNA binding in the PPAR response elements (Mazidi et al., 2016).

Curcumin has been recognized as an activator of PPARγ, although the precise mechanism by which it activates and increases its expression is unknown. However, the activation of this isomer has been shown to reduce the expression of two transcription factors involved in carcinogenes: NF-κβ and signal transducer and activator of transcription 3 (STAT3). In addition, it reduces the expression of transforming growth factor-β (TGF-β), monocyte chemoattractant protein-1 (MCP-1), proinflammatory interleukins (IL-1, IL-2, IL-6, and IL-8), and cyclin D1, among the main ones (Mazidi et al., 2016). The regulation of these genes promotes an antiinflammatory response in vitro and in vivo, and several studies have found that the activation of PPARγ promotes tumor apoptosis and reduces its growth rate in various cancer cell lines in vitro (Prakobwong et al., 2011).

5.5.5 Epigenetic Regulation

Curcumin in various in vitro studies has shown regular epigenetic changes through the inhibition of HDAC enzymes and HATs. The inhibition of these enzymes modifies the state of chromatin, regulating the expression of various genes, including those involved in apoptosis and cell growth (Reuter et al., 2011).

A study conducted by Liu et al. (2005) showed that curcumin induced apoptosis in a cell line of Burkitt's lymphoma, and one of its mechanisms of action was reducing the expression of HDAC 1, 3, and 8, and inhibiting them directly. The inhibition of these HDACs increased the acetylation of H4 histones, which Liu et al. linked to increased expression of proapoptotic genes.

Another study, conducted by Balasubramanyam et al. (2004) in cervical cancer cell lines, found that curcumin induced its apoptosis by inhibiting the acetylation of histones and p53 protein, through the inhibition of HAT p300/CBP.

Curcumin has also been shown to have the ability to inhibit some DNMT enzymes, which is why it is considered a hypomethylating agent. The inhibition of DNMT reduces the methylation of the GpC-rich regions, thus increasing the expression of certain genes. In the case of cancer, regions of DNA that code for proapoptotic genes that are hypermethylated have been identified, so their expression is inhibited (Boyanapalli and Kong, 2015). Link et al. (2013) conducted an in vitro study with three colorectal cancer cell lines that have hypermethylated regions of DNA (HCT116, RKO, and HT29), demonstrating that curcumin culture reduced methylation of specific regions of DNA that were hypermethylated, so that, Link et al. concluded that curcumin is an important chemopreventive agent and that it can play an important role in the treatment of cancer.

5.5.6 Human Studies

The large number of in vitro and in vivo studies on curcumin and its anticarcinogenic effects have shown great potential, which has encouraged the performance of several studies in humans and found promising results. The results of some studies conducted with curcumin in humans are summarized below.

A phase II clinical study, which included 25 patients with advanced pancreatic cancer, gave 8 g of curcumin orally to each patient for 8 weeks. Three of these 25 patients showed important clinical improvements. One patient remained stable for more than 18 weeks and showed slow but constant reduction of the tumor marker CA125, for 1 year, while the size of his tumor remained stable. The second patient had a 73% reduction in his tumor size that lasted for a month, this lesion remained small while secondary lesions grew. The third patient showed stability in his weight and improvement in his well-being for approximately 8 months, during this period he did not show progression of his cancer. Furthermore, no patient showed signs of toxicity or intolerance to curcumin during the study (Dhillon et al., 2008).

Another phase II clinical study evaluated the chemopreventive activity of colorectal cancer of two different doses of curcumin: 2 and 4 g/day orally. The study was carried out with 41 patients who presented with aberrant crypt foci (ACF), considered a premalignant lesion. This study found that the 4 g/day dose significantly reduced the number of ACFs by 40% ($P < .005$), while the 2 g group showed no improvement. Curcumin was well tolerated in both doses and no patients showed toxicity (Carroll et al., 2011).

A pilot clinical study conducted by Hejazi et al. (2013) found that supplementation with 3 g of curcumin orally, in patients with prostate cancer who received radiotherapy, reduced the severity of urinary symptoms related to radiotherapy compared to the placebo group ($P = .011$) (Hejazi et al., 2013). Thus, this study demonstrated that curcumin not only has properties to treat cancer on its own, but also has an important coadjuvant and protective effect against modern medical therapy.

Turmeric and its active metabolite, curcumin, have been shown through various in vitro, in vivo, and clinical studies to be effective as an anticancer agent. Doses between 4 and 8 g a day orally seem to show the best results as a chemopreventive and chemotherapeutic agent. However, it is necessary to consider that, like the drugs used by modern medicine, tumor cells can show resistance to this agent, so only susceptible cells will be affected. Thus, the use of this phytochemical should be considered in combination with other phytochemicals or as a coadjutant of modern medical therapy.

5.5.6.1 *Zingiber officinale*

Zingiber officinale, commonly known as ginger, is a rhizome that has been used historically in different types of medicine for millennia, including Asian, Greek, Roman, Indian, Arabic, and Mediterranean. Over time it has been used to treat different diseases such as hypertension, dementia, fever, infectious diseases, and diabetes and to prevent nausea and/or vomiting in different gastrointestinal disorders. Nowadays, it is recognized that ginger possesses different bioactive metabolites with important antiinflammatory, anticancer, antioxidant, and antiemetic properties (Baliga et al., 2011).

Ginger has a large number of metabolites. However, an important part is volatile, so during cooking or drying this plant volatilizes. Among the nonvolatile bioactive compounds, to which their pharmacological properties are attributed, are gingerols, shogaols, paradols, and zingerone (Baliga et al., 2011; Zhang et al., 2017a).

The gingerols (4-gingerol, 6-gingerol, 8-gingerol, and 10-gingerol) are thermolabile bioactive compounds due to the presence of a β-hydroxy keto group, so that during the dehydration of ginger they are degraded to shogaols. Both compounds, the gingerols and shogaols, present different pharmacological, pharmacokinetic, and bioavailability activities, with the gingerols being more bioactive (Baliga et al., 2011).

For its part, extracts of ginger are rich in bioactive metabolites, which through various in vitro, in vivo, and preclinical studies have shown an antiproliferative, cytotoxic, and antiangiogenic potential. The main anticancer activities of some of the most bioactive metabolites of ginger are summarized below.

5.5.7 10-Gingerol

10-Gingerol (10-G) has been shown to have the most potent anticancer properties compared to the other gingerols. These properties are due to different mechanisms of action, among which are its antioxidant, antiinflammatory capacity, and its ability to modify the genetic expression and the induction of apoptosis of tumor cells. These mechanisms of action participate in an important way in the reduction of tumor initiation, promotion, and progression, being an important agent in chemoprevention and cancer treatment (Zhang et al., 2017a).

10-G has been shown to have antitumor properties against different cancer cell lines in vitro, including ovarian, lung, colon, prostate, and cervical cancer, through the induction of apoptosis and the inhibition of cell proliferation (Zhang et al., 2017a).

10-G exerts its antiproliferative effects by inducing cell cycle arrest in the G1 phase by inactivating phosphatidylinositol 3-kinase (PI3K)/protein kinase B (AKT) and mitogen-activated protein kinase (MAPK), through a reduction in the expression of epidermal growth factor receptors (EGFRs), in the HeLa cervical cancer cell line (Zhang et al., 2017b). EGFR is linked to the activation of signaling pathways mediated by PI3K/AKT and MAPK, which are responsible for increasing the synthesis of enzymes and cyclins that regulate the progression of the cell cycle and the survival of cells. For their part, these signaling pathways are related to the development of different cancers, such as colorectal, melanoma, lung, ovarian, and thyroid (Burotto et al., 2014), therefore 10-G could help in treatment of these cancers.

A study conducted by Ryu and Chung (2015) evaluated the antiproliferative activity of 10-G on the colon cancer cell line HCT-116, finding dose-dependent inhibition. In addition, they found morphological changes indicative of apoptosis, which were related to an increase in the expression of the mitochondrial protein Bax, a protein that regulates the release of mitochondrial cytochrome C and subsequent activation of caspase-mediated apoptosis. This study demonstrates the ability of 10-G to induce apoptosis of tumor cells by regulating the expression of proapoptotic proteins.

A study carried out by Martin et al. (2017) evaluated the anticancer activity of 10-G in vivo and in vitro in triple negative breast cancer models, finding important proapoptotic activity and the inhibition of metastasis. The activity of 10-G in vitro was dose-dependent, finding that at the highest concentration (100 μM) colony formation was completely inhibited and extensive cell death occurred. In addition, they found that, starting at the concentration of

50 μM, potent activation of caspase-3 and caspase-9 occurred, promoting cellular apoptosis. In the in vivo study they used a 4T1Br4 metastatic cancer model in mice, to which they administered 5 mg/kg of 10-G for 23 days (time in which the negative control showed signs of metastasis). At the end of the in vivo study, they found a significant inhibition of growth and tumor metastasis compared to the control group. The researchers concluded that 10-G could be safe and useful for use as a complementary therapy in metastatic breast cancer.

Another in vitro study with the triple negative breast cancer cell line (MDA-MB-231 and MDA-MB-468), conducted by Bernard et al. (2017), compared the anticancer activity of 6-gingerol, 8-gingerol, and 10-gingerol. The findings demonstrated that 10-gingerol was more potent than 6-gingerol and similar to 8-gingerol in inducing apoptosis of this cell line. In addition, they found that 10-gingerol on the MDA-MB-231 cell line increased the permeabilization of the mitochondrial outer membrane, increasing the release of cytochrome C and subsequent caspase activation, inducing cellular apoptosis. In the same way, the studies carried out by Martin et al. (2017) and Bernard et al. (2017) concluded that 10-G is a potent chemotherapeutic agent that could be used as part of a complementary treatment for breast cancer.

5.5.8 6-Gingerol

6-Gingerol (6-G) is another phytochemical found naturally in ginger and, like 10-G, this metabolite has demonstrated an important anticancer potential (Zhang et al., 2017a). Several studies in cancer cell lines have found that 6-G induces cell cycle arrest, apoptosis, and reduces the risk of metastasis, through different mechanisms, which are described below.

6-G induces cell cycle arrest in the G1 phase in colorectal cancer cell lines (HCT-116, SW480, HT-29, LoVo, and Caco-2) by suppressing the expression of cyclin D1 (Lee et al., 2008b). In the case of the LoVo cell line, 6-G induces cell cycle arrest by reducing cyclin D1, in addition to cyclin A, cyclin B1, and cyclin-dependent kinases-1 (CDK1) (Lee et al., 2008a,b; Lin et al., 2012). In addition, an inhibition of the signaling pathway mediated by extracellular signal-regulated kinase (ERK1)/c-jun NH 2-terminal kinase (JNK)/AP-1 has been found in lines SW-480 and HCT-116, thus reducing cell proliferation (Radhakrishnan et al., 2014).

In various studies on colorectal cancer cell lines, 6-G has demonstrated proapoptotic effects by increasing the expression of the nonsteroidal antiinflammatory drug-activated gene-1 (NAG-1). This last gene codes for a cytokine with proapoptotic and antitumorigenic properties. 6-G has also been shown to promote apoptosis by increasing the activation of caspases 3, 8, and 9 (Lee et al., 2008a,b; Radhakrishnan et al., 2014).

A study conducted by Park et al. (2006) evaluated the activity of 6-G in two pancreatic cancer cell lines, BxPC-3 expressing the mutated *p53* gene and HPAC expressing wild-type (wt) *p53*.

The study found that 6-G induced arrest of the cell cycle in both cell lines by reducing the expression of cyclin A and CDKA. In addition, they found that 6-G induced cell death by apoptosis in the cell line with the mutated *p53* gene. Thus, 6-G demonstrates an important potential to treat pancreatic cancer, one of the most aggressive cancers with the lowest survival rates.

6-G has been shown to have a potential inhibitor of metastasis in in vitro and in vivo studies by different mechanisms, including a reduction in the expression of MMP and inhibiting angiogenesis. This last mechanism is not completely understood. However, it is associated with the inhibition of VEGF-induced gene expression in endothelial cells (Kim et al., 2005; Lee et al., 2008a).

5.5.9 Human Studies

Regarding clinical studies on ginger or its bioactive metabolites on cancer, no studies have been found to evaluate its antitumor potential. Rather, its potential has been evaluated with an antiemetic agent to treat the nausea and vomiting induced by the oncological chemotherapy of modern medicine.

A multicenter study that included 576 patients evaluated the activity of ginger consumption to reduce the nausea induced by cancer chemotherapy. The study included three groups with different doses of ginger (0.5, 1, and 1.5 g) and one placebo group. The results showed that all evaluated doses of ginger reduced nausea significantly compared to placebo, finding better results with doses of 0.5 and 1.0 g (P=.017 and P=.036, respectively) (Ryan et al., 2012).

Likewise, a phase II clinical study evaluated the antiemetic activity of 6-G in patients (88 patients in total) with solid tumors receiving moderately and highly emetogenic chemotherapy. The patients in the study all consumed ondansetron, metoclopramide, and dexamethasone. Patients in the treatment group (N=42) received 10 mg of 6-G twice a day for 12 weeks. The results of the study showed that 6-G has a significant antiemetic effect compared to placebo (P=<.001). In addition, the study found that 6-G significantly reduced fatigue (P=.020) and loss of appetite (P=.001) compared to placebo (Konmun et al., 2017).

According to the clinical studies presented, it can be concluded that ginger and its bioactive metabolites have an antiemetic activity and reduce the nausea induced by chemotherapy, significantly. Therefore, the use of 0.5–1.5 g of powdered ginger (divided into two doses per day) or 10 mg of 6-G every 12 h, may be part of complementary cancer therapy to prevent side effects caused by modern oncological chemotherapy.

In summary, ginger and its bioactive metabolites (10-G and 6-G) present an important anticancer potential for different cancers through the induction of apoptosis, arrest of the cell cycle, and inhibition of angiogenesis and metastasis. Clinical studies, despite evaluating its antiemetic activity and not anticancer properties, have shown that ginger and its metabolite,

6-G, have biological activities in humans and do not generate toxicity and/or side effects. Thus, the use of ginger powder or extracts for the complementary treatment of cancer is useful to reduce the side effects of conventional chemotherapy and could represent an important strategy to reduce the initiation, promotion, and/or progression of cancer.

5.5.9.1 *Moringa oleifera*

Moringa oleifera (MO) is an indigenous tree from the north of India, Pakistan, and Nepal, of which all its components (leaves, seeds, flowers, and bark) are considered medicinal. Its medicinal components have been used by traditional Ayurvedic medicine, curators of Siddha, Egyptians, Greeks, and Romans, among the main traditional medicines. Due to its medicinal properties it has been given names like "miracle tree" and "mother's best friend" (Leone et al., 2015).

The different parts of the MO plant have been used for different medicinal purposes, although, in general, the leaves are most commonly used. Its leaves have been used in traditional medicine to treat conditions such as malaria, fever, arthritis, high blood pressure, diabetes, parasitic diseases, skin lesions, and even HIV/AIDS. Its roots have been used to prepare infusions for the treatment of parasites, toothache, and to enhance sexual desire. The flowers are considered aphrodisiac and antiinflammatory, useful for the treatment of hysteria, muscle pain, splenomegaly, and even tumors (Leone et al., 2015).

This section will focus on summarizing the medicinal properties of MO leaves for cancer, since they are the most used part for the treatment of different diseases in traditional medicines. However, if the case arises, a comparison of the biological effect between its different parts will be made.

MO leaves have a large number of bioactive compounds, with the main ones including vitamins, polyphenols, phenolic acids, carotenoids, flavonoids, glucosinolates, alkaloids, tannins, saponins, oxalates, and phytates. The concentration of its bioactive compounds varies depending on the growing region of the plant, since the different environmental conditions modify its synthesis (Leone et al., 2015).

Among its main active metabolites are myricetin, quercetin, kaempferol, isorhamnetin, rutin, caffeic acid, chlorogenic acid, ellagic acid, ferulic acid, gallic acid, N α-L-rhamnopyranosyl vincosamide, 4-(α-L-rhamnopyranosyloxy) phenylacetonitrile (Niazirin), benzyl, and sinalbin (Leone et al., 2015; Rani et al., 2018). However, these are only the main components, as the leaves have a large number of metabolites with potential biological effect. Due to its broad content of metabolites, the extract of the leaves is usually used for different studies and not its metabolites in isolation.

Currently, several in vitro studies have evaluated the potential biological effect of MO extracts, found to possess antiatheroclerotic, immunostimulant, antidiabetic, antihypertensive, antioxidant, antiinflammatory, antimicrobial, and anticancer activities (Rani et al., 2018; Mishra et al., 2015; Kou et al., 2018). However, few studies in humans have been performed, and none in cancer patients.

Some of the in vitro effects of MO extracts on cancer cell lines are summarized below.

A study conducted by Jung (2014) evaluated the anticancer potential and its mechanisms of action of an aqueous extract of MO, in several cancer cell lines, including breast, lung, colon, squamous cell carcinoma, fibrosarcoma, etc. However, Jung only presents the results specifically for the lung cancer cell line A549; while, for the other cell lines, it only mentions that it found a cytotoxic and antiproliferative effect. The cellular mechanism of MO to induce cell death of A549 was through the induction of caspase activation. The antiproliferative effect was probably due to a reduction in the expression of AKT, NF-κβ, ERK, and cyclin D1. Likewise, the MO extract also showed some toxicity in the normal control cell line (COS-7). However, it was much lower than in the cancer cell lines. Thus, the author concluded that the MO extract has a specific cytotoxic and antiproliferative effect against cancer cells and could be an important antitumor agent (Jung, 2014).

Also, another study conducted with lung cancer A549, found that the aqueous extract of MO reduced the intracellular levels of glutathione and the expression of Nrf2 (nuclear transcription factor erythroid 2p45 [NF-E2]-related factor 2), reducing the antioxidant capacity of the cancer cell. Nrf2 regulates the expression of antioxidant enzymes, thus its inhibition reduces the expression of these enzymes. Moreover, the extract also significantly induced an increase in the expression of the p53 protein, caspase-9, and caspase-3/7, thus inducing apoptosis of the cell line (Tiloke et al., 2013).

Another study, conducted with the aqueous extract of MO leaves, found that the extract inhibited cell growth and NF-κβ signaling in different pancreatic cancer lines (Panc-1, p34, and COLO 357). In addition, the extract enhanced the cytotoxic effect of the cisplatin chemotherapeutic agent and induced cell arrest in the G1 phase on the Panc-1 line (Berkovich et al., 2013).

With regard to breast and colon cancer, Al-Asmari et al. (2015) evaluated the anticancer activity of extracts of seeds, leaves, and bark of MO, on cell lines of these cancers (MDA-MB-231 and HCT-8, respectively). The study found that the leaves and bark of MO had important anticancer properties, while the seed extract did not. The results illustrated that leaf and bark extracts inhibited cell growth and induced late apoptosis on both cell lines.

In summary, extracts from the MO leaf have been shown to have significant potential as a chemotherapeutic agent in different cancers, including breast, colon, lung, and pancreas. In addition, according to in vitro studies, MO leaf extract could be an important sensitizing agent to the drug cisplatin for the treatment of pancreatic cancer. Further studies are needed on the mechanisms of action involved in its anticancer potential, since in some cell lines it reduced the expression of antioxidant enzymes, which could represent a risk for normal cells.

In vivo studies and later clinical models are necessary to be able to define effective doses of MO and prevent side effects due to possible toxicity.

5.5.9.2 Azadirachta indica

Azadirachta indica, commonly known as neem, is a tree native to semitropical and tropical climates, and is found in countries such as India, Pakistan, and Bangladesh (Hao et al., 2014). Currently, it is thought that neem is found in at least 30 countries around the world, mainly in Southeast Asia, Africa, Australia, and in South and Central America (Kumar and Navaratnam, 2013). In some parts of Asia it is consumed as a vegetable, although its use is mainly as a medicinal plant to treat bacterial, viral, fungal, and parasitic infections, as well as a herbal medicine for the treatment of diabetes mellitus. Recent in vitro studies have shown that neem possesses anticancer activities such as antiproliferative, proapoptotic, inhibitor of tumor angiogenesis, and immunostimulant, although the mechanisms of action are not yet very clear (Hao et al., 2014).

More than 300 neem phytochemicals have been isolated and identified, so it has a wide therapeutic potential for various diseases. Two main classes of phytochemicals from different parts of the neem have been identified, isoprenoids and nonisoprenoids. Among the isoprenoids are diterpenes, triterpenes, limonoids, vilasinins, and c-secomeliacin. While, in nonisoprenoids, proteins, polysaccharides, sulfur compounds, dihydrochalcone, tannin, and polyphenols have been identified. Likewise, some metabolites derived from neem have been identified that give it its medicinal properties, among them nimbine; this phytochemical possesses biological activities such as antiinflammatory, antipyretic, fungicidal, antiseptic, and antihistamine. Other metabolites derived from neem are nimbolide, azadirachtin, azadirone, azadiradione, and gedunine (Hao et al., 2014; Gupta et al., 2017).

Currently, as far as cancer is concerned, extracts from different parts of neem and one of its main biactive metabolites, known as nimbolide, have been studied. Both the extracts and isolated nimbolide have presented important antiproliferative and proapoptotic activities, in both in vitro and in vivo models. Below are the main anticancer activities of neem extracts and nimbolide, as well as the possible mechanisms of action involved.

5.5.10 Neem Extracts

Studies of neem extracts in breast and prostate cancer cell lines have been shown to have antiproliferative and proapoptotic activities, although their mechanism of action has been poorly studied (Mahapatra et al., 2011; Arumugam et al., 2014).

Among the investigated mechanisms of action it has been found that neem leaf extracts increase the expression of 40 genes in the C4-2B and PC-3M-luc2 cell lines of prostate cancer. Among these genes are antioxidant enzymes (e.g., GCLM and TXNRD1), and those related to cell cycle arrest (e.g., SESN2 and MDM2) (complete list in Mahapatra et al., 2011). Also, among the regulated genes were *AKR1C2*, *AKR1C3*, and *AKR1B10*, which are related to the metabolism of dihydrotestosterone (DHT) in prostate cells. The increase in the latter genes is related to a greater catabolism of DHT and a lower growth of prostate cancer. The extract also reduced the expression of genes related to tumor growth, such as *SMC2*, *SMC3*, *SMC4*, *STAG2*, *TFPI*, and *HELLS*, among others (Mahapatra et al., 2011).

A study conducted by Arumugam et al. (2014) evaluated the antitumorigenic activity of the ethanolic extract of neem leaves against breast cancer tumor progression in a rat model with *N*-methyl-N-nitrosourea-induced tumorigenesis (MNU) and investigated the molecular mechanisms of action. The results of the study showed that the extract of the neem leaf were highly effective in inhibiting tumor progression in rats. Different mechanisms of action were found, including regulation of gene expression and intracellular pathways. There was an increase in the expression of proapoptotic genes, such as p53, B cell lymphoma-2-associated X protein (Bax), Bcl-2 associated death promoter protein (Bad) caspases. Likewise, the expression of the antiapoptotic protein Bcl-2, angiogenic factors (angiopoietin and VEGF-A), regulatory proteins of the cell cycle (cyclin D1, Cdk2, and Cdk4) were reduced. In addition, the activity of the pathways mediated by NF-κβ and MAPK1 were reduced, probably related to the modulation in gene expression. Thus, this study demonstrates the potent anticancer effect of neem leaves extract to inhibit carcinogenesis mediated by MNU.

5.5.11 Nimbolide

Nimbolide (Nb) is one of the main metabolites with anticancer action found in *Azadirachta indica*. Nb is a tetranortriterpenoid with a δ-lactonic ring, which has presented different anticancer activities, such as antiproliferative, proapoptotic, inhibitors of angiogenesis and antimetastases (Elumalai and Arunakaran, 2014).

In vitro, Nb presents cytotoxic and antiproliferative activity against a large number of cancer cell lines, with neuroblastoma (NE-115), osteosarcoma (143B), prostate (MIA PaCa-2, BX-PC3), colorectal cancer (HCT-116, HT-29 and Caco-2), being the main ones (Gupta et al., 2013, Elumalai and Arunakaran, 2014, Kumar et al., 2017; Subramani et al., 2016). Different mechanisms of action have been identified by which Nb exerts these activities, among these is a reduction in the expression of proteins, such as: antiapoptotic (Bcl-2 and Bcl-xL), proliferative (cyclin D1), related to metastasis (MMP-9), and angiogenesis (VEGF), through the inhibition of NF-κβ (Gupta et al., 2013; Elumalai and Arunakaran, 2014; Kumar et al., 2017; Subramani et al., 2016).

Another important mechanism for the development of cancer is EMT (discussed in Section 5.2). A study conducted by Subramani et al. (2016) showed that Nb reduces the EMT of pancreatic cancer cells, which inhibited the migration and invasion of cancer cells. Nb reduced the EMT by an increase in the expression of the intercellular adhesion protein E-cadherin and reduced the expression of Notch-2, vimentin, Snail, Slug, and Zeb, proteins related to the EMT. Thus, Nb plays an important role in the prevention of tumor progression by inhibiting EMT.

5.5.12 Human Studies

Unfortunately, there are no studies in humans that evaluate the anticancer potential of neem. However, a study conducted by Mbah et al. (2007) evaluated the use of an extract of neem leaves in people with HIV/AIDS who had a CD4 cell count of less than 300 cells/μL. This study included 60 patients (between 23 and 50 years) who were given 500 mg of aqueous extract of neem leaves, every 12 h (total 1 g day) for 12 weeks, in order to evaluate its antiretroviral and immunostimulatory effect. The results of the study found that patients had a significant increase in their lymphocyte count (266 cells/μL, +159%) and there were no side effects or alterations in the parameters of liver and kidney function. Thus, this study gives the guideline that the use of 1 g/day of extracts of neem leaves presents a great biological activity and does not represent a danger to health during a period of consumption of 12 weeks.

In summary, the extracts of *Azadirachta indica* and its metabolite, nimbolide, have an important in vitro anticancer activity, through different mechanisms of action that participate in the different stages of carcinogenesis. Thus, its use as part of complementary therapy could represent an important strategy for the prevention and treatment of cancer.

5.6 Conclusion

Medicinal plants have a large number of bioactive metabolites with anticancer effects through various mechanisms of action, which affect the different stages of carcinogenesis, so their use could be effective for chemoprevention and complementary treatment. On the other hand, some mechanisms of action of the metabolites can favor the response to the chemotherapeutic agents of modern medicine and reduce the number or intensity of side effects, thus improving the quality of life of the patient with cancer.

The use of bioactive extracts or metabolites derived from medicinal plants in suitable doses and in an appropriate form could be an important strategy for the prevention and treatment of cancer. Therefore, a greater number of in vitro, in vivo, and clinical investigations are necessary to finish defining the mechanisms of action of the metabolites, which would allow choosing the right dose and time to use them in a patient with cancer.

References

Aggarwal, B.B., Shishodia, S., Takada, Y., Banerjee, S., Newman, R.A., Bueso-Ramos, C.E., Price, J.E., 2005. Curcumin suppresses the paclitaxel-induced nuclear factor-κB pathway in breast cancer cells and inhibits lung metastasis of human breast cancer in nude mice. Clinical Cancer Research 11 (20), 7490–7498. https://doi.org/10.1158/1078-0432.CCR-05-1192.

Al-Asmari, A.K., Albalawi, S.M., Athar, M.T., Khan, A.Q., Al-Shahrani, H., Islam, M., 2015. *Moringa oleifera* as an anti-cancer agent against breast and colorectal cancer cell lines. PLoS One 10 (8), 1–14. https://doi.org/10.1371/journal.pone.0135814.

Arumugam, A., Agullo, P., Boopalan, T., Nandy, S., Lopez, R., Gutierrez, C., et al., 2014. Neem leaf extract inhibits mammary carcinogenesis by altering cell proliferation, apoptosis, and angiogenesis. Cancer Biology and Therapy 15 (1), 26–34. https://doi.org/10.4161/cbt.26604.

Baba, A.I., Câtoi, C., 2007. Comparative Oncology. The Publishing House of the Romanian Academy, Bucharest. Chapter 19, Principles of Anticancer Therapy. Available from: https://www.ncbi.nlm.nih.gov/books/NBK9546/.

Balasubramanian, S., Eckert, R.L., 2007. Curcumin suppresses AP1 transcription factor-dependent differentiation and activates apoptosis in human epidermal keratinocytes. Journal of Biological Chemistry 282 (9), 6707–6715. https://doi.org/10.1074/jbc.M606003200.

Balasubramanyam, K., Varier, R.A., Altaf, M., Swaminathan, V., Siddappa, N.B., Ranga, U., Kundu, T.K., 2004. Curcumin, a novel p300/CREB-binding protein-specific inhibitor of acetyltransferase, represses the acetylation of histone/nonhistone proteins and histone acetyltransferase-dependent chromatin transcription. Journal of Biological Chemistry 279 (49), 51163–51171. https://doi.org/10.1074/jbc.M409024200.

Baliga, M.S., Haniadka, R., Pereira, M.M., D'Souza, J.J., Pallaty, P.L., Bhat, H.P., Popuri, S., 2011. Update on the chemopreventive effects of ginger and its phytochemicals. Critical Reviews in Food Science and Nutrition 51 (6), 499–523. https://doi.org/10.1080/10408391003698669.

Basnet, P., Skalko-Basnet, N., 2011. Curcumin: an anti-inflammatory molecule from a curry spice on the path to cancer treatment. Molecules 16 (6), 4567–4598. MDPI AG. Retrieved from: https://doi.org/10.3390/molecules16064567.

Berkovich, L., Earon, G., Ron, I., Rimmon, A., Vexler, A., Lev-Ari, S., 2013. *Moringa oleifera* aqueous leaf extract down-regulates nuclear factor-kappaB and increases cytotoxic effect of chemotherapy in pancreatic cancer cells. BMC Complementary and Alternative Medicine 13 (212), 1–7. https://doi.org/10.1186/1472-6882-13-212.

Bernard, M.M., McConnery, J.R., Hoskin, D.W., 2017. [10]-Gingerol, a major phenolic constituent of ginger root, induces cell cycle arrest and apoptosis in triple-negative breast cancer cells. Experimental and Molecular Pathology 102 (2), 370–376. https://doi.org/10.1016/j.yexmp.2017.03.006.

Boyanapalli, S.S.S., Kong, A.-N.T., 2015. "Curcumin, the king of spices": epigenetic regulatory mechanisms in the prevention of cancer, neurological, and inflammatory diseases. Current Pharmacology Reports 1 (2), 129–139. https://doi.org/10.1007/s40495-015-0018-x.

Burotto, M., Chiou, V.L., Lee, J.-M., Kohn, E.C., 2014. The MAPK pathway across different malignancies: a new perspective. Cancer 120 (22), 3446–3456. https://doi.org/10.1002/cncr.28864.

Carbone, C., Melisi, D., 2012. NF-κB as a target for pancreatic cancer therapy. Expert Opinion on Therapeutic Targets 16 (Suppl. 2), S1–S10. https://doi.org/10.1517/14728222.2011.645806.

Carroll, R.E., Benya, R.V., Turgeon, D.K., Vareed, S., Neuman, M., Rodriguez, L., et al., 2011. Phase IIa clinical trial of curcumin for the prevention of colorectal neoplasia. Cancer Prevention Research 4 (3), 354–364. https://doi.org/10.1158/1940-6207.CAPR-10-0098.

Dhillon, N., Aggarwal, B.B., Newman, R.A., Wolff, R.A., Kunnumakkara, A.B., Abbruzzese, J.L., Kurzrock, R., 2008. Phase II trial of curcumin in patients with advanced pancreatic cancer. Clinical Cancer Research 14 (14), 4491–4499. https://doi.org/10.1158/1078-0432.CCR-08-0024.

Eferl, R., Wagner, E.F., 2003. AP-1: a double-edged sword in tumorigenesis. Nature Reviews Cancer 3 (11), 859–868. https://doi.org/10.1038/nrc1209.

Elumalai, P., Arunakaran, J., 2014. Review on molecular and chemopreventive potential of nimbolide in cancer. Genomicsa and Informatics 12 (4), 156–164. https://doi.org/10.5808/GI.2014.12.4.156.

Ferlay, J., Soerjomataram, I., Dikshit, R., Eser, S., Mathers, C., Rebelo, M., Bray, F., 2015. Cancer incidence and mortality worldwide: Sources, methods and major patterns in GLOBOCAN 2012. International Journal of Cancer 136 (5), E359–E386. https://doi.org/10.1002/ijc.29210.

Foster, K., Younger, N., Aiken, W., Brady-West, D., Delgoda, R., 2017. Reliance on medicinal plant therapy among cancer patients in Jamaica. Cancer Causes and Control 28 (11), 1349–1356. https://doi.org/10.1007/s10552-017-0924-9.

Francis, P.A., Kris, M.G., Rigas, J.R., Grant, S.C., Miller, V.A., 1995. Paclitaxel (Taxol) and Docetaxel (Taxotere): active chemotherapeutic agents in lung cancer. Lung Cancer 12 (Suppl. 1), S163–S172. https://doi.org/10.1016/0169-5002(95)00432-Z.

Fridlender, M., Kapulnik, Y., Koltai, H., 2015. Plant derived substances with anti-cancer activity: from folklore to practice. Frontiers in Plant Science 6 (October), 1–9. https://doi.org/10.3389/fpls.2015.00799.

Greenwell, M., Rahman, P.K.S.M., 2015. Medicinal Plants: their use in anticancer treatment. International Journal of Pharmaceutical Sciencies and Research 6 (10), 4103–4112. https://doi.org/10.13040/IJPSR.0975-8232.6(10).4103-12.

Gupta, S.C., Prasad, S., Sethumadhavan, D.R., Nair, M.S., Mo, Y.Y., Aggarwal, B.B., 2013. Nimbolide, a limonoid triterpene, inhibits growth of human colorectal cancer xenografts by suppressing the proinflammatory microenvironment. Clinical Cancer Research 19 (16), 4465–4476. https://doi.org/10.1158/1078-0432.CCR-13-0080.

Gupta, S.C., Prasad, S., Tyagi, A.K., Kunnumakkara, A.B., Aggarwal, B.B., 2017. Neem (*Azadirachta indica*): an indian traditional panacea with modern molecular basis. Phytomedicine 34 (May), 14–20. https://doi.org/10.1016/j.phymed.2017.07.001.

Hao, F., Kumar, S., Yadav, N., Chandra, D., 2014. Neem components as potential agents for cancer prevention and treatment. Biochimica et Biophysica Acta 1846 (1), 247–257. https://doi.org/10.1016/j.bbcan.2014.07.002.

Harrington, K.J., 2016. The biology of cancer. Medicine 44 (1), 1–5. https://doi.org/10.1016/j.mpmed.2015.10.005.

Hejazi, J., Rastmanesh, R., Taleban, F.-A., Molana, S.-H., Ehtejab, G., 2013. A Pilot Clinical Trial of Radioprotective Effects of Curcumin Supplementation in Patients with Prostate Cancer. Journal of Cancer Science & Therapy 5 (10), 320–324. https://doi.org/10.4172/1948-5956.1000222.

Henrotin, Y., Priem, F., Mobasheri, A., 2013. Curcumin: a new paradigm and therapeutic opportunity for the treatment of osteoarthritis: curcumin for osteoarthritis management. SpringerPlus 2 (56). https://doi.org/10.1186/2193-1801-2-56.

Hoff, P.M., Kalil, K., 2012. Role of angiogenesis in the pathogenesis of cancer. Cancer Treatment Reviews 38 (7), 825–833. https://doi.org/10.1016/j.ctrv.2012.04.006.

Jaradat, N.A., Shawahna, R., Eid, A.M., Al-Ramahi, R., Asma, M.K., Zaid, A.N., 2016. Herbal remedies use by breast cancer patients in the West Bank of Palestine. Journal of Ethnopharmacology 178, 1–8. https://doi.org/10.1016/j.jep.2015.11.050.

Jones, P.A., Issa, J.P.J., Baylin, S., 2016. Targeting the cancer epigenome for therapy. Nature Reviews Genetics 17 (10), 630–641. https://doi.org/10.1038/nrg.2016.93.

Jung, I.L., 2014. Soluble extract from *Moringa oleifera* leaves with a new anticancer activity. PLoS One 9 (4), 1–10. https://doi.org/10.1371/journal.pone.0095492.

Kalluri, R., Weinberg, R.A., 2009. The basics of epithelial-mesenchymal transition. The Journal of Clinical Investigation 119 (6), 1420–1428. http://doi.org/10.1172/JCI39104.

Kang, H.J., Lee, S.H., Price, J.E., Kim, L.S., 2009. Curcumin suppresses the paclitaxel-induced nuclear factor kappa B in breast cancer cells and potentiates the growth inhibitory effect of paclitaxel in breast cancer nude mice model. The Breast Journal 15 (3), 223–229. https://doi.org/10.1111/j.1524-4741.2009.00709.x.

Kim, E.C., Min, J.K., Kim, T.Y., Lee, S.J., Yang, H.O., Han, S., Kwon, Y.G., 2005. [6]-Gingerol, a pungent ingredient of ginger, inhibits angiogenesis in vitro and in vivo. Biochemical and Biophysical Research Communications 335 (2), 300–308. https://doi.org/10.1016/j.bbrc.2005.07.076.

Konmun, J., Danwilai, K., Ngamphaiboon, N., Sripanidkulchai, B., Sookprasert, A., Subongkot, S., 2017. A phase II randomized double-blind placebo-controlled study of 6-gingerol as an anti-emetic in solid tumor patients receiving moderately to highly emetogenic chemotherapy. Medical Oncology 34 (69), 1–10. https://doi.org/10.1007/s12032-017-0931-4.

Kou, X., Li, B., Olayanju, J., Drake, J., Chen, N., 2018. Nutraceutical or pharmacological potential of *Moringa oleifera* Lam. Nutrients 10 (343), 1–12. https://doi.org/10.3390/nu10030343.

Kumar, S., Inigo, J.R., Kumar, R., Chaudhary, A.K., O'Malley, J., Balachandar, S., Chandra, D., 2017. Nimbolide reduces CD44 positive cell population and induces mitochondrial apoptosis in pancreatic cancer cells. Cancer Letters https://doi.org/10.1016/j.canlet.2017.10.029.

Kumar, V.S., Navaratnam, V., 2013. Neem (*Azadirachta indica*): prehistory to contemporary medicinal uses to humankind. Asian Pacific Journal of Tropical Biomedicine 3 (7), 505–514. https://doi.org/10.1016/S2221-1691(13)60105-7.

Kunnumakkara, A.B., Diagaradjane, P., Guha, S., Deorukhkar, A., Shentu, S., Aggarwal, B.B., Krishnan, S., 2008. Curcumin sensitizes human colorectal cancer xenografts in nude mice to γ-radiation by targeting nuclear factor-κB-regulated gene products. Clinical Cancer Research 14 (7), 2128–2136. https://doi.org/10.1158/1078-0432.CCR-07-4722.

Lamouille, S., Xu, J., Derynck, R., 2014. Molecular mechanisms of epithelial–mesenchymal transition. Nature Reviews Molecular Cell Biology 15 (3), 178–196. https://doi.org/10.1038/nrm3758.

Lee, H.S., Seo, E.Y., Kang, N.E., Kim, W.K., 2008a. [6]-Gingerol inhibits metastasis of MDA-MB-231 human breast cancer cells. Journal of Nutritional Biochemistry 19 (5), 313–319. https://doi.org/10.1016/j.jnutbio.2007.05.008.

Lee, S.-H., Cekanova, M., Baek, S.J., 2008b. Multiple mechanisms are involved in 6-gingerol-induced cell growth arrest and apoptosis in human colorectal cancer cells. Molecular Carcinogenesis 47 (3), 197–208. https://doi.org/10.1002/mc.20374.

Leone, A., Spada, A., Battezzati, A., Schiraldi, A., Aristil, J., Bertoli, S., Cultivation, 2015. Genetic, Ethnopharmacology, Phytochemistry and Pharmacology of *Moringa oleifera* Leaves: An Overview. International Journal of Molecular Sciences 16 (6), 12791–12835. http://doi.org/10.3390/ijms160612791.

Li, L., Aggarwal, B.B., Shishodia, S., Abbruzzese, J., Kurzrock, R., 2004. Nuclear factor-κB and IκB are constitutively active in human pancreatic cells, and their down-regulation by curcumin (Diferuloylmethane) is associated with the suppression of proliferation and the induction of apoptosis. Cancer 101 (10), 2351–2362. https://doi.org/10.1002/cncr.20605.

Li, L., Braiteh, F.S., Kurzrock, R., 2005. Liposome-encapsulated curcumin: in vitro and in vivo effects on proliferation, apoptosis, signaling, and angiogenesis. Cancer 104 (6), 1322–1331. https://doi.org/10.1002/cncr.21300.

Lin, C., Bin, Lin, C.C., Tsay, G.J., 2012. 6-gingerol inhibits growth of colon cancer cell LoVo via induction of G2/M arrest. Evidence-Based Complementary and Alternative Medicine 2012, 1–7. https://doi.org/10.1155/2012/326096.

Link, A., Balaguer, F., Shen, Y., Lozano, J.J., Leung, H.C.E., Boland, C.R., Goel, A., 2013. Curcumin modulates DNA methylation in colorectal cancer cells. PLoS One 8 (2), 1–13. https://doi.org/10.1371/journal.pone.0057709.

Liu, H.L., Chen, Y., Cui, G.H., Zhou, J.F., 2005. Curcumin, a potent anti-tumor reagent, is a novel histone deacetylase inhibitor regulating B-NHL cell line Raji proliferation. Acta Pharmacologica Sinica 26 (5), 603–609. https://doi.org/10.1111/j.1745-7254.2005.00081.x.

Mahapatra, S., Karnes, R.J., Holmes, M.W., Young, C.Y.F., Cheville, J.C., Kohli, M., et al., 2011. Novel molecular targets of *Azadirachta indica* associated with inhibition of tumor growth in prostate cancer. The AAPS Journal 13 (3), 365–377. https://doi.org/10.1208/s12248-011-9279-4.

Marin, Y.E., Wall, B.A., Wang, S., Namkoong, J., Martino, J.J., Suh, J., Ryu, J.H., 2007. Curcumin downregulates the constitutive activity of NF-kappaB and induces apoptosis in novel mouse melanoma cells. Melanoma Research 17 (5), 274–283. https://doi.org/10.1097/CMR.0b013e3282ed3d0e.

Martin, A.C.B.M., Fuzer, A.M., Becceneri, A.B., da Silva, J.A., Tomasin, R., Denoyer, D., et al., 2017. [10]-Gingerol induces apoptosis and inhibits metastatic dissemination of triple negative breast cancer in vivo. Oncotarget 8 (42), 72260–72271. https://doi.org/10.18632/oncotarget.20139.

Mazidi, M., Karimi, E., Meydani, M., Ghayour-Mobarhan, M., Ferns, G.A., 2016. Potential effects of curcumin on peroxisome proliferator-activated receptor-γ in vitro and in vivo. World Journal of Methodology 6 (1), 112–117. https://doi.org/10.5662/wjm.v6.i1.112.

Mbah, A.U., Udeinya, I.J., Shu, E.N., Chijioke, C.P., Nubila, T., Udeinya, F., et al., 2007. Fractionated neem leaf extract is safe and increases CD4+ cell levels in HIV/AIDS patients. American Journal of Therapeutics 14 (4), 369–374. https://doi.org/10.1097/MJT.0b013e3180a72199.

Miozzo, M., Vaira, V., Sirchia, S.M., 2015. Epigenetic alterations in cancer and personalized cancer treatment. Future Oncology 11 (2), 333–348. https://doi.org/10.2217/fon.14.237.

Mishra, A., Kumar, R., Tyagi, A., Kohaar, I., Hedau, S., Bharti, A.C., et al., 2015. Curcumin modulates cellular AP-1, NF-kB, and HPV16 E6 proteins in oral cancer. Ecancermedicalscience 9 (525), 1–12. https://doi.org/10.3332/ecancer.2015.525.

Nishida, N., Yano, H., Nishida, T., Kamura, T., Kojiro, M., 2006. Angiogenesis in cancer. Vascular Health and Risk Management 2 (3), 213–219.

Okada, F., 2014. Inflammation-related carcinogenesis: current findings in epidemiological trends, causes and mechanisms. Yonago Acta Medica 57 (2), 65–72.

Panahi, Y., Darvishi, B., Ghanei, M., Jowzi, N., Beiraghdar, F., Varnamkhasti, B.S., 2016. Molecular mechanisms of curcumins suppressing effects on tumorigenesis, angiogenesis and metastasis, focusing on NF-κB pathway. Cytokine and Growth Factor Reviews 28, 21–29. https://doi.org/10.1016/j.cytogfr.2015.12.004.

Park, Y.J., Wen, J., Bang, S., Park, S.W., Song, S.Y., 2006. [6]-Gingerol induces cell cycle arrest and cell death of mutant p53-expressing pancreatic cancer cells. Yonsei Medical Journal 47 (5), 688–697. https://doi.org/10.3349/ymj.2006.47.5.688.

Petrovska, B.B., 2012. Historical review of medicinal plants' usage. Pharmacognosy Reviews 6 (11), 1–5. https://doi.org/10.4103/0973-7847.95849.

Prakobwong, S., Gupta, S.C., Kim, J.H., Sung, B., Pinlaor, P., Hiraku, Y., et al., 2011. Curcumin suppresses proliferation and induces apoptosis in human biliary cancer cells through modulation of multiple cell signaling pathways. Carcinogenesis 32 (9), 1372–1380. https://doi.org/10.1093/carcin/bgr032.

Radhakrishnan, E.K., Bava, S.V., Narayanan, S.S., Nath, L.R., Thulasidasan, A.K.T., Soniya, E.V., Anto, R.J., 2014. [6]-Gingerol induces caspase-dependent apoptosis and prevents PMA-induced proliferation in colon cancer cells by inhibiting MAPK/AP-1 signaling. PLoS One 9 (8). https://doi.org/10.1371/journal.pone.0104401.

Rani, N.Z.A., Husain, K., Kumolosasi, E., 2018. Moringa genus: a review of phytochemistry and pharmacology. Frontiers in Pharmacology 9 (108), 1–26. https://doi.org/10.3389/fphar.2018.00108.

Reuter, S., Gupta, S.C., Park, B., Goel, A., Aggarwal, B.B., 2011. Epigenetic changes induced by curcumin and other natural compounds. Genes and Nutrition 6 (2), 93–108. https://doi.org/10.1007/s12263-011-0222-1.

Ryan, J.L., Heckler, C.E., Roscoe, J.A., Dakhil, S.R., Kirshner, J., Flynn, P.J., et al., 2012. Ginger (*Zingiber officinale*) reduces acute chemotherapy-induced nausea: a URCC CCOP study of 576 patients. Supportive Care in Cancer 20 (7), 1479–1489. https://doi.org/10.1007/s00520-011-1236-3.

Ryu, M.J., Chung, H.S., 2015. [10]-Gingerol induces mitochondrial apoptosis through activation of MAPK pathway in HCT116 human colon cancer cells. In vitro Cellular and Developmental Biology - Animal 51, 92–101. https://doi.org/10.1007/s11626-014-9806-6.

Sen, T., Samanta, S.K., 2014. Medicinal plants, human health and biodiversity: a broad review Tuhinadri. Advances in Biochemical Engineering/Biotechnology 147 (July 2015), 59–110. https://doi.org/10.1007/10_2014_273.

Shang, L., Wang, M., 2013. Molecular alterations and clinical relevance in esophageal squamous cell carcinoma. Frontiers of Medicine 7 (4), 401–410. https://doi.org/10.1007/s11684-013-0286-y.

Shanmugam, M., Rane, G., Kanchi, M., Arfuso, F., Chinnathambi, A., Zayed, M., Alharbi, S., et al., 2015. The multifaceted role of curcumin in cancer prevention and treatment. Molecules 20 (2), 2728–2769. MDPI AG. Retrieved from https://doi.org/10.3390/molecules20022728.

Sharma, S., Kelly, T.K., Jones, P.A., 2010. Epigenetics in cancer. Carcinogenesis 31 (1), 27–36. https://doi.org/10.1093/carcin/bgp220.

Shrihari, T.G., 2017. Dual role of inflammatory mediators in cancer. Ecancermedicalscience 11, 1–9. https://doi.org/10.3332/ecancer.2017.721.

Siwak, D.R., Shishodia, S., Aggarwal, B.B., Kurzrock, R., 2005. Curcumin-induced antiproliferative and proapoptotic effects in melanoma cells are associated with suppression of IκB kinase and nuclear factor κB activity and are independent of the B-Raf/mitogen-activated/extracellular signal-regulated protein kinase pathway and the Akt pathway. Cancer 104 (4), 879–890. https://doi.org/10.1002/cncr.21216.

Subramani, R., Gonzalez, E., Arumugam, A., Nandy, S., Gonzalez, V., Medel, J., et al., 2016. Nimbolide inhibits pancreatic cancer growth and metastasis through ROS-mediated apoptosis and inhibition of epithelial-to-mesenchymal transition. Scientific Reports 6 (August 2015), 1–12. https://doi.org/10.1038/srep19819.

Tanaka, T., Shimizu, M., Kochi, T., Moriwaki, H., 2013. Chemical-induced carcinogenesis. Journal of Experimental and Clinical Medicine 5 (6), 203–209. https://doi.org/10.1016/j.jecm.2013.10.009.

Thompson, S.L., Compton, D.A., 2011. Chromosomes and cancer cells. Chromosome Research 19 (3), 433–444. https://doi.org/10.1007/s10577-010-9179-y.

Tiloke, C., Phulukdaree, A., Chuturgoon, A.A., 2013. The antiproliferative effect of *Moringa oleifera* crude aqueous leaf extract on cancerous human alveolar epithelial cells. BMC Complementary and Alternative Medicine 13 (226), 1–8. https://doi.org/10.1186/1472-6882-13-226.

Torre, L.A., Siegel, R.L., Ward, E.M., Jemal, A., 2016. Global cancer incidence and mortality rates and trends–an update. Cancer Epidemiology Biomarkers and Prevention 25 (1), 16–27. https://doi.org/10.1158/1055-9965.EPI-15-0578.

Vogelstein, B., Papadopoulos, N., Velculescu, V.E., Zhou, S., Diaz, L. a, Kinzler, K.W., 2013. Cancer genome lanscapes. Science 339 (6127), 1546–1558. https://doi.org/10.1126/science.1235122.

Weston, A., Harris, C.C., 2003. Multistage carcinogenesis. In: Kufe, D.W., Pollock, R.E., Weichselbaum, R.R., et al. (Eds.), Holland-Frei Cancer Medicine, sixth ed. BC Decker, Hamilton, ON. Available from: https://www.ncbi.nlm.nih.gov/books/NBK13982/.

Xia, Y., Shen, S., Verma, I.M., 2014. NF-κB, an active player in human cancers. Cancer Immunology Research 9 (2), 823–830. https://doi.org/10.1158/2326-6066.CIR-14-0112.

Yadav, L., Puri, N., Rastogi, V., Satpute, P., Sharma, V., 2015. Tumour angiogenesis and angiogenic inhibitors: a review. Journal of Clinical and Diagnostic Research 9 (6), XE01–XE05. https://doi.org/10.7860/JCDR/2015/12016.6135.

Zhang, F., Thakur, K., Hu, F., Zhang, J.G., Wei, Z.J., 2017a. 10-Gingerol, a phytochemical derivative from "tongling white ginger", inhibits cervical cancer: insights into the molecular mechanism and inhibitory targets. Journal of Agricultural and Food Chemistry 65. https://doi.org/10.1021/acs.jafc.7b00095.

Zhang, F., Thakur, K., Hu, F., Zhang, J.-G., Wei, Z.-J., 2017b. Cross-talk between 10-gingerol and its anti-cancerous potential: a recent update. Food and Function 8 (8), 2635–2649. https://doi.org/10.1039/C7FO00844A.

PART 5

Bioavailability of Bioactive Compounds In Vitro and In Vivo Models

CHAPTER 6

Phenolic Compound Bioavailability Using In Vitro and In Vivo Models

Helena Dias de Freitas Queiroz Barros, Mário Roberto Maróstica Junior
Department of Food and Nutrition, Faculty of Food Engineering, University of Campinas, Campinas, Brazil

6.1 Introduction

Recently, the relationship between food intake and health has been the main point of scientific investigations, with the intention of identifying the specific components that impart beneficial effects. Among these compounds, polyphenols should be highlighted. Polyphenols belong to a diverse group of molecules that are consumed in all diets. They originate from plant-based foods and have been termed as non-nutrients, plant secondary metabolites, phytonutrients, and dietary bioactives (Williamson, 2017). They are the most abundant antioxidants in our diet and are widespread constituents of fruits, vegetables, cereals, olives, dry legumes, chocolate, and beverages, such as tea, coffee, and wine (D'archivio et al., 2007). Some examples of polyphenol compounds in food are shown in Table 6.1.

Phenols are important compounds because of their contribution to human health and their multiple biological effects, such as antioxidant activity, antimutagenic and/or anticarcinogenic activities, and antiinflammatory action (Karakaya, 2004). These compounds range from simple, low-molecular-weight, single aromatic-ring compounds to the large and complex tannins and derived polyphenols (Crozier et al., 2009). Depending on the number of these phenol rings and on the structural elements bound to them, polyphenols are classified into different groups, namely the flavonoids, phenolic acids, phenolic alcohols, stilbenes, and lignans (Tatullo et al., 2016).

Despite their wide distribution, the health effects of dietary polyphenols depend on how they are extracted from food and on their intestinal absorption, metabolism, and biological action with target tissues (Tressera-Rimbau et al., 2017). This chapter aims to explain the bioavailability of phenolic compounds, how these compounds are ready to be absorbed, and consequently exert the beneficial effects observed.

Bioactive Compounds. https://doi.org/10.1016/B978-0-12-814774-0.00006-2

Table 6.1: Polyphenolic Compounds in Foods

	Source	Polyphenol Content (mg/kg Fresh wt or mg/L)	References
Hydroxycinnamic acids	Kiwi (100 g)	600–1000	Clifford (1999)
Caffeic acid	Aubergine (200 g)	600–660	
Ferulic acid	Apple (200 g)	50–600	
Sinapic acid	Pear (200 g)	15–600	
Anthocyanins			
Cyanidin	Blackberry (100 g)	1000–4000	Clifford (2000)
Peonidin	Blueberry- (100 g)	250–5000	
Delphinidin	Black grape (200 g)	300–7500	
Malvidin	Cherry (200 g)	350–4500	
Flavonols			
Kaempferol	Leek (200 g)	30–225	Hertog et al. (1992)
Myricetin	Cherry tomato (200 g)	15–200	Crozier et al. (1997)
Flavones			
Apigenin	Celery (200 g)	20–140	Crozier et al. (1997)
Flavanones			
Hesperetin	Grapefruit juice (200 mL)	100–650	Robards et al. (1997)
Naringenin	Lemon juice (200 mL)	50–300	

6.2 Bioaccessibility × Bioavailability

The amount and bioavailability are key to elucidating the importance of dietary polyphenols in health. Bioavailability is the proportion of the nutrient that is digested, absorbed, and metabolized through normal pathways (McGhie and Walton, 2007). Only polyphenols released from the food matrix by the action of digestive enzymes (small intestine) and microbiota (large intestine) are bioaccessible in the gut and therefore potentially bioavailable (Hithamani; Srinivasan, 2014).

Bioaccessibility is defined as the amount of a food constituent, present in the gut as a result of its release from the solid food matrix, which might be able to pass through the intestinal barrier. Additionally, bioavailability of phytochemicals is dependent on their stability, their release from the food matrix, and the efficiency of their transepithelial passage (Rubió et al., 2014).

The polyphenols' beneficial effects can only be truly effective if they reach the relevant tissues and exert their action at a sufficient concentration to have a biological effect (Rueda et al., 2017). Knowledge of the bioavailability and metabolism of the various polyphenols is necessary to evaluate their biological activity within target tissues (Manach et al., 2004).

Generally, the most abundant phenolic compounds in the human diet are not necessarily the most active in vivo, either because they have a lower intrinsic activity or because they are poorly absorbed from the intestine, highly metabolized, or rapidly eliminated (Dias et al., 2017).

6.3 Mechanisms Associated With Bioavailability of Polyphenols

6.3.1 Release and Absorption of Polyphenols

Polyphenols are classified as flavonoids and nonflavonoids and are found in two major structural formats: attached to sugars, which increase its solubility, known as glycosides, or as a single compound known as aglycones (Santhakumar et al., 2018). The chemical structure of polyphenols, more than the concentration, determines the rate and extent of absorption and the nature of the metabolites circulating in the plasma (D'archivio et al., 2007). Most polyphenols are stored as glycosylated, acylated forms or decorated with rhamnoside moieties (Santino et al., 2017). Generally, aglycones can be absorbed from the small intestine; however, most polyphenols are present in food in the form of esters, glycosides, or polymers that cannot be absorbed in the native form (D'archivio et al., 2007).

When phenolic compounds are consumed in the diet, they are released from the matrix after mastication (Aura, 2008). The mechanical action of mastication determines the breaking of cells with the release of polyphenols linked more weakly to the cell wall structure and those contained in vacuoles. The polyphenols linked more closely to the cell wall, especially in skin cells, are released during the digestive gastro-pancreatic phase as a consequence of the action of the acidic environment of the stomach and of the alkaline environment of the intestine (Tagliazucchi et al., 2010).

The stomach reduces the particle size of food, which further enhances the release of phenolic compounds (Scalbert et al., 2002). The gastrointestinal tract (GIT) may be considered as an efficient extractor, where part of the phytochemicals contained in food matrices is extracted and becomes available for uptake in the intestine (Bouayed et al., 2011).

In the GIT, the biotransformation of the conjugated forms unmodified in the oral cavity occurs through the beta-hydrolysis of sugar moieties in the O-glycoside flavonoids through the phase I (oxidation, reduction, and hydrolysis) and phase II (conjugation) enzymatic detoxification pathways, resulting in various water-soluble conjugate metabolites capable of crossing the enteric barrier for further distribution to organs and finally excreted in urine (Santhakumar et al., 2018). The hydrolysis of the glycoside moiety is a requisite step for absorption. The type of sugar attached to the molecule is the most important determinant of the site and extent of absorption, but the position of the sugar affects the mechanisms involved in intestinal uptake (Donovan et al., 2006).

After intake, a small amount of the ingested polyphenols is absorbed from the intestine and can be converted to glucuronidated and sulfated conjugates by the enzymes of intestinal tissue and can enter the bloodstream (Vetrani et al., 2016). This absorption step is mediated by the lactase phlorizin hydrolase (LPH), an enzyme is present on the luminal side of the brush border in the small intestine and can act on dietary glycosides before absorption (Day et al., 2000). LPH exhibits broad substrate specificity for flavonoid-O-β-D-glucosides, and the released aglycone may then enter the epithelial cells by passive diffusion as a result of its increased lipophilicity and its proximity to the cellular membrane (Del Rio et al., 2013).

The rapid transfer into the enterocyte of the released aglycone is due to an efficient transport mechanism in parallel with the cytosolic β-glucosidase (CBG) activity. In the intestine, LPH protrudes into an unstirred boundary layer and is positioned in close proximity to the sodium-dependent glucose transporter, SGLT1. The aglycone released by the action of LPH may result in an increased local concentration, stimulating diffusion across the brush border. Rapid conjugation within the enterocyte to glucuronide and sulfate, and blood flow to remove the conjugates, will maintain a concentration gradient of the aglycone with the gut lumen (Day et al., 2000).

Thus, there are two possible routes by which the glycoside conjugates are hydrolyzed, and the resultant aglycones appear in the epithelial cells, namely LPH/diffusion and transport/CBG (Del Rio et al., 2013). The relative contributions of "LPH/diffusion" and "transport/CBG" depend on the position of glycosylation (Day et al., 1998).

Only 5%–10% of the amount of polyphenols ingested is absorbed in the small intestine. Of this amount, 90%–95% enters in the circulation as conjugated compounds produced by a combination of methylation, sulfate conjugation, glucuronide conjugation and, in the case of some phenolic acids, also by glycine conjugation (Clifford, 2004). The path taken by polyphenols after ingestion can be observed in Fig. 6.1.

6.3.2 Metabolism and Elimination of Polyphenols

Once absorbed by the intestinal epithelium and before entering the systemic circulation, polyphenols suffer a certain degree of phase II enzymatic detoxification with the production of different conjugated products, such as sulfates, glucuronides, and methylated derivatives through the action of sulfotransferases (SULTs), uridine-5′-diphosphate glucuronosyltransferases (UGT), and catechol-O-methyltransferase (COMTs), respectively (Del Rio et al., 2013; Manach et al., 2004). There is also efflux of at least some of the metabolites back into the lumen of the small intestine and this is thought to involve members of the adenosine triphosphate (ATP)-binding cassette (ABC) family of transporters including multidrug resistance protein (MRP) and P-glycoprotein (P-gp) (Crozier et al., 2009).

In the portal bloodstream, metabolites are bound to albumin and transported to the liver, specifically in the Golgi apparatus and possibly also in the peroxisome in hepatocytes, where they can be oxidatively degraded and subjected to further phase II metabolism, enterohepatic transport may result in some recycling back to the small intestine through bile excretion (Viskupicová et al., 2008).

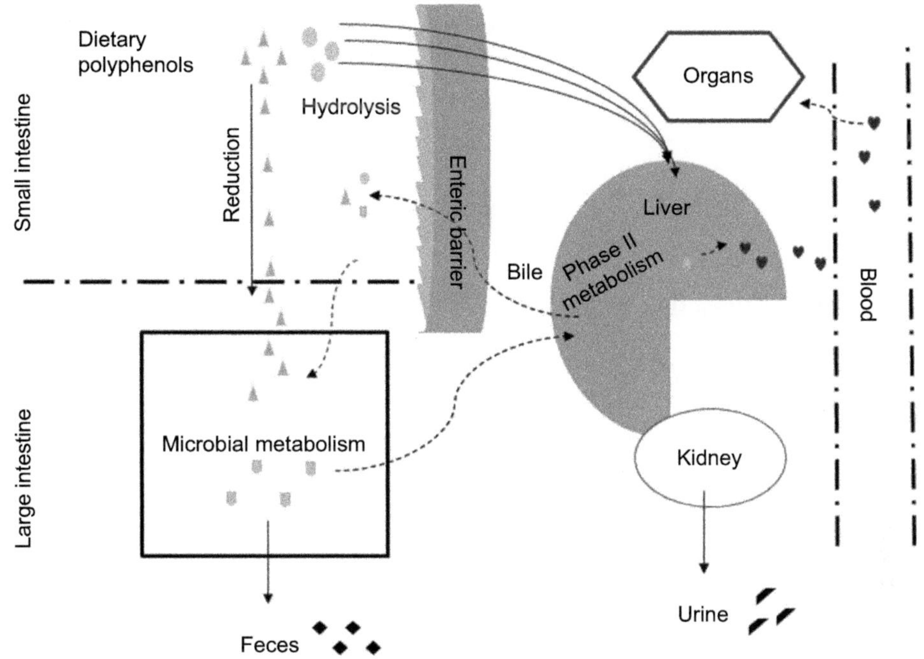

Figure 6.1

The path of polyphenols after ingestion. *Figure based on Del Rio et al., 2013; Viskupicová, J., Ondrejovic, M., Sturdík, E., 2008. Bioavailability and metabolism of flavonoids. Journal of Food and Nutrition Research 47 (4), 151–162; Crozier, A., Jaganath, I.B., Clifford, M.N., 2009. Dietary phenolics: chemistry, bioavailability and effects on health. Natural Product Reports 26 (8), 965–1096; Donovan, J.L., Manach, C., Faulks, R.M., Kroon, P.A., 2006. Absorption and metabolism of dietary secondary metabolites. In: Crozier, A., Clifford, M.N., Ashihara, H. (Eds.), Plant Secondary Metabolites. Occurrence, Structure and Role in the Human Diet. Blackwell Publishing, Oxford, New York, pp. 303–351.*

The remaining unmodified polyphenols (~90%–95%) that are resistant to the action of LPH/CBG and the conjugated forms which reach the small intestine through the bile, cross the intestinal tract and accumulate in the large intestine, where they further undergo a bout of gut microbiota enzymatic action consequently resulting in the production of a variety of metabolites exhibiting different physiological effects (Del Rio et al., 2013; Santhakumar et al., 2018).

The gut microbiota acts as a powerful bioreactor able to break the complex structures of polyphenols into different low-molecular-weight molecules, which are more easily absorbable and exert diverse biological functions (Santhakumar et al., 2018). The microbiota hydrolyzes glycosides into aglycones and extensively metabolizes the aglycones into various aromatic acids (Manach et al., 2004) that are well absorbed through the colonic barrier. These aromatic acids are phenylvaleric, phenylpropionic, phenylacetic, and benzoic acids (Scalbert et al., 2002). The hydrolysis of glycosides results in metabolites that are potentially more biologically active than the parent compounds (Selma et al., 2009).

Apart from the interindividual variation in daily intake of polyphenols, interindividual differences in the composition of the gut microbiota may lead to differences in bioavailability and bioefficacy of polyphenols and their metabolites (Cardona et al., 2013). Metabolites produced in the large intestine subsequently undergo further phase II metabolism, locally and/or in the liver after absorption. They then enter the blood compartment, reach peripheral tissues, and are finally excreted in the urine in substantial amounts, largely exceeding the excretion of phenolic metabolites formed in the upper gastrointestinal tract (Zanotti et al., 2015).

Metabolites of polyphenols may follow two pathways of excretion: (1) via the biliary or (2) the urinary route. Large, extensively conjugated metabolites are more likely to be eliminated in the bile, whereas small conjugates such as monosulfates are preferentially excreted in urine (Manach et al., 2004; Crespy et al., 2003). Intestinal bacteria possess β-glucuronidases that are able to release free aglycones from conjugated metabolites secreted in bile. Aglycones can be reabsorbed, which results in enterohepatic cycling (Manach et al., 2004).

After consumption of flavonoids, only a very small fraction of the dose is typically recovered in urine as forms containing the intact flavonoid ring. Indirect evidence of elimination by bile in humans, along with animal models, supports the theory that elimination in bile is quantitatively the most important route of elimination for some or most flavonoids (Donovan et al., 2006).

Studies deal with the determination of how much of the polyphenols can be absorbed into the circulation system after they get into the body. Evidences suggest that polyphenols are absorbed in a relatively low amount. A study developed by Wu et al. (2002) evaluated that the absorption rate of anthocyanins from berries is less than 1%. The absorption rates of flavan-3-ol and procyanins are less than 5% (Zhu et al., 2000); isoflavones less than 1% (Cassidy et al., 2006); and flavonols (quercetin) between 3% and 7% (Day et al., 2001). Naturally low aqueous solubility, poor gastrointestinal stability, passive diffusion, and active efflux of phenolic phytochemicals in the GIT result in low absorption (Li et al., 2015).

The absorption of isoflavones and gallic acid is best, followed by catechins, flavanones, and quercetin glucosides. Proanthocyanidins, galloylated tea catechins, and anthocyanins are the least well absorbed (Manach et al., 2004). Studies developed for Czank et al. (2013) showed that anthocyanins are more bioavailable than previously described and that their metabolites are still present in the circulation for 48 h after ingestion. Table 6.2 presents the bioavailability and metabolism of polyphenols.

6.3.3 Effect of Polyphenols in Microbiota

The interaction between microbiota and polyphenols is a reciprocal correspondence since different microbial groups are able to change the structure of polyphenols in more absorbable metabolized forms, and polyphenols can modulate the microbiota composition (Santino et al., 2017). The metabolism of polyphenols by microbiota involves the cleavage of glycosidic

Table 6.2: Bioavailability and Metabolism of Polyphenols

Compound	Bioavailability	Relevant Human Metabolities
Phenolic acid	Well absorbed in the upper part of gut. Maximal plasma concentration is reached 30 min after ingestion	Dihydrocaffeic and dihydroferrulic acids, feruloylglycine, hydroxibenzoic
Flavonols	Limited amount in aglycone structure is absorbed in small intestine, the remaining have rearrangements by microbiota	Aglycones: phase II metabolites in liver. Remaining compounds are biotransformed in hidroxyphenyl-acetic derivates, procatechuic and propionic acids
Anthocyanins	Limited amount of glycosylated anthocyanins is directly absorbed, the remaining have rearrangements by microbiota	Methyl, glucoronide, and sulfate conjugates of anthocyanins
Flavan-3-ol and proanthocyanidins	Small amount of flavan-3-ol is bioaccessible in small intestine the remaining monomers and proanthocyanins have rearrangements by microbiota	Aglycones in small intestine: phase II metabolites in liver. Larger fractions are biotransformed to phenylpropionic, phenilacetic, and benzoic acids by microbiota
Ellagitannins	Are not directly absorbed and suffer hydrolysis and rearrangements by microbiota. Free ellagic acid can be absorbed in the small intestine	Urolothin A and B (free form, glucoronide and sulfate)

Based on Zanotti, I., Dall'asta, M., Mena, P., Mele, L., Bruni, R., Ray, S., Del Rio, D., 2015. Atheroprotective effects of (poly)phenols: a focus on cell cholesterol metabolism. Food Function *13 (6).*

linkages and breakdown of the polyphenols' heterocycle. Glycans, which are products of glycosidic cleavage, are important for the establishment and survival of colonic organisms, allowing its paper on nutrient metabolism (Ratmanesh, 2011). *Firmicutes* possess a disproportionately smaller number of glycan-degrading enzymes when compared to *Bacteroidetes*, suggesting, therefore, that the ingestion of different phenolic compounds could reshape the gut microbiota (Cardona et al., 2013).

As a metabolic organ, cellular composition of the human gut microbiota is determined by a dynamic process of selection and competition. Prevalence of the *Bacteroidetes* community following regular wine vinegar ingestion or polyphenol-rich fruits and green tea, due to having more glycan-degrading enzymes, indicates an important role of the polyphenol on the intestinal microbiota composition modulation (Ratmanesh, 2011). Anthocyanins have been shown to stimulate the growth of lactic acid bacteria and increase malolactic fermentation. These bacteria can cleave the anthocyanin molecules and use the sugar moiety as a carbohydrate source (Duda-Chodak et al., 2015).

Polyphenols may modulate the microbiota balance through their more growth-promoting effects on *Bacteroides*, and may exert an effect through suppressing growth on *Firmicutes*. Phenolic compounds alter gut microbiota, and consequently alter the *Bacteroides/Firmicutes* balance (Ratmanesh, 2011).

The effect of phenolic compounds on gut microbiota modulation has gained much attention in recent years, but the influence of polyphenols on specific gut bacteria is still not clear. One of the main limitations in previous studies is that most phenolic fractions and pure phenolic compounds have been analyzed without considering the bioavailability and chemistry of phenolic compounds in the colon (Ozdal et al., 2016).

Another limitation is that the information obtained from in vitro studies about the role of individual phenolic compounds on gut microbiota cannot be directly extrapolated to what occurs in the physiological context of the gut ecosystem. Human and animal intervention studies involve very high doses of individual phenolic compounds, or high amounts of foods rich in phenolic content, neither of which represents the regular diet. Therefore, there is a lack of adequate in vivo studies (Ozdal et al., 2016).

6.3.4 Interactions Between Polyphenols and Other Compounds

Lipids, proteins, and carbohydrates are compounds often found in the environment surrounding phenolic compounds. They can come into contact with phenolic compounds and interact with them. Interactions between polyphenols and molecules from food were mostly based on different noncovalent hydrophobic interactions (Jakobek, 2015). The antioxidant effects of dietary polyphenols prevent the oxidation of vitamins and other nutrients (lipids, proteins, cholesterol), thus preserving a higher quality of nutrient intake (Tomás-Barberána and

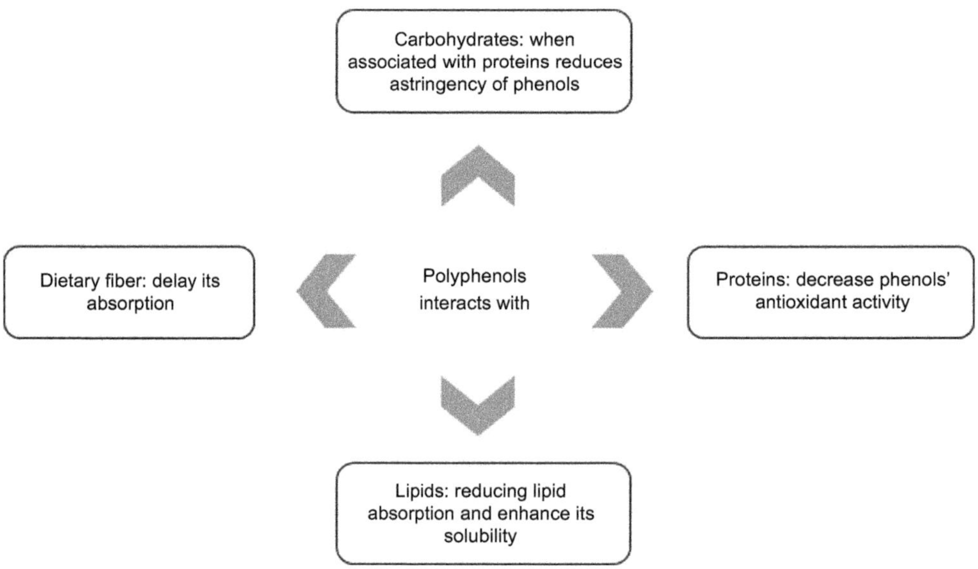

Figure 6.2
Interaction between polyphenols and macronutrients.

Andrés-Lacueva, 2012). Fig. 6.2 show the interaction between polyphenols and macronutrients.

Carbohydrates are frequently used in the food industry as food colloids (gums) and are naturally present in several food products, thereby affecting their astringent features. The use of a complex of carbohydrates–polyphenols–protein was able to inhibit the interaction and precipitation of salivary proteins with tannins and thus influence the perceived astringency of some food products. The extent and mechanism by which this inhibition occurs are related to the carbohydrate structure and in the last instance to the protein structure (Soares et al., 2012). Phenolic compounds interact with starch, inhibiting its hydrolysis by action of α-amylase and amyloglucosidase (Kandil et al., 2012).

Dietary fibers can restrict the diffusion of the enzymes to their substrate, which may allow polyphenols to be carried to the colon (Palafox-Carlos et al., 2011). Recent findings confirm that the fibers have a role as a carrier of dietary antioxidants to the colon due to the cumulative synergistic antioxidant power of polyphenols and other minor constituents (Saura-Calixto, 2011). In addition, polyphenols associated with dietary fibers can enhance the excretion of lipids, protein, water, and total fecal output; they can have positive effects on lipid metabolism, total cholesterol, LDL-cholesterol, and triacylglycerides; and they can increase the antioxidant activity in the large intestine (Saura-Calixto, 2011).

Proteins affect the antioxidant activity of polyphenols mostly in a negative way. They usually decrease the antioxidant activity due to their strong binding affinity to polyphenols. Protein–phenolic interactions influence the structure, functional and nutritional properties, and digestibility of proteins. The presence of phenolic compounds also affects protein solubility, which is an important factor in protein functionality, because protein insolubility also hinders other protein functional properties. The increase in the content of phenolic compounds decreases the solubility of proteins (Ozdal et al., 2013).

Interaction between lipids and polyphenols is related with a decrease in lipase activity and consequently to reduction of fat absorption by the organism, which is because lipase is related to the lipolysis process. Also, it seems that lipids protected polyphenols in their passage through the gastrointestinal tract, delivering them to lower parts of the GIT where they will be metabolized and absorbed to exert the bioactive properties (Jakobek, 2015).

Thereby, the bioavailability of phenolic compounds can be affected by molecular interactions between these potentially bioactive compounds and the food matrix. These interactions could be either beneficial or detrimental for the bioactivities associated with phenolic compounds (Tomas et al., 2018), giving polyphenols a very different role. They could protect polyphenols from oxidation during their passage through the gastrointestinal tract and deliver them to the colon more intact where they can be metabolized under the influence of microbiota, diminishing the content of lipid absorbed by the organism, and also reducing the astringency of food (Jakobek, 2015).

6.4 In vitro × in vivo Bioavailability Methods

In vitro gastrointestinal models have been developed to simulate the digestion and biotransformation of dietary components throughout the human GIT. Most bioaccessibility studies have used a static model that does not consider the dynamics of transit during digestion or the varying microbial and digestive conditions in different segments of the GIT. On the other hand, multistage, dynamic in vitro models, comprised of stomach, small intestine, and the three colonic compartments that simulate both upper and lower gastrointestinal digestion, have been developed and validated (Ekbatan et al., 2016).

There are principally two in vitro methods for measuring bioaccessibility and/or bioavailability of polyphenols: a gastrointestinal model for bioaccessibility, and the Caco-2 models for bioavailability (Etcheverry et al., 2012). Various digestion models have been proposed, often impeding the possibility to compare results across research teams. For example, a large variety of enzymes from different sources, such as of porcine, rabbit, or human origin, have been used, differing in their activity and characterization. Differences in pH, mineral type, ionic strength, and digestion time, alter enzyme activity and other phenomena, and may also considerably alter the results. Considering these, a standard protocol was developed by Minekus et al. (2014).

Simulated digestion methods typically include the oral, gastric, and small-intestinal phases, and occasionally large-intestinal fermentation. These methods try to mimic physiological conditions in vivo, taking into account the presence of digestive enzymes and their concentrations, pH, digestion time, and salt concentrations, among other factors (Minekus et al., 2014). Depending on the physical state of matter (solid or liquid), the oral phase is required. Solid samples should be mixed with salivary amylase at pH 7.0 for 2 min, this step is optional for liquid samples. After the oral phase, samples are submitted for the gastric step where pepsin is added and the pH is adjusted to 2.0 for 2 h. Then, samples follow to the intestinal phase, where the pH is adjusted to 7.0 and pancreatin and bile salts are added. It is important to put samples in a shaking incubator in all steps and with a temperature control of 37°C to simulate the body temperature (Minekus et al., 2014).

Caco-2 cells belong to a human epithelial cell line derived from a human colonic adenocarcinoma. Following the gastric digestion of the food, pancreatin/bile is added and the digest is added to the cells. In vivo, cellular integrity is maintained through the presence of an intestinal mucus layer. However, in vitro, some of the following methods must be used to prevent enzymatic degradation of the cells (Etcheverry et al., 2012).

One method is the introduction of a dialysis membrane secured with a silicone O-ring to a plastic insert, which is placed on top of the cell monolayer. The intestinal digest is placed on top of the dialysis membrane, thus preventing the enzymes from reaching the cells. Other methods involve inactivation of the enzymes by acidifying the intestinal digests to pH 2 or by lowering the temperature of the digests and subsequently filtering the samples (Etcheverry et al., 2012).

A high in vitro activity is not always translated into comparable activity in vivo, thus, it is very important to determine the bioaccessibility and bioavailability of the compound of

Table 6.3: Advantages and Limitations of in vitro and in vivo Studies

Method	Purpose	Advantages	Limitations
Gastrointestinal models	Measures bioaccessibility. However, when coupled to intestinal cells, bioavailability can also be measured	• Incorporates digestion parameters • Allows the analysis of digestion contents in various steps	Omission of colon action in digestion and absorption process
Caco-2 cell model	Measures bioavailability	• Allows the study of nutrient or food component competition at the site of absorption	Requires trained personnel with knowledge of cell culture methods
In vivo (animal and human studies)	Measures bioavailability	• Allows the identification of metabolite compounds in plasma • Allows the knowledge of physiological occurrence of gut microbiota	• Expensive • Requires trained personnel

interest. In general, in vitro methods are somewhat limited due to the omitted role of the colon in digestion and absorption (Swieca et al., 2017). In these cases, in vivo methods can be applied, which can be done using animals (normally mice or rats) and humans. Normally, in vivo animal models are used before human trials and every in vivo study must be approved by the research ethics committee prior to its inception. Animal models contribute to better understanding the mechanisms and biological effects that could be likely to happen in the human body. The metabolism of polyphenols has been the subject of numerous animal studies (mostly in rodents) (Dueñas et al., 2014). Table 6.3 show the advantages and limitations of in vitro and in vivo methods for determination of polyphenol metabolites.

For the evaluation of bioavailability of polyphenols, human and animals studies are performed with administration of an oral dose of the compounds of interest, after administration, blood samples are collected in predetermined periods that vary according to the maximum plasma levels reached by the compounds or their metabolites. A great challenge in these studies is the identification of the compounds in plasma, once the metabolite amounts rarely exceed nM concentrations and the metabolites are rapidly absorbed and metabolized (Zanotti et al., 2015).

6.5 Conclusion

Polyphenols comprise a large group of compounds consumed with health benefits that are already well documented. However, the beneficial effects on the body are related to bioavailability, that is, what actually reaches the blood circulation after ingestion and also for its capacity in modulating gut microbiota favoring, in this way, the metabolism of these compounds and consequently guaranteeing the beneficial effects resulting from their consumption. For this to happen these compounds are subjected to several processes since their release from the food matrix, through hydrolysis, phase II metabolism, intestinal microbiota action, and absorption. All these steps will guarantee that the compound can in fact be absorbed by the organism. In addition, the bioavailability of phenolic compounds can be affected by their interaction with the food matrix causing, or not, benefits to their bioactivity. Much information regarding the metabolism and absorption of phenolic compounds still needs to be elucidated for a better understanding of the beneficial effects caused by their consumption. For this, much research still needs to be done to facilitate the construction of knowledge between food chemistry and human health.

References

Aura, A.M., 2008. Microbial metabolism of dietary phenolic compounds in the colon. Phytochemistry Reviews 7, 407–429.

Bouayed, J., Hoffmann, L., Bohn, T., 2011. Total phenolics, flavonoids, anthocyanins and antioxidant activity following simulated gastro-intestinal digestion and dialysis of apple varieties: bioaccessibility and potential uptake. Food Chemistry 128, 14–21.

Cardona, F., Andrés-Lacueva, C., Tulipani, S., Tinahones, F.J., Queipo-Ortuño, M.I., 2013. Benefits of polyphenols on gut microbiota and implications in human health. Journal of Nutritional Biochemistry 24, 1415–1422.

Cassidy, A., Brown, J.E., Hawdon, A., Faughnan, M.S., King, L.J., Millward, J., Zimmer-Nechemias, L., Wolfe, B., Setchelly, K.D.R., 2006. Factors affecting the bioavailability of soy isoflavones in humans after ingestion of physiologically relevant levels from different soy foods. The Journal of Nutrition 136 (1), 45–51.

Clifford, M.N., 1999. Chlorogenic acids and other cinnamates –nature, occurrence and dietary burden. Journal of the Science of Food and Agriculture 79, 362–372.

Clifford, M.N., 2000. Anthocyanins – nature, occurrence and dietaryburden. Journal of the Science of Food and Agriculture 80, 1063–1072.

Clifford, M.N., 2004. Diet-derived phenols in plasma and tissues and their implications for health. Planta Medica 70, 1103–1114.

Crespy, V., Morand, C., Besson, C., Cotelle, N., Vézin, H., Demigné, C., Rémésy, C., 2003. The splanchnic metabolism of flavonoids highly differed according to the nature of the compound. American Journal of Physiology - Gastrointestinal and Liver Physiology 284, G980–G988.

Crozier, A., Lean, M.E.J., McDonald, M.S., Black, C., 1997. Quantitative analysis of the flavonoid content of commercial tomatoes, onions, lettuce, and celery. Journal of Agriculture and Food Chemistry 45, 590–595.

Crozier, A., Jaganath, I.B., Clifford, M.N., 2009. Dietary phenolics: chemistry, bioavailability and effects on health. Natural Product Reports 26 (8), 965–1096.

Czank, C., Cassidy, A., Zhang, Q., Morrison, D.J., Preston, T., Kroon, P.A., Botting, N.P., Kay, C.D., 2013. Human metabolism and elimination of the anthocyanin, cyanidin-3-glucoside: a 13C-tracer study. The American Journal of Clinical Nutrition 97, 995–1003.

D'archivio, M., Filesi, C., Benedetto, R., Gargiulo, R., Giovannini, C., Masella, R., 2007. Polyphenols, dietary sources and bioavailability. Annali dell'Istituto Superiore di Sanita 43 (4), 348–361.

Day, A.J., DuPont, M.S., Ridley, S., Rhodes, M., Rhodes, M.J.C., Morgan, M.R.A., Williamson, G., 1998. Deglycosylation of flavonoid and isoflavonoid glycosides by human small intestine and liver L-glucosidase activity. FEBS Letters 436, 71–75.

Day, A.J., Cañada, F.J., Díaz, J.C., Kroon, P.A., Mclauchlan, R., Faulds, C.B., Plumb, G.W., Morgan, M.R.A., Williamson, G., 2000. Dietary flavonoid and isoflavone glycosides are hydrolysed by the lactase site of lactase phlorizin hydrolase. FEBS Letters 468, 166–170.

Day, A.J., Mellon, F., Barron, D., Sarrazin, G., Morgan, M.R.A., Williamson, G., 2001. Human metabolism of dietary flavonoids: identification of plasma metabolites of quercetin. Free Radical Research 35 (6), 941–952.

Del Rio, D., Rodriguez-Mateos, A., Spencer, J.P.E., Tognolini, M., Borges, G., Crozier, A., 2013. Dietary (Poly) phenolics in Human Health:Structures, Bioavailability, and Evidence of Protective Effects Against Chronic Diseases. Antioxidants & Redox Signaling 18 (14).

Dias, T.R., ALves, M.G., Casal, S., Oliveira, P.F., Silva, B.M., 2017. Promising potential of dietary (poly)phenolic compounds in the prevention and treatment of diabetes mellitus. Current Medicinal Chemistry 24, 334–354.

Donovan, J.L., Manach, C., Faulks, R.M., Kroon, P.A., 2006. Absorption and metabolism of dietary secondary metabolites. In: Crozier, A., Clifford, M.N., Ashihara, H. (Eds.), Plant Secondary Metabolites. Occurrence, Structure and Role in the Human Diet. Blackwell Publishing, Oxford, New York, pp. 303–351.

Duda-Chodak, A., Tarko, T., Satora, P., Sroka, P., 2015. Interaction of dietary compounds, especially polyphenols, with the intestinal microbiota: a review. European Journal of Nutrition 54, 325–341.

Dueñas, M., Muñoz-González, I., Cueva, C., Jiménez-Girón, A., Sánchez-Patán, F., Santos-Buelga, C., Moreno-Arribas, M.V., Bartolomé, B., 2014. A survey of modulation of gut microbiota by dietary polyphenols. BioMed Research International 2015.

Ekbatan, S.S., Sleno, L., Sabally, K., Khairallah, J., Azadi, B., Rodes, L., Prakash, S., Donnelly, D.J., Kubow, S., 2016. Biotransformation of polyphenols in a dynamic multistage gastrointestinal model. Food Chemistry 204, 453–462.

Etcheverry, P., Grusak, M.A., Fleige, L.E., 2012. Application of *in vitro* bioaccessibility and bioavailability methods for calcium, carotenoids, folate, iron, magnesium, polyphenols, zinc, and vitamins B6, B12, D, and E. Frontiers in Physiology 3.

Hertog, M.G.L., Hollman, P.C.H., Katan, M.B., 1992. Content of potentially anticarcinogenic flavonoids of 28 vegetables and 9 fruits commonly consumed in The Netherlands. Journal of Agriculture and Food Chemistry 40, 2379–2383.

Hithamani, G., Srinivasan, K., 2014. Effect of domestic processing on the polyphenol content and bioaccessibility in finger millet (Eleusine coracana) and pearl millet (*Pennisetum glaucum*). Food Chemistry 164, 55–62.

Jakobek, L., 2015. Interactions of polyphenols with carbohydrates, lipids and proteins. Food Chemistry 175, 556–567.

Kandil, A., Li, J., Vasanthan, T., Bressler, D.C., 2012. Phenolic acids in some cereal grains and their inhibitory effect on starch liquefaction and saccharification. Journal of Agricultural and Food Chemistry 60, 8444–8449.

Karakaya, S., 2004. Bioavailability of phenolic compounds. Critical Reviews in Food Science and Nutrition 44 (6), 453–464.

Li, Z., Jiang, H., Xu, C., Gu, L., 2015. A review: using nanoparticles to enhance absorption and bioavailability of phenolic phytochemicals. Food Hydrocolloids 43, 153–164.

Manach, C., Scalbert, A., Morand, C., Rémésy, C., Jiménez, L., 2004. Polyphenols: food sources and bioavailability. The American Journal of Clinical Nutrition 79, 727–747.

McGhie, T.K., Walton, M.C., 2007. The bioavailability and absorption of anthocyanins: towards a better understanding. Molecular Nutrition and Food Research 51, 702–713.

Minekus, M., Alminger, M., Alvito, P., Ballance, S., Bohn, T., Bourlieu, C., Carrière, F., Boutrou, R., Corredig, M., Dupont, D., Dufort, C., Egger, L., Golding, M., Karakaya, S., Kirkhus, B., Feunteun, S.L., Lesmes, U., Macierzanka, A., Mackie, A., Marze, S., McClements, D.J., Ménard, O., Recio, I., Santos, C.N., Singh, R.P., Vegarud, G.E., Wickham, M.S.J., Weitschies, W., Brodkorb, A., 2014. A standardised static *in vitro* digestion method suitable for food – an international consensus. Food Functional 5 (1113).

Ozdal, T., Capanoglu, E., Altay, F., 2013. A review on protein–phenolic interactions and associated changes. Food Research International 51, 954–970.

Ozdal, T., Sela, D.A., Xiao, J., Boyacioglu, D., Chen, F., Capanoglu, E., 2016. The reciprocal interactions between polyphenols and gut microbiota and effects on bioaccessibility. Nutrients 78 (8).

Palafox-Carlos, H., Ayala-Zavala, J.F., González-Aguilar, G.A., 2011. The role of dietary fiber in the bioaccessibility and bioavailability of fruit and vegetable antioxidants. Journal of Food Science 76 (1).

Ratmanesh, R., 2011. High polyphenol, low probiotic diet for weight loss because of intestinal microbiota interaction. Chemico-Biological Interactions 189, 1–8.

Robards, K., Li, X., Antolovich, M., Boyd, S., 1997. Characterisation of citrus by chromatographic analysis of flavonoids. Journal of the Science of Food and Agriculture 75, 87–101.

Rubió, L., Macià, A., Castell-Auví, A., Pinent, M., Blay, M.T., Ardévol, A., Romero, M.P., Motilva, M.J., 2014. Effect of the co-occurring olive oil and thyme extracts on the phenolic bioaccesability and bioavailability assessed by *in vitro* digestion and cell models. Food Chemistry 149, 277–284.

Rueda, A., Cantarero, S., Seiquer, I., Cabrera-Vique, C., Olalla, M., 2017. Bioaccessibility of individual phenolic compounds in extra virgin argan oil after simulated gastrointestinal process. Lebensmittel-Wissenschaft und -Technologie- Food Science and Technology 75, 466–472.

Santhakumar, A.B., Battino, M., Alvarez-Suarez, J.M., 2018. Dietary polyphenols: structures, bioavailability and protective effects against atherosclerosis. Food and Chemical Toxicology 113, 49–65.

Santino, A., Scarano, A., Santis, S., Benedictis, M., Giovinazzo, G., Chieppa, M., 2017. Gut microbiota modulation and anti-inflammatory properties of dietary polyphenols in IBD: new and consolidated perspectives. Current Pharmaceutical Design 23, 2344–2351.

Saura-Calixto, F., 2011. Dietary fiber as a Carrier of dietary antioxidants: an essential physiological function. Journal of Agricultural and Food Chemistry 59 (1).

Scalbert, A., Morand, C., Manach, C., Rémésy, C., 2002. Absorption and metabolism of polyphenols in the gut and impact on health. Biomedicine and Pharmacotherapy 56, 276–282.

Selma, M.V., Espín, J.C., Tomás-Barberán, F.A., 2009. Interaction between phenolics and gut microbiota: role in human health. Journal of Agricultural and Food Chemistry 57, 6485–6501.

Soares, S., Mateus, N., Freitas, V., 2012. Carbohydrates inhibit salivary proteins precipitation by condensed tannins. Journal of Agricultural and Food Chemistry 60, 3966–3972.

Swieca, M., Gawlik-Dziki, U., Dziki, D., Baraniak, B., 2017. Wheat bread enriched with green coffee – *In vitro* bioaccessibility and bioavailability of phenolics and antioxidant activity. Food Chemistry 221, 1451–1457.

Tagliazucchi, D., Verzelloni, E., Bertolini, D., Conte, A., 2010. In vitro bio-accessibility and antioxidant activity of grape polyphenols. Food Chemistry 120, 599–606.

Tatullo, M., Simone, G.M., Tarullo, F., Irlandese, G., Vito, D., Marrelli, M., Santacroce, L., Cocco, T., Ballini, A., Scacco, S., 2016. Antioxidant and antitumor activity of a bioactive polyphenolic fraction isolated from the brewing process. Nature Scientific Reports 6 (36042).

Tomas, M., Beekwilver, J., Hall, R.D., Simon, C.D., Sagdic, O., Capanoglu, E., 2018. Effect of dietary fiber (inulin) addition on phenolics and *in vitro* bioaccessibility of tomato sauce. Food Research International 106, 129–135.

Tomás-Barberán, F.A., Andrés-Lacueva, C., 2012. Polyphenols and health: current state and progress. Journal of Agricultural and Food Chemistry (Symposium introduction).

Tressera-Rimbau, A., Arranz, S., Eder, M., Vallverdú-Queralt, A., 2017. Dietary polyphenols in the prevention of stroke. Oxidative Medicine and Cellular Longevity,10. Article ID 7467962.

Vetrani, C., Rivellese, A.A., Annuzzi, G., Adiels, M., Borén, J., Mattila, I., Oresic, M., Aura, A.M., 2016. Metabolic transformations of dietary polyphenols: comparison between *in vitro* colonic and hepatic models and *in vivo* urinary metabolites. Journal of Nutritional Biochemistry 33, 111–118.

Viskupicová, J., Ondrejovic, M., Sturdík, E., 2008. Bioavailability and metabolism of flavonoids. Journal of Food and Nutrition Research 47 (4), 151–162.

Williamson, G., 2017. The role of polyphenols in modern nutrition. Nutrition Bulletin 42, 226–235.

Wu, X., Cao, G., Prior, R.L., 2002. Absorption and metabolism of anthocyanins in elderly women after consumption of elderberry or blueberry. The Journal of Nutrition 132 (7), 1865–1871.

Zanotti, I., Dall'asta, M., Mena, P., Mele, L., Bruni, R., Ray, S., Del Rio, D., 2015. Atheroprotective effects of (poly)phenols: a focus on cell cholesterol metabolism. Food Function 13 (6).

Zhu, M., Chen, Y., Li, R.C., 2000. Oral absorption and bioavailability of tea catechins. Planta Medica 66 (5), 444–447.

PART 6

Bioactive Compounds as Ingredients of Functional Foods

CHAPTER 7

Bioactive Compounds as Ingredients of Functional Foods: Polyphenols, Carotenoids, Peptides From Animal and Plant Sources New

Sibele Santos Fernandes, Michele Silveira Coelho, Myriam de las Mercedes Salas-Mellado

Laboratory of Food Technology, School of Chemistry and Food, Federal University of Rio Grande, Rio Grande, Brazil

7.1 Introduction

From the 21st century, society has been seeking a healthy lifestyle and food habits. The increase in life expectancy and changes in eating habits, preferably in the reduction of caloric content and balance of complementary diets, help consumers to prevent serious diseases, such as type II diabetes, obesity, osteoporosis, cardiovascular diseases, Alzheimer's disease, and Parkinson's disease, among others. In view of this, there is pressure in several sectors for the development of products that meet this demand, such as in the food industries, researchers, health professionals, and regulatory authorities (Prakash et al., 2017). In this context, functional foods have great potential. Functional foods correspond to a portion of the human diet and provide health benefits, reducing the risk of chronic diseases, in addition to those provided by adequate nutrition.

The term "functional foods" came from Japan in the late 1980s where the food industries responded to a Japanese government appeal that was concerned about the increase in the incidence of noncommunicable diseases in the Japanese elderly population. In 1991, functional foods were regulated under Japanese law under the name of "Foods for Specified Health Use" (FOSHU) (European Commission, 2010). Subsequently, exhaustive research studies were carried out to identify and elucidate functional food ingredients. The most potent sources of fiber, protein, energy, minerals, vitamins, and antioxidants, which are recognized as functional food ingredients, are grains, legumes, and cereals (Prakash et al., 2017). Currently, the focus of many bioactive compound extraction researches focuses on the use of food process residues. Many researchers are discovering new alternative uses for such "wastes" as potential value-added ingredients.

Bioactive Compounds. https://doi.org/10.1016/B978-0-12-814774-0.00007-4

In this context, this chapter aims to elucidate the characteristics of bioactive compounds present as natural constituents or as fortifiers in foods with the potential to provide health benefits beyond the basic nutritional value of the product. In addition, we evidence the use of this concept in the market.

7.2 Bioactive Ingredients

A bioactive compound is any compound present in foods, animals, or plants that has an effect on the body that consumes it. In that direction, bioactive ingredients are those that, when inserted in foods, provide some effect of improvement to health. Such ingredients may be present in the usual foods with naturally occurring bioactive substances, e.g., dietary fiber; in foods supplemented with bioactive substances, e.g., probiotics and antioxidants; or fortified in usual, e.g., prebiotics (Grajek et al., 2005).

Bioactive ingredients, often referred to as functional ingredients, are compounds extracted from a source food, such as fruits, cereals, vegetables, and food processing residues, which preserve their characteristics even after extraction. One difference that must be considered is that bioactive ingredients are not food additives. Certain food ingredients used in the food industry to impart or enhance sensory characteristics are not functional ingredients but only additives.

Public health recommendations that can be made on bioactive foods and ingredients that have health benefits are currently under discussion. Many conferences, meetings, and congresses aim to discuss the evaluation of bioactive components for public health recommendations. The Office of Dietary Supplements at the Institute of Medicine establishes the requirements for the essential nutrients, but this is not to be used for the bioactive components of health, since this process has not yet been defined for bioactive foods and ingredients (Weaver, 2014).

The main bioactive ingredients include prebiotics, probiotics, amino acids, peptides, proteins, omega-3, structured lipids, phytochemicals and plant extracts, minerals, vitamins, fibers, special carbohydrates, carotenoids, and antioxidants, all of which can be obtained from various sources and are served in the form of functional foods, beverages, personal care products, and supplements to consumers.

Bioactive components, such as chitosan, polyunsaturated fatty acids, and astaxanthin from macroalgae, microalgae, industrial waste from fish, and other marine animals have excellent potential as functional food ingredients, since they have advantageous physiological effects and benefits for the health, such as anticancer or antiinflammatory activity. Among all bioactive compounds from marine origins, the most abundant among polyphenols is the phlorotannins, which can be active ingredients in nutraceuticals due to their antioxidant activity (Suleria et al., 2015).

Identifying the characteristics of bioactive ingredients is of paramount importance for their application in any industry sector. Yuanzhi (*Radix palygalae*), a traditional Chinese medicinal herb, is often used as a soothing nerve medication or as an expectorant. Wang et al. (2014) evaluated the drying performance of this compound in order to increase the efficiency of use of the bioactive ingredients present.

Bioavailability is a term related to the availability of bioactive ingredients in food formulations. For the most part, bioavailability incorporates a range of parts, clutch absorption, distribution, metabolism, and excretion, including biochemical and physiological effects. Bioactive ingredients can be administered by many routes. Administration through formulated food products is by the oral route. However, the entry of bioactive ingredients into the systemic circulation is restricted by several barriers, such as acidic conditions of the stomach, brush border membrane, and proteases in the gut lumen, metabolism by liver enzymes (the "first-pass effect") and tightly bound intestinal epithelial cells (enterocytes) (Jafari et al., 2017).

The bioavailability of functional food components and the levels required in humans are critical factors needed to optimize health benefits. Current information in this regard is insufficient and hazy, because of this it is necessary to provide consumers with more information to effectively guide them to make broader choices of diets that contain optimal levels of functional food components that promote health (Abuajah et al., 2015).

7.3 Identification of Bioactive Ingredients

The identification of a bioactive food or ingredient is carried out through its connection with health or a specific disease through diet. From scientific identification, claims can be formulated by food manufacturers and communicated directly to the consumer or for government approval (Weaver, 2014).

To claim the benefits of a bioactive ingredient, processes, spanning several sectors, must be developed. The interaction between scientists and policymakers is iterative because the process needs to be framed to know what evidence should be collected and the quality of the evidence should be evaluated by policymakers before public health recommendations can be communicated to healthcare providers, food manufacturers, and finally, consumers (Weaver, 2014).

The development of good biomarkers for exposure and for the effect of bioactive components and their relation to health is of paramount importance to aggregate evidence of the same, since the bioactivity of a food may not be attributable to a single constituent. If several constituents constitute the bioactive effect and multiple tissues respond to this effect, then monitoring the causal link between a bioactive source and health is indeed a challenge (Weaver, 2014).

7.3.1 Of Natural Origin

Plants have been used for nutritional purposes by people since the beginning of mankind. However, after discovering its medicinal properties, flora has become a useful source of compounds with important roles in the prevention and/or treatment of diseases, promoting better health. In fact, the ancestral use of herbal plants can be considered as the basis for the use of naturally bioactive molecules depending on traditional medicine as primary health care, mainly through the use of plant extracts and their bioactive compounds (Azmir et al., 2013).

Bioactive compounds from plant materials can be extracted by various extraction techniques, and most of these techniques are based on the extracting power of different solvents in use and the application of heat and/or mixing. To obtain bioactive compounds of plants, the classic techniques are the extraction of Soxhlet, maceration, and hydrodistillation (Azmir et al., 2013).

The bioactive compounds may be isolated from their natural environment and then incorporated into the lipid phase of the emulsion delivery systems. Alternatively, bioactive agents may be left in their natural environment, e.g., fruits or vegetables, and then ingested with emulsion-based excipient systems, which may not have inherent health benefits, but enhance the biological activity of bioactive ingredients coingested with it, altering its bioaccessibility, absorption, and/or chemical transformation (Varzakas et al., 2016).

Due to the vast biodiversity, the marine world is a rich natural resource of many biologically active compounds, such as polyunsaturated fatty acids (PUFAs), sterols, proteins, polysaccharides, antioxidants, and pigments, that can be derived from a vast array of sources, including marine plants, microorganisms, and sponges, all of which contain their own unique set of biomolecules (Rasmussen and Morrissey, 2007; Lordan et al., 2011). In view of this, the marine-derived nutrients and other marine bioactive components have excellent potential as functional food ingredients, because they have advantageous physiological effects with medicinal characteristics and added health benefits such as anticancer or antiinflammatory activity (Lordan et al., 2011).

In recent times, fibers have been being encouraged for use as an ingredient with specific functions in food production. Due to the insoluble and soluble properties of fibers, they have a range of technological attributes, such as binding to water, gelling, and building structures and can be used as a fat substitute. Fiber production studies for its use as a bioactive ingredient are concentrated primarily on fruits such as apple, grape, lemon, mango, orange, and peach, and vegetables such as carrot, cauliflower, onion, potato, and tomato (Shea et al., 2012).

7.3.2 From Byproducts

Many researches seek to use as much of their product as possible, by studying alternatives to their waste as functional or financial benefits to their industries. Food residues can be highly nutritional and functional food ingredients, including polysaccharides, vitamins, minerals, dietary fibers, and bioactives such as flavonoids and lycopene (Helkar et al., 2016). The market in this field is competitive and the development of new byproduct-based ingredients is a challenge for the food industry. Therefore, there is an increasing interest in considering not only the nutritional quality of these ingredients, but also its distribution, cost, and other additional benefits, since the use of these materials as food ingredients would make them added-value products (Bensadón et al., 2010).

Fruits, such as mango, banana, and the citrus family, generate a lot of waste such as bark, pulp, seeds, and stones, which represent a major disposal problem due to the lack of infrastructure to deal with these large quantities of available biomass, lack of processing facilities, and high processing costs, especially in developing countries. Such residues are of great importance due to the presence of phenolic compounds, which provide nutraceutical properties to fruit residues, and to the biological properties, such as anticancer, antimutagenicity, antiallergenicity, and antiaging activity, which have been reported both for natural and synthetic antioxidants (Varzakas et al., 2016).

Helkar et al. (2016) presented a literature review of bioactive ingredients that can be extracted from food processing residues. Fruit processing generates byproducts with excellent compounds that can be used as functional food ingredients, instead of being discarded. Bioactive compounds have already been extracted from xoconostle pells (Morales et al., 2015), pomegranate peels and seeds (Kaderides et al., 2015), apple pomace (Rana et al., 2015), and orange pomace (Shea et al., 2012), e.g., to obtain bioactive ingredients with antioxidant properties, primarily, antihypertensive, anticancer, antidiabetic, and hypolipidemic activities; and functional, such as increasing fiber content and improved water/oil-holding and binding properties.

Gamma-oryzanol is a phytosterol from rice bran, a byproduct obtained from the rice-polishing process. This phytosterol is a mixture of ferulic acid (FA) esters of triterpene alcohols such as cycloartenol and 24-methylene cycloartenol. These triterpenic alcohols are now widely used as bioactive ingredients in food and as a precursor of FA in the production of biovanilin. FA derivatives are active ingredients for the development of therapies for the treatment and prevention of cancer and metabolic disorders (Varzakas et al., 2016). The byproducts of brewers are utilized as raw materials for the extraction of compounds such as sugars, proteins, acids, and antioxidants, which can be used as bioactive ingredients (Aliyu and Bala, 2011).

Seeds, as agro-industrial residues, can be used as a source of macronutrients and/or raw material for extraction of vegetable oils, since they present great quantities of bioactive compounds (Varzakas et al., 2016).

The development and utilization of seafood byproducts as functional food ingredients have been accelerated due to the growing health benefits knowledge of seafood diets/products. The seafood-derived functional food ingredients may create a positive contribution to health and wellbeing. These days, consumers prefer foods that have a significant potential to improve health, increase longevity, and reduce the risk of or delay the onset of diseases (Pal and Suresh, 2016).

7.4 Claims of Bioactive Ingredients

For the claim of functional food ingredients, the aspects to be considered are that the functional effect should be different from normal nutrition, it must be demonstrated satisfactorily, and their benefits may lead to improved physiological function or reduced risk of developing a pathological process. However, not all bioactive compounds present these aspects, which means that some are not considered to be functional food ingredients.

The European Food Safety Authority (EFSA) considers three criteria for scientific reasoning for a health claim: (1) definition and characterization of the food or functional ingredient; (2) definition of the claimed (physiological beneficial) effect; and (3) establishing a cause and effect relationship between food consumption or the functional ingredient and the claimed effect (Boer et al., 2016). Therefore, the European Union makes available through the positive list in the Annex to Regulation 432/2012, the specific conditions of use which focus on different aspects, such as the amount of the food required to be consumed containing the active substance to see the effect, the source of the active ingredient, as well as the matrix in which the active ingredient is presented (European Commission, 2010).

Table 7.1 presents some examples of bioactive ingredients that have been claimed according to the European Union.

In the USA, the Food and Drug Administration has grouped together several compounds that show potentiability as new bioactive food ingredients. These new bioactive ingredients are

Table 7.1: Examples of Bioactive Ingredients With Claims by the European Union

Font	Bioactive Ingredient	Effect	Condition
Olive oil polyphenols	Hydroxytyrosol and its derivatives	Contributes to the protection of blood lipids from oxidative stress	Olive oil that contains at least 5 mg of hydroxytyrosol and its derivatives per 20 g of olive oil
Cocoa flavanols	Monomeric (mainly epicatechin) and oligomeric flavanols (procyanidins)	Maintains the elasticity of blood vessels, which contributes to normal blood flow	Containing 34%–37% monomeric flavanols (in their total flavanol content)
Oats or barley	β-Glucans	Contributes to the reduction of the glucose rise after a meal	Food must contain ≥4 g of β-glucans from oats or barley for each 30 g of available carbohydrates in a quantified portion
		Lowers/reduces blood cholesterol. Blood cholesterol lowering may reduce the risk of (coronary) heart disease	Daily intake of 3 g of oat or barley β-glucan

Data adapted from Boer, A., Urlings, M.J.E., Bast, A., 2016. Active ingredients leading in health claims on functional foods. Journal of Functional Foods 20, 587–593.

generally considered as safe (GRAS), and include vegetable oil, sterol esters, phytostanol esters, lactoferin, fish oil concentrate, tuna oil, diacylglycerol, and inulin, among others (Burdock et al., 2006).

The bioavailability of bioactive ingredients after consumption, their mechanism and mode of action, as well as the biological effects in vivo and in vitro and the structure–function relationship and the physiological mechanism by which the health benefits are manifested, are still not well understood. Therefore, research on these aspects needs more attention.

To prove that naturally occurring bioactive substances provide some health benefit and can be used as bioactive ingredients, there is a dilemma with regard to nutritional research, as investigating the preventive activity can be difficult when the effect is only moderate. This means that the effect of the compounds on the human body may be very small in relatively short periods but can contribute significantly to health when consumed throughout life as part of the daily diet (Biesalski et al., 2009).

The use of biomarkers, that is a chemical or biological test result in an analyzed biological material related to a certain exposure, susceptibility, or biological effect, may be an essential element for the bioactive substance efficacy test and nutritional science in general. Using biomarkers, the moderate effects of natural bioactive ingredients in the human body can be studied more easily (Biesalski et al., 2009).

7.5 Preservation of Bioactive Ingredients

In addition to consumer awareness, another major challenge facing the successful use of bioactive ingredients is determining the appropriate food processing and storage conditions (temperature, exposure to oxygen and light) where they will be inserted and also finding ways to sustain stability of the ingredients in the gastrointestinal environment (pH, digestion enzymes, and long transit time) (Vos et al., 2010). In this context, one technique that allows the preservation of bioactive ingredients is encapsulation.

Besides that, as the incorporation of bioactive molecules into commercial food products is a challenging task due to its poor stability and low solubility rate, novel techniques to increase the solubility of bioactive poorly water-soluble natural products are being developed. Recently, this problem has been solved using nanotechnology, which is an innovative approach for substantial improvement of the solubility and bioavailability of bioactive ingredients (Recharla et al., 2017).

Encapsulation is a process used to trap one substance (called a core material or active agent) into another (coating or support material/wall). Encapsulation techniques can prolong the shelf-life of products by protecting the active components against degradation, as well as masking unwanted taste or odor. The choice of encapsulation method and wall material depends on multiple factors, including required particle size, physical and chemical properties of the core and wall, application of the final product, desired release mechanisms, the scale of production, and cost (Ré, 1998).

Regardless of the encapsulation technique, the physical properties of the encapsulation systems, such as size, shape, surface charge, and decorations, and their mechanical properties, play an important role in the fate of ingredients encapsulated in the bloodstream, including biodistribution, vascular dynamics, absorption, debugging, kinetic release, and degradation (Jafari et al., 2017).

The most recent studies have focused on nanoencapsulation techniques, due to the advantages when compared with microencapsulation techniques, because of the increase in the stability and viability of the bioactives, as well as other applications to improve the quality of food products. Although nanoencapsulation can occur in compounds such as carbohydrates, proteins, or lipids, lipid-based nanoparticles have advantages, such as biocompatibility, enhanced encapsulation efficiency, and targeted effect. That is why recent research has focused on lipid-based transporters to be applied in the pharmaceutical and food industries (Katouzian et al., 2017).

In addition to the various factors mentioned above and which can be mitigated or solved by encapsulation, the various digestion barriers can also be overcome by the application of a variety of nanoencapsulation systems such as nanoemulsions, nanoliposomes, and polymeric mineral nanoparticles (NPs). The nanoencapsulated ingredients can pass through the intestinal epithelium either between the enterocytes (paracellular route) or through the cells via the transcytosis mechanism. Nonloaded ingredients can also pass through the epithelium by diffusion through the cells (transcellular route) (Jafari et al., 2017).

Situ et al. (2017) studied to develop liposomes as a nano-delivery system for bioactive ingredients (phosphatidylcholine, melatonin, and cholesterol) using supercritical carbon dioxide (SC–CO_2). Results from in vitro release experiments showed that the melatonin liposomes were resistant to degradation in a simulated gastric environment and improved the bioavailability of melatonin via controlled release in a simulated small intestine environment. In addition, the particle size was nanometer scale.

7.6 Development of Products With Bioactive Ingredients

The development of foods containing bioactive ingredients is a branch of research that is gaining more prominence after numerous studies of obtaining these compounds, since the effectiveness of the functional and nutraceutical products in the prevention of diseases depends on the preservation of the stability, bioactivity, and bioavailability of active ingredients (Fang and Bhandari, 2010). This represents a considerable challenge because only a small proportion of molecules remains available after oral administration, generally due to insufficient gastric residence time, low permeability, and/or solubility in the intestine, as well as instability under conditions encountered in food processing or in the gastrointestinal tract (Leonard, 2000; Silva et al., 2016).

In order to develop potential innovative functional food products, the following investigations are required: (1) the identification and quantification of bioactive compounds, (2) the establishment of appropriate dosage and delivery systems to incorporate bioactive compounds into foods, (3) the analysis of absorption and bioavailability of the incorporated ingredient, (4) testing the safety of bioactive compounds incorporated into foods, (5) stability studies of product storage, and (6) investigation of possible interactions between active ingredients and other food components (Recharla et al., 2017).

The way a food is processed affects its functional components. Some food processing techniques increase the concentration of these functional components in food, while others reduce it (Abuajah et al., 2015). Therefore, it is of paramount importance to study how the bioactive compound will behave during food processing, as well as its bioavailability when inserted in the food matrix. However, there are not many studies on the application of bioactive compounds as food ingredients. Most studies are concentrated on obtaining these compounds, and in the evaluation of their properties and biodisponibility.

Mildner-Szkudlarz et al. (2013) developed biscuits by adding fiber residue-rich grains with significant amounts of bioactive compounds, such as polyphenols and carotenoids. This authors concluded, based on sensory profile analysis, that biscuits could be incorporated at a level of up to 10% of white grape pomace in order to obtain sensorially acceptable products. At 10% white grape pomace enhancement, the biscuits had 64.86 g of total dietary fiber contents per kg dry matter and 30.51 mg phenolic compounds per 100 g dry matter. For these reasons, grape byproducts can be used as ingredients in a novel biscuit formulation as an alternative source of dietary fiber and phenols, and thereby become a functional food.

Kaderides et al. (2015) incorporated pomegranate peel extract in hazelnut paste through encapsulation to evaluate the stabilizing efficiency of the extract against oxidative deterioration. The encapsulated phenolic extract was found to be efficient in improving the shelf life of hazelnut paste.

Marsanasco et al. (2015) developed a functional food in pasteurized chocolate milk from the addition of essential fatty acids linolenic acid (omega-3) and linoleic acid (omega-6) contained in soy phosphatidylcholine (SPC) which is a natural lipid. Liposomes showed significant stability of all parameters and a protective effect over thermolabile fatty acids, with the remaining half of this vitamin encapsulated. The incorporation of SPC did not change the acceptability of commercial milk. Also, when the potential consumer beforehand knew about the addition of nutritional bioactive compounds, it generated a positive effect on acceptability for all the formulations.

Lagos et al. (2015) conducted a survey of patents on the application of bioactive compounds in different food industries from 2011 to 2015. The study showed that patents for foods containing bioactive ingredients are concentrated in the beverage industry, the others mentioned were chocolate and as a sweetener.

7.7 Bioactive Ingredients and Their Market Insertion

The food industries have shown a positive trend towards the development of food products fortified with bioactive ingredients. In fact, a number of functional food products such as dietary supplements, medical foods, and food additives are already available in the market (Varzakas et al., 2016).

The interest in bioactive food and ingredients is high among consumers. The International Food and Nutrition Council has conducted research with consumers of functional foods for 15 years. Among 1005 participants in the 2013 Functional Food Consumer Survey, 45% said they were very interested and 86% said they were very or interested in learning more about foods that have benefits beyond basic nutrition (Weaver, 2014).

The markets for bioactive products and ingredients are interdependent on one another, as any change in one will have a direct impact on the other. Growing awareness and reliability in bioactive products is providing a strong consumer base for the marketplace (Marketsandmarkets, 2017).

According to a new report by Grand View Research (2016), increased income coupled with increased consumer health awareness will boost growth in the global bioactive ingredients market to USD 51.71 billion by 2024. In 2015, fibers were the major product segment and accounted for more than 25% of global revenue, due to increased awareness of its benefits, and Asia Pacific was the main consumer and accounted for more than 35% of total revenue.

Among the research, a global perspective was made regarding the types and applications of bioactive ingredients between 2014 and 2024. In this period, the most used ingredients will be fiber, vitamins, omega-3 fatty acids, plant extracts, minerals, carotenoids, antioxidants, probiotics, and others. In relation to applications, the focus will be on functional foods and beverages, dietary supplements, clinical nutrition, personal care, and other products (Grand View Research, 2016). According to the Marketsandmarkets (2017) report, a global market research and consulting company, global trends and forecasts for the bioactive ingredients market for 2018 were projected to grow from $23.8 billion in 2013 to $33.6 billion in 2018, with a compound annual growth rate of 7.2%.

The global market for bioactive ingredients is highly competitive, with the presence of small and large suppliers. The Asia–Pacific region is the dominant market for bioactive ingredients due to increasing population and urbanization in several developing countries of the region.

However, India, China, Brazil, and Russia are the four nations that are most potent in the bioactive industry (Marketsandmarkets, 2017).

There are already companies specialized in the development and production of high-efficacy bioactive ingredients. For the cosmetics industry, among the companies that can be cited as examples are Assessa, in Brazil, with bioactive ingredients obtained from algae for application in cosmetics; and Montpellier, France, which comes from exclusive oil eco-extraction technology producing 100% natural, sustainable, and clinically proven bioactive compounds for cosmetic applications, including antiaging qualities, reduction of dark spots, and skin whitening.

The leading supplier of bioactive ingredients on the market is DuPont (USA), followed by Royal DSM (Netherlands), Cargill (USA), ADM (USA), and BASF (Germany). These five companies cover 35.2% of the market, while the other suppliers participate in the dominant share of 64.8% (Marketsandmarkets, 2017).

7.8 New Bioactive Ingredients

Rapid growth in innovation is the result of the identification of new bioactive compounds, where the industry focuses its efforts on the development of more technically and economically feasible processes to guarantee the sensory and functional and/or nutritional properties of these in food products. In this sense, the use and incorporation of antioxidants, lipophilic compounds, and microbiological strains will be a great alternative for the production of supplements and foods that will meet human and animal needs (Lagos et al., 2015).

Consumers have created a strong interest in novel bioactive compounds as ingredients in functional foods from marine natural resources. Among them, an example studied is fucoidians, due to their antioxidant, antiinflammatory, antiallergic, antitumor, antiobesity, anticoagulant, antiviral, antihepatopathy, antiuropathy, and antirenalpathy effects. Fucoidans are a complex series of sulfated polysaccharides found widely in the cell walls of brown algae. These special properties of fucoidans have supported it being applied in functional foods for disease prevention and health promotion (Vo and Kim, 2012).

Chia can be considered as a functional food because, apart from contributing to human nutrition, it would be an excellent ingredient to consider in the formulation of functional foods, such as bakery products, cereal bars, etc. Therefore, chia would be an excellent ingredient to consider in the formulation of functional foods. From these seeds, various bioactive compounds can be obtained, with emphasis on the mucilage of chia, essential fatty acids, and proteins, and used in formulations of various foods to promote health.

Chia mucilage, which results from the addition of water to chia seeds and is rich in soluble fibers, has excellent emulsifying properties. Fernandes and Salas-Mellado (2017) replaced 25%, 50%, 75%, and 100% of fat in breads and cakes with chia mucilage, aiming to reduce the lipid content and increase the fiber content and thus promote health. Chia seeds are also potential sources of biologically active peptides. Chia protein hydrolysates with improved biological activity may prove to be an effective functional ingredient in a wide range of foods (Segura-Campos et al., 2013). In relation to polyunsaturated fatty acids from chia seeds, Nadeem et al. (2017) developed margarine containing omega-3 fatty acids from chia oil.

Among the fruits, pomegranate is the one that is currently gaining prominence, because it has the highest antioxidant activity, with anthocyanins, ellagic acid derivatives, and hydrolysable tannins responsible for this activity. Due to the rising value of pomegranate, pomegranate juices are now included in many food products, such as other juices, yogurts, and jams, in order to increase their nutritive values. In addition, three fermented products, i.e., pomegranate wine, pomegranate vinegar, and pomegranate sour concentrate or molasses are produced directly from pomegranate juices. The visibility of these products in markets and their consumption are increasing. During the fermentation process, some bioactive ingredients decrease, and others form, so that the antioxidant activities of fermented pomegranate products have been found to be high in recent studies (Kalaycıoğlu and Erim, 2017).

Melanoidins are Maillard reaction products that result from the heat processing of foods containing reducing sugars and free amino groups. These products are responsible for the specific aroma and color formations that give processed foods their typical appearance and greatly improve acceptance by consumers. Melanoidins present biological properties, such as antioxidant, antimicrobial, and antiinflammatory activity, that can be considered as potential food ingredients in the search for healthier and tasty foods. Future investigations could be focused on the development of tailor-made melanoidins in foods to fulfill the technological or functional needs of consumers (Mesías and Delgado-Andrade, 2017).

Taofiq et al. (2017) studied the incorporation of medicinal mushroom (*G. lucidum*) ethanolic extracts in a cosmetic base cream. These authors verified the presence of different bioactive molecules in the classes of phenolic compounds, terpenes, especially triterpenes, can be correlated with the exhibited bioactivity. Further studies should be performed with the extract for application as an ingredient in food.

References

Abuajah, C.I., Ogbonna, A.C., Osuji, C.M., 2015. Functional components and medicinal properties of food: a review. Journal of Food Science and Technology 52 (5), 2522–2529.

Aliyu, S., Bala, M., 2011. Brewer's spent grain: a review of its potentials and applications. African Journal of Biotechnology 10 (3), 324–331.

Available, 2017. Marketsandmarkets. https://www.marketsandmarkets.com/PressReleases/bioactive-ingredients.asp.

Azmir, J., Zaidul, I.S.M., Rahmana, M.M., Sharif, K.M., Mohameda, A., Sahena, F., Jahurul, M.H.A., Ghafoor, K., Norulaini, N.A.N., Omar, A.K.M., 2013. Techniques for extraction of bioactive compounds from plant materials: a review. Journal of Food Engineering 117 (4), 426–436.

Bensadón, S., Hervert-Hernández, D., Sáyago-Ayerdi, S.G., Goñi, I., 2010. By-products of *Opuntia ficus*-indica as a source of antioxidant dietary fiber. Plant Foods for Human Nutrition 65 (3), 210–216.

Biesalski, H.K., Dragsted, L.O., Elmadfa, I., Grossklaus, R., Muller, M., Schrenk, D., Walter, P., Weber, P., 2009. Bioactive compounds: definition and assessment of activity. Nutrition 25 (11–12), 1202–1205.

Boer, A., Urlings, M.J.E., Bast, A., 2016. Active ingredients leading in health claims on functional foods. Journal of Functional Foods 20, 587–593.

Burdock, G.A., Carabin, I.G., Griffiths, J.C., 2006. The importance of GRAS to the functional food and nutraceutical industries. Toxicology 221 (1), 17–27.

European Commission, 2010. Functional Foods. Food, Agriculture & Fisheries & Biotechnology.

Fang, Z., Bhandari, B., 2010. Encapsulation of polyphenols - a review. Trends in Food Science and Technology 21 (10), 510–523.

Fernandes, S.S., Salas-Mellado, M.M., 2017. Addition of chia seed mucilage for reduction of fat content in bread and cakes. Food Chemistry 227, 237–244.

Grajek, W., Olejnik, A., Sip, A., 2005. Probiotics, prebiotics and antioxidants as functional foods. Acta Biochimica Polonica 52 (3), 665–671.

Grand View Research, 2016. Available from: http://www.grandviewresearch.com/press-release/global-bioactive-ingredients-market.

Helkar, P.B., Sahoo, A.K., Patil, N.J., 2016. Review: food industry by-products used as a functional food ingredients. International Journal of Waste Resources 6 (3), 1–6.

Jafari, S.M., Esfanjani, A.F., Katouzian, I., Assadpour, E., 2017. Release, characterization of nanoencapsulated food ingredients. In: Jafari, S.M. (Ed.), Nanoencapsulation of Food Bioactive Ingredients Principles and Applications. Elsevier, India, p. 401.

Kaderides, K., Goula, A.M., Adamopoulos, K.G., 2015. A process for turning pomegranate peels into a valuable food ingredient using ultrasound-assisted extraction and encapsulation. Innovative Food Science and Emerging Technologies 31, 204–215.

Kalaycıoğlu, Z., Erim, F.B., 2017. Total phenolic contents, antioxidant activities, and bioactive ingredients of juices from pomegranate cultivars worldwide. Food Chemistry 221, 496–507.

Katouzian, I., Esfanjani, A.F., Jafari, S.M., Akhavan, S., 2017. Formulation and application of a new generation of lipid nano-carriers for the food bioactive ingredients. Trends in Food Science and Technology 68, 14–25.

Lagos, J.B., Vargas, F.C., Oliveira, T.G., Makishi, G.L.A., Sobral, P.J.A., 2015. Recent patents on the application of bioactive compounds in food: a short review. Current Opinion in Food Science 5, 1–7.

Leonard, N.B., 2000. Stability testing of nutraceuticals and functional foods. *Hand-book of Nutraceuticals and Functional Foods*. CRC Press, Boca Raton, FL.

Lordan, S., Ross, R.P., Stanton, C., 2011. Marine bioactives as functional food ingredients: potential to reduce the incidence of chronic diseases. Marine Drugs 9 (6), 1056–1100.

Marsanasco, M., Márquez, A.L., Wagner, J.R., Chiaramoni, N.S., Alonso, S.V., 2015. Bioactive compounds as functional food ingredients: characterization in model system and sensory evaluation in chocolate milk. Journal of Food Engineering 166, 55–63.

Mesías, M., Delgado-Andrade, C., 2017. Melanoidins as a potential functional food ingredient. Current Opinion in Food Science 14, 37–42.

Mildner-Szkudlarz, S., Bajerska, J., Zawirska-Wojtasiaka, R., Górecka, D., 2013. White grape pomace as a source of dietary fibre and polyphenols and its effect on physical and nutraceutical characteristics of wheat biscuits. Journal of the Science of Food and Agriculture 93 (2), 389–395.

Morales, P., Barros, L., Ramírez-Moreno, E., Santos-Buelga, C., Ferreira, I.C.F.R., 2015. Xoconostle fruit (*Opuntia matudae* Scheinvar cv. Rosa) by-products as potential functional ingredientes. Food Chemistry 185, 289–297.

Nadeem, M., Imran, M., Taj, I., Ajmal, M., Junaid, M., 2017. Omega-3 fatty acids, phenolic compounds and antioxidant characteristics of chia oil supplemented margarine. Lipids in Health and Disease 102, 1–12.

Pal, G.K., Suresh, P.V., 2016. Sustainable valorisation of seafood by-products: recovery of collagen and development of collagen-based novel functional food ingredients. Innovative Food Science and Emerging Technologies 37 (Part B), 201–215.

Prakash, B., Kujur, A., Singh, P.P., Kumar, A., Yadav, A., 2017. Plants-Derived bioactive compounds as functional food ingredients and food preservative. Nutrition and Food Science 2 (5), 1–7.

Rana, S., Guptad, S., Rana, A., Bhushan, S., 2015. Functional properties, phenolic constituents and antioxidant potential of industrial apple pomace for utilization as active food ingredient. Food Science and Human Wellness 4, 180–187.

Rasmussen, R.S., Morrissey, M.T., 2007. Marine biotechnology for production of food ingredients. Advances in Food & Nutrition Research 52 (6), 237–292.

Ré, M.I., 1998. Microencapsulation by spray drying. Drying Technology 16, 1195–1236.

Recharla, N., Riaz, M., Ko, S., Park, S., 2017. Novel technologies to enhance solubility of food-derived bioactive compounds: a review. Journal of Functional Foods 39, 63–73.

Segura-Campos, M.R., Salazar-Vega, I.M., Chel-Guerrero, L.A., Betancur-Ancona, D.A., 2013. Biological potential of chia (*Salvia hispanica* L.) protein hydrolysates and their incorporation into functional foods. LWT - Food Science and Technology 50 (2), 723–731.

Shea, N.O., Arendt, E.K., Gallagher, E., 2012. Dietary fibre and phytochemical characteristics of fruit and vegetable by-products and their recent applications as novel ingredients in food products. Innovative Food Science and Emerging Technologies 16, 1–10.

Silva, B.V., Barreira, J.C.M., Oliveira, M.B.P.P., 2016. Natural phytochemicals and probiotics as bioactive ingredients for functional foods: extraction, biochemistry and protected-delivery technologies. Trends in Food Science and Technology 50, 144–158.

Situ, W., Song, X., Luo, S., Liang, Y., 2017. A nano-delivery system for bioactive ingredients using supercritical carbon dioxide and its release behaviors. Food Chemistry 228, 219–225.

Suleria, H.A.R., Osborne, S., Masci, P., Gobe, G., 2015. Marine-based nutraceuticals: an innovative trend in the food and supplement industries. Marine Drugs 13 (10), 6336–6351.

Taofiq, O., Heleno, S.A., Calhelha, R.C., Alves, M.J., Barros, L., González-Paramás, A.M., Barreiro, M.F., Ferreira, I.C.F.R., 2017. The potential of Ganoderma lucidum extracts as bioactive ingredients in topical formulations, beyond its nutritional benefits. Food and Chemical Toxicology 108, 139–147.

Varzakas, T., Zakynthinos, G., Verpoort, F., 2016. Functional foods. Foods 5, 1–32.

Vo, T., Kim, S., 2012. Fucoidans as a natural bioactive ingredient for functional foods. Journal of Functional Foods 5 (1), 16–27.

Vos, P., Faas, M.M., Spasojevic, M., Sikkema, J., 2010. Encapsulation of nano beta-glucan for preservation of functionality and targeted delivery of bioactive food component. International Dairy Journal 20, 292–302.

Wang, Y., Li, Y., Zhao, Y., Chang, L., 2014. Effects of drying process of Yuanzhi (Radix Palygalae) on its bioactive ingredients. Journal of Traditional Chinese Medicine 34 (2), 206–213.

Weaver, C.M., 2014. Bioactive foods and ingredients for health. Advances in Nutrition (Bethesda, Md.) 5 (3), 306S–311S.

CHAPTER 8

Bioactive Compounds and Their Potential Use as Ingredients for Food and Its Application in Food Packaging

Betty Jarma Arroyo[1], Andrelina Pinheiro Santos[2], Enayde de Almeida de Melo[1], Anderson Campos[1], Lara Lins[1], Luis Carlos Boyano-Orozco[3]

[1]*Department of Domestic Science, Program of Food Science and Technology, Universidade Federal Rural de Pernambuco, Recife, Brazil;* [2]*Department of Chemical Engineering, Universidade Federal de Pernambuco, Recife, Brazil;* [3]*Department of Biochemical Engineering of the Escuela Nacional de Ciencias Biológicas, Instituto Politécnico Nacional, Mexico City, Mexico*

8.1 Introduction

In recent years, interest in the consumption of bioactive compounds has grown (Lähteenmäki, 2013), since there are several benefits that have been attributed to the consumption of these compounds, among which the prevention of cardiovascular diseases, cancer, diabetes, and Alzheimer's disease can be highlighted (Liu, 2013). This represents a great opportunity for the food industry, since the development of food products enriched or fortified with bioactive compounds could be a profitable business (Annunziata and Vecchio, 2013).

Among the most studied bioactive compounds are the secondary metabolites, such as antibiotics, mycotoxins, alkaloids, and phenolic compounds (Martins et al., 2011; Kepekçi et al., 2013). The latter have a great diversity of structures from simple molecules, such as vanillin, gallic acid, and caffeic acid to more complex molecules such as stilbenes, flavonoids, and polymers derived from various groups (Cheynier, 2012). The most common sources of bioactive compounds are plant materials such as fruits, vegetables, legumes, whole grains, and algae (Nirmala et al., 2014; Zakaria and Kamal, 2016). However, the large amount of waste generated by the food industry also represents a striking source of this type of compound. By giving added value to what was previously considered waste, it becomes a potential raw material, obtaining new foods (functional foods) in this way the environmental impact is reduced, and the production chains are optimized.

On the other hand, in the 20th century, advancements in packaging technology appeared as intelligent or smart packaging and active packaging (oxygen scavengers, antimicrobial agents, respiration controllers, and aroma/odor absorbers) (Brody et al., 2008). The first is capable of

Bioactive Compounds. https://doi.org/10.1016/B978-0-12-814774-0.00008-6

detecting, sensing, recording, tracing, or communicating information about the quality and/or state of the product during the whole food chain (Yam et al., 2005). The second is based on the concept of the incorporation of certain components into packaging systems that release or absorb substances from or into the packed food or the surrounding environment, so as to prolong the shelf life and sustain the quality, safety, and sensory characteristics of the food (Camo et al., 2008; Vermeiren et al., 1999).

In this chapter we will focus on the bioactive compounds that currently have greater relevance in the food industry, the advantages of adding them to food matrices, and which of them are used in the bioactive packaging industry, including the most promising types of packaging, and the impact it causes both in food and in the food chain.

8.2 Utilization of Phenolic Compounds as Ingredients in the Development of Functional Foods

It is known as functional food when it has a similar appearance to conventional food and it is consumed in a habitual diet, aside from fulfilling the basic function of nourishment, it can provide physiological benefits such as health-promoting properties or disease prevention (Kaur and Das, 2011). One of the main techniques for developing functional foods is through fortification, which consists of adding one or more components in order to correct or improve a potential biological activity of the food product (Świeca et al., 2014). With the intention of developing functional foods, different investigations have focused on incorporating bioactive compounds into different dietary matrices of high importance such as bread, yogurt, juices, and jams.

There are several plant sources of phenolic compounds, among these, green coffee and tea have become quite popular, since both have been related to the prevention of different diseases, such as obesity, due to the content of chlorogenic acids, quercetin-3-glucoside, among other compounds (Sato et al., 2011; Porto-Figueira et al., 2015; Craig et al., 2016; Choi et al., 2016; Pan et al., 2016).

Grapes, for example, are considered to be one of the most important sources of phenolic compounds for the human diet (Toaldo et al., 2016), and when they are processed to obtain products such as wine, juices, and jams, a significant amount of waste is generated (seeds and skins) rich in flavanols like oligomeric and polymeric compounds and anthocyanins whose potential as natural antioxidants is highly prominent (Rockenbach et al., 2011). Researchers have been interested in the use of phenolic compounds from this fruit as ingredients for the elaboration of different food products.

Toaldo et al. (2013) enriched juices of three grape varieties (Concord, Isabel, and Bordo) with grape seeds (*Vitis labrusca* L.); the enriched juices presented values of phenolic compounds expressed as mg equivalents of gallic acid/L of 293.8 ± 11.3, 973.6 ± 38.6, and 484.1 ± 24 for the

Concord, Isabel, and Bordo varieties, respectively. Likewise, the antioxidant capacity increased significantly in the three juices, where the Bordo variety obtained the highest values, and also observed an increase in the levels of some essential minerals (Ca, Mg, Na, K, Mn, Zn).

On the other hand, Corrêa et al. (2014) elaborated guava jams enriched with 30% grape juice, the enriched jam presented an amount greater than 200% of the amount of phenolic compounds initially present in the product and an antioxidant capacity value 18% higher than the original formulation, therefore, they classified this new jam as a product that could provide health benefits. It is important to emphasize that the antioxidant capacity of the products derived from the grape does not depend exclusively on the total content of phenolic compounds, but rather on the type of phenolic compound present in such products (Burin et al., 2014).

Foods such as bread, which for many years have continued to be part of the diet of a large percentage of the world population, have attracted attention to be studied as a model food in the incorporation of different bioactive compounds. Mildner-Szkudlarz et al. (2011) enriched sourdough mixed rye bread with 10% grape byproducts; these authors obtained increases of 39% and 37% of soluble and insoluble dietary fiber, respectively, and showed increases in antioxidant capacity due to the presence of compounds such as procyanidin B1 and B2, catechin, epicatechin, caffeic acid, and myricetin, however, increasing the proportion of the byproducts also increased the hardness, gumminess, and sharpness of the bread, so it is important to determine the sensory acceptance of this type of product.

Other research, such as that carried out in 2013 by Mildner-Szkudlarz and Bajerska, determined through a study in rats that rye bread fortified with grape byproducts (dry skin powder and lyophilized extract) generates lower susceptibility to the effect of oxidative stress caused by a diet with high cholesterol levels, it also increases the high-density lipoprotein cholesterol and prevents the accumulation of visceral fat, which is well known to increase the risk of atherosclerotic diseases and is associated with cardiometabolic risk (Shin et al., 2012; Tanaka et al., 2016).

There are several studies which make reference to the introduction of phenolic compounds in bread. Dziki et al. (2014) reported different investigations which studied the addition of plant materials rich in phenolic compounds in wheat bread and the increase in antioxidant capacity obtained as a result of that addition.

The addition of phenolic compounds to the yogurt manufacturing process has also been studied; Guiné et al. (2016) enriched two types of yogurt (fruit flavor, cinnamon flavor) with wine, they reported increases in antioxidant capacity as the amount of wine added increased, nevertheless there was no effect on the acidity in any of the proportions of wine, the sensory analysis showed that the product has great potential to be successful in the market since its acceptance was high.

In another research, Amirdivani and Baba (2015) observed a similar behavior when they incorporated green tea before the fermentation stage in the elaboration of a yogurt, these authors reported an increase in antioxidant capacity both at the beginning and end of the process, and they explained that the presence of green tea increases the growth of lactic acid bacteria (LAB) which could subsequently increase the antioxidant activity of yogurt during the fermentation.

The extract of pomegranate peel has also been studied as an ingredient in the preparation of yogurt, it has been illustrated that increasing the concentration of this extract significantly increases the content of phenolic compounds, flavonoids, and antioxidant capacity by including it both before and after the inoculation, the compounds increase more markedly when included before inoculation because the compounds could be broken down into smaller forms that are more extractable or stable during inoculation (El-Said et al., 2014). Thus, if phenolic compounds are incorporated into yogurt, there may be an increase of these in the finished product due to the decomposition of the compounds derived during the fermentation of the milk (Amirdivani and Baba, 2015).

All the above-mentioned investigations obtained foods with greater antioxidant capacity, high quality, and with promising effects for health. To catalog a food as functional, it is necessary to determine the effect caused by the addition of bioactive compounds of such a product, and test the real effect on the health of it. For this, it is necessary to carry out in vitro and in vivo studies, such as bioavailability studies where it is possible to know the absorption capacity of these compounds and their availability within the human body (Carbonell-Capella et al., 2014).

8.3 The Food Packaging Industry and Current Challenges

Traditional packaging has as a fundamental premise: "to be as inert as possible," i.e., there should be a minimum of interaction between the food and packaging, but at the same time it must be able to provide a protective barrier to food from external influences that ensure food handling and also can guarantee the preservation of food quality for a certain time.

However, as society is becoming increasingly complex, users (food producers, food processors, logistic operators, retailers, and consumers) continuously demand innovative and creative food packaging to guarantee food safety, quality, and traceability (Vanderroost et al., 2014); the necessary major current and future challenges to fast-moving consumer goods packaging include legislation, global markets, longer shelf-life, convenience, safer and healthier food, environmental concerns, authenticity, and food waste (Kerry, 2014; Realini and Begonya, 2014).

Therefore, food packaging technology is continuously evolving in response to growing challenges from a modern society, including changes in the use of materials that are usually derived from petroleum products and cause problems in waste disposal (Avella et al., 2005; Othman, 2014).

The development of new techniques for food storage includes edible coatings and films, which present as potential to replace conventional packaging, contributing to increase the shelf-life of food products, as well as acting in reducing gas exchange, respiration, moisture migration, and solute in the case of vegetables, and also reducing metabolic disturbances and lowering the rates of oxidative reaction (Azeredo et al., 2011; Ooi et al., 2012; Galus and Lenart, 2013; Maran et al., 2013; Pascall and Lin, 2013).

8.4 Biopolymer Packaging: Edible Films and Coatings

Edible films and coatings are layers of materials applied to food products that have a positive impact on the conservation, distribution, and commercialization of these foods because they protect these products from physical, chemical, and microbiological damage. These films may contribute with the addition of antioxidants, antimicrobials, nutraceuticals, flavoring agents, and various additives that improve the quality, handling, and integrity of the packaged item (Tavassoli-Krafani et al., 2016). Among the materials used in film production are biopolymers and their blends, which are viable substitutions to the use of traditional plastic packaging (Brito et al., 2011).

When preparing packaging, certain characteristics relating to raw materials, such as being inert, nontoxic, and impermeable to microorganisms, must be considered in order to fulfill their preservation and to preserve it from degradation as well as contact with external factors (Sung et al., 2013).

Different parts of plants and fruits from many agricultural crops have been found to be a viable source as fillers for composite industries (Majeed et al., 2013). In addition, some agricultural waste available in nature can also be used as a source of raw material for renewable materials (Azeredo, 2009). Biopolymers such as polysaccharides, proteins, and lipids can be used for the formation of edible films and coatings (Azeredo, 2009; Atarés and Chiralt, 2016), partly because of their biodegradable properties (Kammouna et al., 2013), and the advantage of being adaptable in terms of controlled release of substances, where the package is no longer a passive barrier, but actively contributes to the preservation of food (Martínez-Abad et al., 2014; Fabra et al., 2016).

The main difference between an edible coating and a film is that the first is applied in liquid form on the food, usually by spraying or immersing the product in the solution of edible material, and the second is molded as solid sheets, this is because its passes through a drying process and is subsequently applied to the food surface shaped as bags, packages, or capsules (Falguera et al., 2011), as shown in Fig. 8.1.

The use of films and edible coatings has the advantage of incorporating several actives in the polymer matrix, which increases the functional attributes of these packages.
In addition, they can be consumed along with food (Rojas-Graü et al., 2009). According to

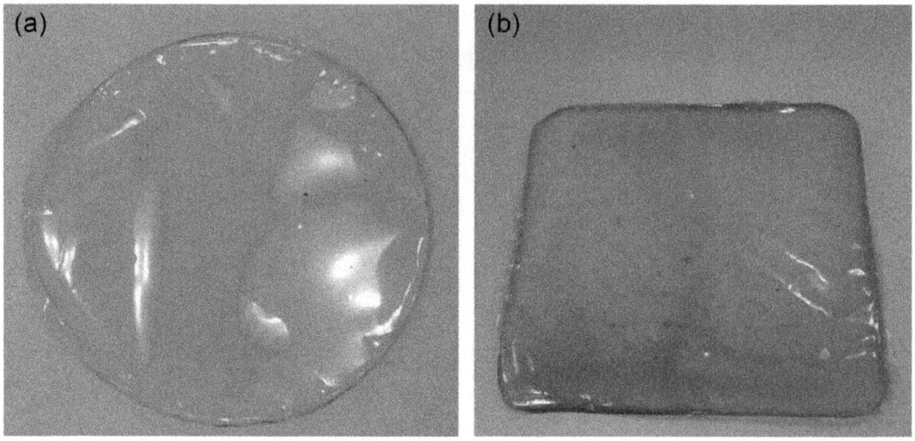

Figure 8.1
Examples of film shapes according to the mold they were made. (a) Pure chitosan film, (b) chitosan film with mint extract. *Source: Personal files (2017).*

Fakhoury et al. (2012), the elaboration of films is made from a combination of components with specific purposes. One or more film-forming agents (matrices), the solvent, the plasticizer, and other additives are added, and the pH-adjusting agent may also be added.

However, some hydrogels formed from natural polymers do not possess mechanical properties appropriate for use in controlled-release systems (García et al., 2017), for example, protein and carbohydrate packaging films are generally good barriers against oxygen at low to intermediate relative humidity and have good mechanical properties; however, their barrier against water vapor is weak due to their hydrophilic nature (Rhim et al., 2013). That's why researchers are constantly seeking different combinations by adding or blending several biopolymers and other biocompounds in order to obtain better matrices, enhancing their thermal, mechanical, and barrier properties.

One way to improve the structure of biodegradable films is by the addition of plasticizers. The incorporation of plasticizers in the filmogenic solution may lead to a decrease in the elongation stress and modulus of elasticity, as well as an increase in elongation at the break in the films (Santana and Kieckbusch, 2013). The plasticizers act by altering the mechanical property of the film, generating modifications to its barrier properties (Espitia et al., 2014).

Santos (2012) showed that the use of plasticizers in the production of biopolymers generates an increase in the free volume of the system, as well as improving the extensibility and flexibility of the films by means of the reduction of the intermolecular forces and the expansion of the polymer chains, reducing the fissures in the film and improving its strength.

The matrices of biopolymers are quite broad and can be divided into different categories based on the origin of the raw materials and their manufacturing processes which include:

1. Natural biopolymers such as plant carbohydrates (starch, cellulose, chitosan, alginate, agar, carrageenan, etc.), and animal or plant origin proteins (soy protein, wheat gluten, gelatin, collagen, whey protein, casein, etc.);
2. Synthetic biodegradable polymers;
3. Biopolymers produced by microbial fermentation such as microbial polyesters and microbial polysaccharides (Bordes et al., 2009; Clarinval and Halleux, 2005).

As mentioned above, the poor mechanical and barrier properties that most of them present currently limit their industrial use, this is one of the reasons why some components and substances besides plasticizers such as antioxidant agents, antimicrobials, texture modifiers (Osorio et al., 2011), nutrients, flavorings, colorants, spices, surfactants, among others (Kang et al., 2013), are also incorporated into these matrices. Therefore, we have selected some of the most studied blends used for the elaboration of this type of packaging.

8.5 Antioxidant Properties in Biopolymer Packages

Of the various types of active packaging, those that exert an antioxidant effect are among those that are most important for the industry, mainly food. Oxidation is one of the main degradation reactions that occur in food, limiting its conservation (Dantas et al., 2015).

Specifically, antioxidant active packaging seeks to prevent or slow down the oxidation of certain food components, like lipids and proteins, which leads to the deterioration of physical characteristics (such as flavor and color) of those food products. This active material approach requires the intentional incorporation of antioxidants within the packaging materials and their further migration to those foods (Robertson, 2012).

Oxygen has a degrading effect on the quality of a wide variety of food products. Foods rich in polyunsaturated fatty acids, such as vegetable oils, tend to undergo the oxidation process. The application in these products of incorporated films and coatings of antioxidants in their formulation represents an opportunity to solve this problem (Berton-Carabin et al., 2013; Bonilla and Sobral., 2012).

The incorporation of antioxidant agents into the polymeric matrix of films is one of the main foci of packaging development. The effectiveness of the antioxidant action of biodegradable films has been tested by different methods, such as: 2,2-diphenyl-1-picryl-hydralized radical (DPPH) and total phenolic compounds. Radical capture methods are commonly employed, such as DPPH, DPD, FRAP, and ABTS, and determine the ability of an antioxidant agent present in the biopolymer matrix of the film to intercept free radicals. Other methods involve the quantification of specific compounds with recognized antioxidant activity, as in the case of phenolic compounds (Bonilla and Sobral., 2012).

Several studies have investigated the effect of the addition of antioxidant agents in biodegradable films. Ramos et al. (2014) investigated the effect of thymol addition on DPPH-modified polylactic acid films and found that the developed biocomposites have the potential to be applied as antioxidant films. Films added with natural extracts, such as grape marc, in chitosan films resulted in films with antioxidant activity verified by different methods, such as ABTS and DPPH (Ferreira et al., 2014). The methods described evaluate the antioxidant activity of the films themselves. However, other methodologies may be employed together in order to deepen the understanding of the action of the active films in protecting the food of interest, in which the film is applied and stays in contact with foods that tend to undergo the oxidation process, as an example we have puff pastry. In this case, the antioxidant action of the films is investigated through tests carried out on the food itself, in order to evaluate its quality and shelf-life and whether the packaging has had an effect in protecting the food from oxidation.

Some studies have investigated the oxidative stability of packaged vegetable oils in active biodegradable films. Reis et al. (2015) developed films of starch with extract of yerba mate and mango pulp that acted against the oxidation of palm oil, an effect attributed to the content of phenolic compounds or flavonoids in the formulation of the film. The incorporation of astaxanthin-rich chrysanthemum extract was studied in polylactic acid films (Samsudin et al., 2014) and polyethylene (Colín-Chávez et al., 2013), and both retarded the oxidation of soybean oil.

Dorménech-Asensi et al. (2013) used tomato paste in the production of mortadella and verified an increase of the lycopene content in the samples and a reduction of the lipid oxidation in the product as a function of the concentration of tomato paste added. Barbosa-Pereira et al. (2014) developed films with the addition of a commercial rosemary extract and two synthetic antioxidants (BHT and propyl gallate) to package meat and verified that the active films in contact with the meat samples reduced (approximately 60%) the TBRAS (thiobarbituric acid reactive substances, method used to determine the lipid oxidation of the samples) after 9 days of storage in relation to the control.

Vargas (2015) used native Brazilian plant extracts (guaraná, pitanga, and rosemary) to retard the oxidation and deterioration of chilled ground meat, verifying that the use of pitanga extract increased the stability in relation to the lipid oxidation of the product and presented antimicrobial action by 6 days against psychotrophic bacteria.

Some authors have suggested the use of extracts of leaves of mint and curry as an alternative to antioxidants in cuts of pork (Biswas et al., 2012), carqueja in eggs (Lázaro et al., 2013), clove extracts, rosemary, cassia bark, licorice, nutmeg, and cardamom in pork (Kong et al., 2010), among others.

8.6 Nanotechnology and the Food Packaging Industry

The increased perception of consumers toward products in which chemical preservatives have been removed has shifted the focus to using natural active agents (Demitri et al., 2016). However, there is a need to incorporate into packaging materials, active agents that can maintain but mostly improve the quality of the mechanical and barrier properties.

The basic underlying principle behind the use of active packaging depends on the incorporation of particular components inside the polymer and intrinsic characteristics of the polymer itself used as a packaging vehicle (Gontard, 2000).

Recent advances in nanotechnology, particularly the ability to produce nanoparticles in different shapes and sizes, have led to the creation of a wide range of nanostructured compounds with antimicrobial properties (Sogvar et al., 2016). The application of nanotechnology to polymer matrices may open new possibilities for improving not only the properties but also the cost–price efficiency (Sorrentino et al., 2007), since because of their size, nanoparticles have a proportionally larger surface area and consequently more surface atoms than their microscale counterparts (Azeredo, 2009).

It has already been proven that low loading of nanofillers (5 wt%) into bio-nanocomposite technology can produce new biomaterials with specific properties and high performances for packaging applications (Reddy et al., 2013; Rhim et al., 2013).

There are several types of nanostructure responsible for multiple functions, sometimes providing active or "smart" properties to the packaging system such as antimicrobial activity, enzyme immobilization, and biosensing (Azeredo 2009).

The nanosized fillers can be either organic or inorganic, natural biopolymers, natural antimicrobial agents, metal, and metal oxides (Othman, 2014), that may improve not only the biopolymers' mechanical and barrier properties but also offer other functions and applications in food packaging such as antimicrobial agent, biosensor, and oxygen scavenger (Azeredo et al., 2011; Rhim et al., 2013).

One of the most common types of metal studied to produce bio-nanocomposite materials is ZnO, which has been found to have many applications in daily life, such as in drug delivery, cosmetics, and medical devices (Yan et al., 2009), due to its strong antimicrobial effect on a board spectrum of microorganisms (Jones et al., 2008). Moreover, it has been listed as generally recognized as safe by the US Food and Drug Administration (21CFR182.8991; Xie et al., 2011; Sogvar et al., 2016).

Nanofillers also have the ability to exhibit other desired functions and applications in food packaging such as as biosensors and oxygen scavengers (Othman, 2014). Then, nanofillers can not only protect food against environmental factors, but also incorporate properties to the packaging materials, allowing several potential options to create and innovative in the food packaging industry.

References

Amirdivani, S., Baba, A.S.H., 2015. Green tea yogurt: major phenolic compounds and microbial growth. Journal of Food Science and Technology 52, 4652–4660.

Annunziata, A., Vecchio, R., 2013. Consumer perception of functional foods: a conjoint analysis with probiotics. Food Quality and Preference 28, 348–355.

Atares, L., Chiralt, A., 2016. Essential oils as additives in biodegradable films and coatings for active food packaging. Trends in Food Science and Technology 48, 51–62.

Avella, M., Vlieger, J.J.D., Errico, M.E., Fischer, S., Vacca, P., Volpe, M.G., 2005. Biodegradable Starch/Clay Nanocomposite Films for Food Packaging Applications. Food Chemistry 93, 467–474.

Azeredo, H.M.C.D., Mattoso, L.H.C., Mchugh, T.H., 2011. In: Boreddy Reddy (Ed.), Nanocomposites in Food Packaging –A Review, Advances in Diverse Industrial Applications of Nanocomposites. InTech. Available from: https://www.intechopen.com/books/advances-in-diverse-industrial-applications-ofnanocomposites/nanocomposites-in-food-packaging-a-review.

Azeredo, H.M.C., 2009. Nanocomposites for food packaging applications. Food Research International 42, 1240–1253.

Barbosa-Pereira, L., Angulo, I., Lagarón, J.M., Paseiro-Losada, P., Cruz, J.M., 2014. Development of new active packaging films coated with natural phenolic compounds to improve the oxidative stability of beef. Meat Science 97, 249–254.

Berton-Carabin, C.C., Genot, C., Gaillard, D., Guilbert, Ropers, M.H., 2013. Design of interfacial films to control lipid oxidation in oil-in- water emulsions. Food Hydrocolloids 33 (1), 99–105.

Biswas, A.K., Chatli, M.K., Sahoo, J., 2012. Antioxidant potential of curry (*Murraya koeniggi* L.) and mint (*Mentha spicata*) leaf extracts and their effect on colour and oxidative stability of raw ground pork meat during refrigeration storage. Food Chemisty 133, 467–472.

Bonilla, J., Sobral, P.J.A., 2012. Edible films and coatings to prevent the detrimental effect of oxygen on food quality: possibilities and limitations. Journal of Food Engineering 110, 208–213.

Bordes, P., Pollet, E., Avérous, L., 2009. Nano-biocomposites: biodegradable polyester/nanoclay systems. Progress in Polymer Science 20, 25–55.

Brito, G.F., Agrawal, P., Araújo, E.M., Mélo, T.J.A., 2011. Biopolímeros, Polímeros Biodegradáveis e Polímeros Verdes. Revista Eletrônica de Materiais e Processos 6, 127–139.

Brody, A.L., Bugusu, B., Han, J.H., Koelsch Sand, C., Mchugh, T.H., 2008. Innovative food packaging solutions. Journal of Food Science 73, 107–116.

Burin, V.M., Ferreira-Lima, N.E., Panceri, C.P., Bordignon-Luiz, M.T., 2014. Bioactive compounds and antioxidant activity of *Vitis vinifera* and *Vitis labrusca* grapes: evaluation of different extraction methods. Microchemical Journal 114, 155–163.

Camo, J., Antonio Beltrán, J., Roncalés, P., 2008. Extension of the display life of lamb with an antioxidant active packaging. Meat Science 80 (4), 1086–1091.

Carbonell-Capella, J.M., Buniowska, M., Barba, F.J., Esteve, M.J., Frigola, A., 2014. Analytical methods for determining bioavailability and bioaccessibility of bioactive compounds from fruits and vegetables: a review. Comprehensive Reviews in Food Science and Food Safety 13, 155–171.

Cheynier, V., 2012. Phenolic compounds: From plants to foods. Phytochemistry Reviews 11 (2–3), 153–177. https://doi.org/10.1007/s11101-012-9242-8.

Choi, B.K., Park, S.B., Lee, D.R., Lee, H.J., Jin, Y.Y., Yang, S.H., Suh, J.W., 2016. Green coffee bean extract improves obesity by decreasing body fat in high-fat diet-induced obese mice. Asian Pacific Journal of Tropical Medicine 9, 635–643.

Clarinval, Am, Halleux, J., 2005. Classification of biodegradable polymers. In: Smith, R. (Ed.), Biodegradable Polymers for Industrial Applications. Woodhead Publishing Ltd., Cambridge, UK, pp. 3–31.

Colín-Chávez, C., Soto-Valdez, H., Peralta, E., Lizardi-Mendoza, J., Balandrán-Quintana, R.R., 2013. Fabrication and properties of antioxidant polyethylene-based films containing marigold (Tagetes erecta) extract and application on soybean oil srability. Packaing Technology and Science 26, 267–280.

Corrêa, R.C.G., Haminiuk, C.W.I., Sora, G.T.S., Bergamasco, R., Vieira, A.M.S., 2014. Antioxidant and rheological properties of guava jam with added concentrated grape juice. Journal of the Science of Food and Agriculture 94, 146–152.

Craig, A.P., Fields, C., Liang, N., Kitts, D., Erickson, A., 2016. Performance review of a fast HPLC-UV method for the quantification of chlorogenic acids in green coffee bean extracts. Talanta 154, 481–485.

Dantas, E.A., Costa, S.S., Cruz, L.S., Bramont, W.B., Costa, A.S., Padilha, F.F., Druzian, J.I., Machado, B.A.S., 2015. Characterization and evaluation of the antioxidant properties of biodegradable films incorporated with tropical fruit pulps. Ciência Rural 45, (1), 142–148.

Demitri, C., De Benedictis, V.M., Madaghiele, M., Corcione, C.E., Maffezzoli, A., 2016. Nanostructured active chitosan-based films for food packaging applications: effect of graphene stacks on mechanical properties. Measurement 90, 418–423.

Doménech-Asensi, G., et al., 2013. Effect of the addition of tomato paste on the nutritional and sensory properties of mortadella. Meat Science 93, 213–219.

Dziki, D., Różyło, R., Gawlik-Dziki, U., Świeca, M., 2014. Current trends in the enhancement of antioxidant activity of wheat bread by the addition of plant materials rich in phenolic compounds. Trends in Food Science and Technology 40, 48–61.

El-Said, M.M., Haggag, H.F., Fakhr El-Din, H.M., Gad, A.S., Farahat, A.M., 2014. Antioxidant activities and physical properties of stirred yoghurt fortified with pomegranate peel extracts. Annals of Agricultural Sciences 59, 207–212.

Espitia, P.J.P., Du, W., Avena-bustillos, R., Soares, J., 2014. Edible films from pectin: physical-mechanical and antimicrobial properties – a review. Food Hydrocolloids 35, 287–296.

Fabra, M.J., López-Rubio, A., Lagaron, J.M., 2016. Use of the electrohydrodynamic process to develop active/bioactive bilayer films for food packaging applications. Food Hydrocolloids 55, 11–18.

Fakhoury, F.M., Martelli, S.M., Bertan, L.C., Yamashita, F., Mei, I.L.H., Queiroz, F.P.C., 2012. Edible films made from blends of manioc starch and gelatin–Influence of different types of plasticizer and different levels of macromolecules on their properties. Food Science and Technology 49 (1), 149–154.

Falguera, V., Quintero, J.P., Jiménez, A., Muñoz, J.A., Ibarz, A., 2011. Edible films and coatings: structures, active functions and trends in their use. Trends in Food Science and Technology 22 (6), 292–303.

Ferreira, A.S., Nunes, C., Castro, A., Ferreira, P., Coimbra, M.A., 2014. Influence of grape pomace extract incorporation on chitosan films properties. Carbohydrate Polymers 113, 490–499.

Galus, S., Lenart, A., 2013. Development and characterization of composite edible films based on sodium alginate and pectin. Journal of Food Engeneering 115, 459–465.

García, J., Ruiz-Durántez, E., Valderrutena, N.E., 2017. Interpenetrating polymer networks hydrogels of chitosan and poly (2-hydroxyethyl methacrylate) for controlled release of quetiapine. Reactive and Functional Polymers 117, 52–59.

Gontard, N., 2000. Panorama des emballages alimentaires actifs. In: Gontard N. (Cord.) Lesemballages actifs. Tec & Doc, p. 1.

Guiné, R.P.F., Rodrigues, A.P., Ferreira, S.M., Gonçalves, F.J., 2016. Development of yogurts enriched with antioxidants from wine. Journal of Culinary Science and Technology 14 (3), 263–275.

Jones, N., Ray, B., Ranjit, K.T., Manna, A.C., 2008. Antibacterial activity of ZnO nanoparticle suspensions on a broad spectrum of microorganisms. Microbiology Letters 279 (1), 71–76.

Kammouna, M., Haddarc, M., Kallelc, T.K., Dammakc, M., Sayari, A., 2013. Biological properties and biodegradation studies of chitosan biofilms plasticized with PEG and glycerol. International Journal of Biological Macromolecules 62, 433–438.

Kang, H., Kim, S., You, Y., Lacroix, M., Han, J., 2013. Inhibitory effect of soy protein coating formulations on walnut (Julgans regia L.) kernels against lipid oxidation. Food Science and Technology 51, 393–396.

Kaur, S., Das, M., 2011. Functional foods: an overview. Food Science and Biotechnology 20, 861–875.

Kepekçi, R.A., Polat, S., Çelik, A., Bayat, N., Saygideger, S.D., 2013. Protective effect of *Spirulina platensis* enriched in phenolic compounds against hepatotoxicity induced by CCl4. Food Chemistry 141, 1972–1979.

Kerry, J.P., 2014. New Packaging Technologies, Materials and Formats for Fast-Moving Consumer Products Author Links Open Overlay Panel. Innovations in Food Packaging (Second Edition), 549–584.

Kong, B., Zhang, H., Xiong, Y.L., 2010. Antioxidant activity of spice extracts in a liposome system and in cooked pork patties and the possible mode of action. Meat Science 85, 772–778.

Lähteenmäki, L., 2013. Claiming health in food products. Food Quality and Preference 27, 196–201.

Lázaro, S.F., Fonseca, L.D., Martins, E.R., de Oliveira, N.J., Duarte, E.R., 2013. Effect of arqueous extracts of baccharis trimera on development and hatching of *Rhipicephalus microplus* (Acaridae) eggs. Veterinary Parasitology 194, 79–82.

Liu, R.H., 2013. Dietary bioactive compounds and their health implications. Journal of Food Science 78, A18–A25.

Majeed, K., Jawaid, M., Hassan, A., Abu Bakar, A., Abdul Khalil, H.P.S., Salema, A.A., Inuwa, I., 2013. Potential materials for food packaging from nanoclay/natural fibres filled hybrid composites. Materials and Design 46, 391–410.

Maran, J.P., Sivakumar, V., Sridhar, R., Immanuel, V.P., 2013. Development of model for mechanical properties of tapioca starch based edible films. Industrial Crops and Products 42, 159–168.

Martínez-Abad, A., Ocio, M.J., Lagaron, J.M., 2014. Morphology, physical properties, silver release, and antimicrobial capacity of ionic silver-loaded poly (L-lactide) films of interest in food-coating applications. Journal of Applied Polymer Science 131 (21), 41001.

Martins, S., Mussatto, S.I., Martínez-Avila, G., Montañez-Saenz, J., Aguilar, C.N., Teixeira, J.A., 2011. Bioactive phenolic compounds: production and extraction by solid-state fermentation. A review. Biotechnology Advances 29, 365–373.

Mildner-Szkudlarz, S., Bajerska, J., 2013. Protective effect of grape by-product-fortified breads against cholesterol/cholic acid diet-induced hypercholesterolaemia in rats. Journal of the Science of Food and Agriculture 93 (13), 3271–3278.

Mildner-Szkudlarz, S., Zawirska-Wojtasiak, R., Szwengiel, A., Pacyński, M., 2011. Use of grape by-product as a source of dietary fibre and phenolic compounds in sourdough mixed rye bread. International Journal of Food Science and Technology 46, 1485–1493.

Nirmala, C., Bisht, M.S., Laishram, M., 2014. Bioactive compounds in bamboo shoots: health benefits and prospects for developing functional foods. International Journal of Food Science and Technology 49, 1425–1431.

Ooi, Z.X., Ismail, H., Bakar, A.A., Aziz, N.A.A., 2012. Properties of the crosslinked plasticized biodegradable poly(vinyl alcohol)/rambutan skin waste flour blends. Journal of Applied Polymer Science 125, 1127–1135.

Osorio, F.A., Molina, P., Matiacevich, S., Enrione, J., Skurtys, O., 2011. Characteristics of hydroxy propyl methyl cellulose (HPMC) based edible film developed for blueberry coatings. Procedia Food Science 1, 287–293.

Othman, S.H., 2014. Bio-nanocomposite materials for food packaging applications: types of biopolymer and nano-sized filler. Agriculture and Agricultural Science Procedia 2, 296–303.

Pan, M.H., Tung, Y.C., Yang, G., Li, S., Ho, C.T., 2016. Molecular mechanisms of the anti-obesity effect of bioactive compounds in tea and coffee. Food and Function 7 (11), 4481–4491.

Pascall, M.A., Lin, S.J., 2013. The application of edible polymeric films and coatings in the food industry. Food Processing and Technology 4, 116.

Porto-Figueira, P., Figueira, J.A., Pereira, J.A., Câmara, J.S., 2015. A fast and innovative microextraction technique, µSPEed, followed by ultrahigh performance liquid chromatography for the analysis of phenolic compounds in teas. Journal of Chromatography A 11 (1424), 1–9.

Ramos, M., Jiménez, A., Peltzer, M., Garrigós, M.C., 2014. Development of novel nano-biocomposite antioxidant films based on poly (lactic acid) and thymol for active packaging. Food Chemistry 6, 149–155.

Realini, C.E., Begonya, M., 2014. Active and intelligent packaging systems for a modern society. Meat Science 98, 404–419.

Reddy, M.M., Vivekanandhan, S., Misra, M., Bhati, S.K., Mohanty, A.K., 2013. Biobased plastics and bionanocomposites: current status and future opportu-nities. Progress in Polymer Science 38, 1653–1689.

Reis, L.C.B., De Souza, C.O., Da Silva, J.B.A., Martins, A.C., Nunes, I.L., Druzian, J.I., 2015. Active biocomposites of cassava starch: the effect of yerba mate extract and mango pulp as antioxidant additives on the properties and the stability of a packaged product. Food and Bioproducts Processing 94, 382–391.

Rhim, J.W., Park, H.M., Ha, C.S., 2013. Bio-nanocomposites for food packaging applications. Progress in Polymer Science 38, 1629–1652.

Robertson, G.L., 2012. Food Packaging: Principles and Practice, third ed. CRC Press, United States of America.

Rockenbach, I.I., Gonzaga, L.V., Rizelio, V.M., Gonçalves, A.S., Genovese, M.I., Fett, R., 2011. Phenolic compounds and antioxidant activity of seed and skin extracts of red grape (*Vitis vinifera* and *Vitis labrusca*) pomace from Brazilian winemaking. Food Research International 44, 897–901.

Rojas-graü, M.A., Soliva-Fortuny, R., Martín-Belloso, O., 2009. Edible coatings to incorporate active ingredients to fresh-cut fruits: a review. Trends in Food Science and Technology 20, 438–447.

Samsudin, H., Soto-valdez, H., Auras, R., 2014. Poly (lactic acid) film incorporated with marigold flower extract (*Tagetes erecta*) intended for fatty-food application. Food Control 46, 55–66.

Santana, A.A., Kieckbusch, T.G., 2013. Physical evaluation of biodegradable films of calcium alginate plasticized with polyols. Brazilian Journal of Chemical Engineering 30, 835–845.

Santos, E.M.C., 2012. Filmes biodegradáveis de galactomanana: uso na conservação de frutos (Ph.D. thesis). Universidade Federal do Ceará, Centro de Ciências, Ceará. 126 p.

Sato, Y., Itagaki, S., Kurokawa, T., Ogura, J., Kobayashi, M., Hirano, T., Sugawara, M., Iseki, K., 2011. In vitro and in vivo antioxidant properties of chlorogenic acid and caffeic acid. International Journal of Pharmaceutics 403 (1–2), 136–138.

Shin, H.-Y., Lee, D.C., Chu, S.H., Jeon, J.Y., Lee, M.K., Im, J.A., Lee, J.W., 2012. Chemerin levels are positively correlated with abdominal visceral fat accumulation. Clinical Endocrinology 77, 47–50.

Sogvar, O.B., Saba, M.K., Emamifar, A., Hallaj, R., 2016. Influence of nano-ZnO on microbial growth, bioactive content and postharvest quality of strawberries during storage. Innovative Food Science and Emerging Technologies 35, 168–176.

Sorrentino, A., Gorrasi, G., Vittoria, V., 2007. Potential perspectives of bionanocomposites for food packaging applications. Trends in Food Science and Technology 18 (2), 84–95.

Sung, S.-Y., Sin, L.T., Tee, T.-T., Bee, S.-T., Rahmat, A.R., WaWA, R., Tan, A.-C., Vikhraman, M., 2013. Antimicrobial agents for food packaging applications. Trends in Food Science and Technology 33, 110–123.

Świeca, M., Sęczyk, Ł., Gawlik-Dziki, U., Dziki, D., 2014. Bread enriched with quinoa leaves – the influence of protein–phenolics interactions on the nutritional and antioxidant quality. Food Chemistry 162, 54–62.

Tanaka, K.I., Kanazawa, I., Sugimoto, T., 2016. Reduced muscle mass and accumulation of visceral fat are independently associated with increased arterial stiffness in postmenopausal women with type 2 diabetes mellitus. Diabetes Research and Clinical Practice 122, 141–147.

Tavassoli-Krafani, E., Sekarchizadeh, H., Masoudpour-Ehabadi, M., 2016. Development of edible films and coatings from alginates and carrageeens. Carbohyrate Polymers 137, 360–374.

Toaldo, I.M., Fogolari, O., Pimentel, G.C., de Gois, J.S., Borges, D.L.G., Caliari, V., Bordignon-Luiz, M., 2013. Effect of grape seeds on the polyphenol bioactive content and elemental composition by ICP-MS of grape juices from *Vitis labrusca* L. Food Science and Technology 53, 1–8.

Toaldo, I.M., Cruz, F.A., da Silva, E.L., Bordignon-Luiz, M.T., 2016. Acute consumption of organic and conventional tropical grape juices (*Vitis labrusca* L.) increases antioxidants in plasma and erythrocytes, but not glucose and uric acid levels, in healthy individuals. Nutrition Research 36 (8), 808–817.

Vanderroost, M., Ragaert, P., Devlieghere, F., De Meulenaer, B., 2014. Intelligent food packaging: the next generation. Trends in Food Science and Technology 39, 47–62.

Vargas, F.C., 2015. Characterization of aqueous extracts of Pitanga (Eugênia uniflora L), Guaraná (Paullinina cupana Kunth) and Rosemary (Rosmarinus officinalis) and its application to promote the stability of refrigerated beef. Thesis (PhD in Animal Quality and Productivity) - Faculty of Animal Science and Food Engineering, University of São Paulo, Pirassununga, 2015. https://doi.org/10.11606/T.74.2016.tde-03022016-102107.

Vermeiren, L., Devlieghere, F., Van Beest, M., De Kruijf, N., Debevere, J., 1999. Developments in the active packaging of foods. Trends in Food Science and Technology 10 (3), 77–86.

Xie, Y., He, Y., Irwin, P.L., Jin, T., Shi, X., 2011. Antibacterial activity and mechanism of action of zinc oxide nanoparticles against *Campylobacter jejuni*. Applied and Environmental Microbiology 77 (7), 2325–2331.

Yam, K.L., Takhistov, P.T., Miltz, J., 2005. Intelligent packaging: concepts and applications. Journal of Food Science 70 (1), 1–10.

Yan, S., Salley, S.O., Ng, K.Y.S., 2009. Simultaneous transesterification and esterification of unrefined or waste oils over ZnO–La2O3 catalysts. Applied Catalysis A: General 353 (2), 203–212.

Zakaria, S.M., Kamal, S.M.M., 2016. Subcritical water extraction of bioactive compounds from plants and algae: applications in pharmaceutical and food ingredients. Food Engineering Reviews 8, 23–34.

Further Reading

Abugoch, L., Tapia, C., Plasencia, D., Pastor, A., Castro-Mandujano, O., López, L., Escalona, V.H., 2016. Shelf-life of fresh blueberries coated with quinoa protein/chitosan/sunflower oil edible film. Journal of the Science of Food and Agriculture 96, 619–626.

Alves, V.D., Mali, S., Beléia, A., Grossmann, M.V.E., 2007. Effect of glycerol and amylose enrichment on cassava starch film properties. Journal of Food Engineering 78, 941–946.

Andrade, R.M.S., Ferreira, M.S.L., Gonçalves, E.C.B.A., 2016. Development and characterization of edible films based on fruit and vegetable residues. Journal of Food Science 81 (2), 412–418.

Elsabee, M.Z., Abdou, E.S., 2013. Chitosan based edible films and coatings: a review. Materials Science and Engineering 33, 1819–1841.

Gutiérrez, T.J., Tapia, M.S., Pérez, E., Famá, L., 2015. Structural and Mechanical properties of edible films made from native and modified cush-cush yam and cassava starch. Food Hydrocolloids 45, 211–217.

Jayakumar, R., Prabaharan, M., Nair, S.V., Tokura, S., Selvamurugan, N., 2010. Novel carboxymethyl derivatives of chitin and chitosan materials and their biomedical applications. Progress in Materials Science 55, 675–709.

Jo, W.S., Song, H.-Y., Song, N.B., Lee, J.H., Min, S.C., Song, K.B., 2014. Quality and microbial safety of "Fuji" apples coated with carnauba-shellac wax containing lemongrass oil. Food Science and Technology 55, 490–497.

Olivas, G.I., Barbosa-cánovas, G.V., 2004. Alginate-Calcium Films: Water Vapor Permeability and Mechanical Properties as Affected by Plastizicerplasticizer and Relative Humidity. IFT Annual Meeting, Las Vegas.

Otoni, C.G., Moura, M.R., Aouada, F.A., Camilloto, G.P., Cruz, R.S., Lorevice, M.V., Soares, N.F., Mattoso, L.H., 2014. Antimicrobial and physical-mechanical properties of pectin/papaya puree/cinnamaldehyde nanoemulsion edible composite films. Food Hydrocolloids 41, 188–194.

Paula, G.A., Benevides, N.M.B., Cunha, A.P., Oliveira, A.V., de Pinto, A.M.B., Morais, J.P.S., de Azeredo, H.M.C., 2015. Development and characterization of edible films from mixtures of k-carrageenan, t-carrageenan and alginate. Food Hydrocolloids 47, 140–145.

Rodríguez, M., Osés, J., Ziani, K., Maté, J.I., 2006. Combined effect of plasticizers and surfactants on the physical properties of starch based edible films. Food Research International 39, 840–846.

Suyatma, N.E., Tighzert, L., Copinet, A., 2005. Effects of hydrophilic plasticizers on mechanical, thermal and surface properties of chitosan films. Journal of Agricultural and Food Chemistry 53, 3950–3957.

Wu, Y., Luo, X., Li, W., Song, R., Li, J., Li, Y., Li, B., Liu, S., 2016. Green and biodegradable composite films with novel antimicrobial performance based on cellulose. Food Chemistry 197, 250–257.

Zhang, H.Y., Li, R.P., Liu, W.M., 2011. Effects of chitin and its derivative chitosan on postharvest decay of fruits: a review. International Journal of Molecular Sciences 12, 917–934.

PART 7

Health Benefits of Bioactive Compounds

CHAPTER 9

Association Between Diet, Health, and the Presence of Bioactive Compounds in Foods

Michele Silveira Coelho, Sibele Santos Fernandes, Myriam de las Mercedes Salas-Mellado

Laboratory of Food Technology, School of Chemistry and Food, Federal University of Rio Grande, Rio Grande, Brazil

Bioactive compounds occur in small amounts in foods and are considered as non-nutritional but vital ingredients for the maintenance of human health (Patil et al., 2009). Research on bioactive ingredients for human consumption has increased in the recent past due to consumer awareness of their associated benefits with health maintenance and well-being. Among these compounds, those related to antioxidant activity are the subject of various research studies, due to their enormous importance in human health.

This chapter lists the bioactive compounds considered to be potentially functional, as well as the effect on health attributed to them.

9.1 Fruits

Consumption of fruits is universally recommended due to the supply of nutrients and phytochemicals in the human diet and is linked to a lower incidence of various diseases (Slavin and Lloyd, 2012). Asia is the largest-producing region for tropical fruits, followed by Latin America, the Caribbean, Africa, and Oceania (FAO, 2011). Brazil possesses a geographical region with suitable climatic conditions for a large number of native fruits that may possess excellent agro-industrial potential, thus representing an interesting economic income for local growers. The evaluation of their bioactive properties will thus strengthen their position in the market, reaching specific markets created by consumer demand for new products able to maintain health and preventing diseases, as well as the growing market of functional ingredients.

Regarding the compounds contained in these fruits that could potentially lead to health benefits, polyphenols are present as major compounds (Schreckinger et al., 2010). Polyphenols are a chemically heterogeneous group with approximately 10,000 compounds (Andersen and Markham, 2006), which according to their structure (number of phenol rings and the type and number of structural elements binding) are grouped into nonflavonoids and flavonoids. The nonflavonoid compounds are phenolic acids and stilbenes, while flavonoids

Bioactive Compounds. https://doi.org/10.1016/B978-0-12-814774-0.00009-8

are divided into anthocyanins, flavanols, flavonols, flavones, flavanones, and isoflavones (Wightman and Heuberger, 2015).

Antioxidant activity is usually related to the presence of phenolic compounds; these exhibit specific common structures that allow them to act as reducing agents, hydrogen donators, and singlet oxygen quenchers, among other reaction mechanisms (Pietta, 2000). At a cellular level, various antioxidant compounds are known to be able to stabilize or deactivate free radicals, thus preventing damage to cell structures. Their significance within human health has been extensively described, playing such diverse roles as protection against cardiovascular diseases (by reducing chronic inflammation and improving endothelium function), certain forms of cancer and cytotoxic effects, among others (Proestos et al., 2006; Robles-Sánchez et al., 2011). Besides their antioxidant capacity, phenolic compounds present in plants may also possess antimicrobial properties (Chakraborty and Mitra, 2008; Rauha et al., 2000). Although there are several studies on the antioxidant capacity of fruits, their antimicrobial properties are scarcely screened. The antimicrobial capacity of a fruit (or its extracts) is of utmost importance, because despite the large amount of preservation techniques available nowadays, the spoilage and deterioration of food products by microorganisms is still a problem that has not yet been completely controlled. Most foodborne illnesses are caused by microbial pathogens present in the food due to contamination during the process from farm to fork, or caused by toxins produced by those contaminants. Certain species, including specific strains of *Escherichia coli*, *Staphylococcus aureus*, and *Listeria monocytogenes*, can even cause fatal infections in humans (Šiler et al., 2014).

Therefore, many food products need to be protected through addition of preservatives, but the growing tendency to avoid chemical compounds in foods is leading to the search and development of alternative natural substances able to simultaneously increase the shelf-life of foods and provide a high degree of safety regarding foodborne pathogens, as well as present reduced hypersensitivity reactions.

In this context, native fruits to South America have received special attention from the scientific community in recent years due to their nutritional and functional properties (Da Silva et al., 2014; Denardin et al., 2015; Oliveira et al., 2012).

Table 9.1 shows the common name, bioactive compounds, and main health benefits of tropical berries. These fruits do not occur in all tropical countries due to varying climate and soils. However, depending on the time of year, it is possible to find these fruits at fairs and specialized markets, attended by people interested in typical tropical foods. These fruits are usually seasonal and grow in a tropical climate, where the harvest occurs mainly during the warmer months of the year (Costa et al., 2013).

Several publications demonstrate that juçara fruit contains excellent nutritional characteristics and bioactive compounds such as anthocyanins, flavonoids, and phenolic acids, which are

Table 9.1: Health Benefits From Fruits (Costa et al., 2013)

Fruit	Bioactive Compounds	Health Benefits
Açai, Asai palm, Azaí Huasaí, Manaca palm	Anthocyanins, flavonoids, phenolic acids, procyanidin, lignans, stilbenes	Increase plasma antioxidant capacity (Mertens-Talcott et al., 2008); decrease of oxidative stress (Noratto et al., 2011; Petruck et al., 2017); antiinflammatory effects (Kang et al., 2011; Noratto et al., 2011); improvement of endothelial function (Michalska et al., 2010; Rocha et al., 2007); antioxidant potential and protected human neuron-like cells (Paz et al., 2015; Torma et al., 2017; Carvalho et al., 2017); platelet aggregation (Michalska et al., 2010); ameliorating properties over metabolic syndrome (Udani et al., 2011); antiallergic (Horiguchi et al., 2011); hypocholesterolemic effects (Feio et al., 2012; De Souza et al., 2010); and anticancer properties (Del Pozo-Insfran et al., 2006)
Pitanga, Brazilian cherry, Ñangapirí	Anthocyanins, carotenoids, flavonols	Antidiarrheic, diuretic, antirheumatic, antifebrile, and antidiabetic (Oliveira et al., 2006); antimicrobial activity against *S. aureus*, *L. monocytogenes*, *C. lipolytica,* and *C. guilliermondii* (Victoria et al., 2012); antitrypanosoma (Santos et al., 2012); β-adrenergic-induced hypotension in rat heart (Consolini and Sarubbio, 2002)
Jaboticaba, Guapurú, Uva de árbol, Brazilian grape tree	Anthocyanins, ellagic and gallic acid, carotenoids, depsides, tannins, rutin, vitamin C	Antioxidant potential increase in rats plasma (Leite et al., 2011); antiinflammatory, against asthma, and antidiarrhea (Lima et al., 2008; Reynertson et al., 2006); inhibition of IL-8 production (Reynertson et al., 2006); antiproliferative effects against tumor cell lines (Leite et al., 2012); protective effect in cardiovascular disease and type II diabetes mellitus (Lenquiste et al., 2012)
Camu-camu, Cacari, Camocamo	Anthocyanins, ellagic acid, flavan-3-ols, vitamin C	High antioxidant capacity (Rufino et al., 2010); inhibition of LPS-induced NO release in RA W 264.7 cells (Yazawa et al., 2011); decrease of oxidative stress and antiinflammatory (Inoue et al., 2008)
Jambolão, black plum	Anthocyanins, ellagic acid, quercetin, rutin, vitamin C	Antiscorbutic and diuretic features (Benherlal and Arumughan, 2007; Gordon et al., 2011); numerous pharmacological features (Baliga et al., 2011); antidiabetic effects (Baliga et al., 2011; Benherlal and Arumughan, 2007; De Bona et al., 2011; Gordon et al., 2011; Teixeira et al., 1997; Teixeira et al., 2000)

related to potent antioxidant activity, and it has been suggested that juçara fruits may exert antioxidant effects in vitro (Bicudo et al., 2014; Borges et al., 2011, 2013; Cardoso et al., 2015a; Da Silva et al., 2013; Rufino et al., 2010; Schulz et al., 2015). Studies in vivo with healthy individuals (Cardoso et al., 2015a), animal models (Cardoso et al., 2015b; Oyama et al., 2016), and cell cultures (Borges et al., 2013) have suggested that juçara fruit may exert a positive effect on the antioxidant status. Due to its exceptionally high antioxidant capacity, juçara fruit has been recently described as a "super food" (Cunha Júnior et al., 2016; Cunha Junio et al., 2015; Felzenszwalb et al., 2013).

In apple, grape, orange, papaya, banana, strawberry, and cherry, iron concentrations vary between 0.8–2 mg/100 g of dry matter (Hui, 2006). Although the bioavailability of iron is lower in plant foods, the juçara fruit intake could be encouraged to contribute to the treatment of iron deficiency, a common nutritional disorder worldwide, that may affect up to 40% of the global population (De Benoist et al., 2008; WHO, 2007a,b). Another element worth mentioning is calcium, which is also among the nutrients with high frequency of inadequacy, mainly in children (Carvalho et al., 2015). This element is found in levels of 349.4–596.7 mg/100 g of dry matter during juçara fruit ripening (Schulz et al., 2015).

Flavonoids are known to act as both antioxidants and antiinflammatory molecules, thus these compounds are also related to the capacity of modulating the signaling pathways involved in inflammation, cell survival, neurotransmission, and enhancing neuroplasticity (Hwang et al., 2012; Rendeiro et al., 2015). A study of the neuroprotective effects was described by Wong et al. (2013), which examined whether açai (*Euterpe oleracea*) extract afforded protection against β-amyloid (Aβ)-mediated loss of cell viability and oxidative stress associated with antifibrillary effects. Pretreatment with açai extract significantly improved the viability of a cell line derived from a pheochromocytoma of the rat adrenal medulla (PC12)and inhibited thioflavin T (ThT), which emits fluorescence when bound to β-amyloid fibrils. Açai extract also disrupted Aβ1–42 fibril and aggregate morphology. These results may underlie a neuroprotective effect of the açai extract, since neurotoxicity in Alzheimer's disease, for example, is associated with increased levels of β-amyloid (Aβ) peptide in the brain, manifest via Aβ oligomer and fibril formation.

9.2 Polyunsaturated Fatty Acids (PUFA)

Cardiovascular disease (CVD) is a leading cause of death in the United States. Nearly 1 in 3 deaths (800,000) each year are from CVD. Total direct medical expenditures on CVD have averaged about $193.1 billion annually and are projected to soar to more than $818 billion by the year 2030 (Heidenreich et al., 2011).

The increasing interest in incorporating omega-3 PUFA into the diet has been driven by the extensive literature indicating that these dietary PUFA promote health and disease prevention. This literature includes in vitro cell-based and animal-feeding studies, as well as observational studies and randomized controlled trials in humans. Omega-3 PUFA are naturally enriched in a number of common foods such as fatty fish like salmon or tuna as well as in several common plant-derived oils like canola, soybean, and chia oils. Widely available dietary supplements, such as fish oils and flax seed oil, are also rich sources of omega-3 PUFA, while other foods including eggs, dairy products, and baked goods that have been supplemented with omega-3 PUFA are now increasingly available (Whelan and Rust, 2006).

The lipid profile of fruits has great nutritional importance because the predominance of unsaturated fatty acids, especially monounsaturated fatty acids, can have a positive effect on health, especially for the risk of cardiovascular disease, as the dietary fatty acid composition regulates lipids and lipoprotein metabolism (Cheng et al., 2016; Ooi et al., 2015). A study by Novello et al. (2015) aimed to evaluate the freeze-dried juçara (*Euterpe edulis*) extract regarding the reduction of cardiovascular risk in mice with apolipoprotein E (ApoE−/−) knockout. Mice groups which received juçara extract presented decreased values of total cholesterol, low-density lipoprotein (LDL), as well as the ratios of total cholesterol/high-density lipoprotein (HDL) and LDL/HDL. Feio et al. (2012) and De Souza et al. (2010) also observed hypocholesterolemic effects for açai in rabbits and rats, respectively.

The Food and Agriculture Organization of the United Nations (FAO, 2010) recommends a minimum daily intake of 2.5% of energy for linoleic fatty acid and 0.5% for linolenic fatty acid. Considering a diet of 2000 kcal, an intake portion of 100 g of juçara fruits could provide up to 2.6% linoleic acid and 0.1% linolenic acid. According to the values of fatty acids published by Morais et al. (2016) and Bunea et al. (2012), an intake of 100 g of other fruits such as avocado, pineapple, banana, papaya, passion fruit, watermelon, and melon could provide up to 0.1% linoleic acid and 0.02% linolenic acid for a diet of 2000 kcal. For blueberries and bilberries, these values are 0.1% for both essential fatty acids.

In recent decades, different oilseeds, specifically of the medicinal plants, have attracted much attention as vegetable seed oils could be the main source of dietary ingredients related to their fatty acid composition and bioactive compounds. The fatty acids content, particularly essential fatty acids, linoleic, and linolenic acids from the source seed oil are very important because the human body cannot produce them. In a review article (De Falco et al., 2017) the authors discussed research on chia oil. Chia seed oil ranges from 25% to 50% and contains high concentrations of polyunsaturated fatty acids. Research demonstrated that oil extracted from chia seeds also contain several phenolic compounds such as tocopherols, phytosterols, and carotenoids with their related antioxidant activity that plays a very important role in the deterioration of the oil due to lipid oxidation. A consumption of 7.3 g of chia seed per day provides 100% of the recommended intake of omega-3 (ω-3) fatty acids, which help to prevent chronic diseases related to diet. It was widely demonstrated that in *S. hispanica* seeds ω-3 is the most abundant component among fatty acids, in particular, the content of α-linolenic acid (C18:3) is over 50% of all fatty acids. Therefore, chia seed can be considered as a natural source of ω-3 which plays a very important role in human nutrition and in human health due to its antiiflammatory, antiarrhythmic, and antithrombotic activity.

Marineli et al. (2014) characterized chia seed oil from Chile using the positive ion easy ambient sonic-spray ionization mass spectrometry technique and reported ranks of fatty acid abundance in the following order: α-linolenic acid (62.8%), linoleic acid (18.23%), palmitic acid (7.07%), oleic acid (7.04%), and stearic acid (3.36%). Amato et al. (2015) reported the first data on the quality of chia seeds produced in Europe, from an experiment conducted in Basilicata (south Italy), in particular the oil extracted from Italian chia seeds was not significantly different from those grown in the traditional area (Peru) and a new area

(Australia). However, the oil extracted in Italy was richer in chlorophyll, carotenoids, and α-linolenic acid, but showed higher free acidity and peroxides. As mentioned previously, chemical composition and oil yield can be affected by several factors, such as extraction technique and geographical area. For example, Ixtaina et al. (2011) used two extraction techniques to obtain oil from chia seeds purchased from different sources, Argentina and Guatemala. In both seeds, the oil yield was much lower in pressing than in solvent extraction (20.3% and 24.8% compared to 26.7% and 33.6%, respectively). This finding is in agreement with that reported by Dabrowski et al. (2016), who also evaluated the influence of the extraction method on the composition of chia seed oil. In fact, the recovery of oil was reported lower by pressing than by extraction methods.

Total fat content in white lupin seeds is from about 8% (Andrzejewska et al., 2016) to 11.5% (Sujak et al., 2006). In the subcontinental climate white lupin seeds are characterized by an approximately 8% lower fat content than in the Mediterranean climate (Annicchiarico et al., 2014). White lupin seeds are an interesting source of favorable ratios of important fatty acids used in the prophylaxis of circulatory system diseases (Simopoulos, 2003). In their development, people have consumed fats in which the ratio of ω-6 to ω-3 acids was similar, whereas now in the diets of highly developed countries it is up to 15 to 1. Excess ω-6 acids in the diet constitute a risk factor, while the 2:1 ratio, which is observed in white lupin seeds, has an enormous effect and decreases mortality associated with circulatory system diseases.

Meat and meat products are important sources of fat in the human diet; however, the natural concentration of polyunsaturated fatty acids (PUFA) in red meat, especially the ω-3 family, is relatively low. Enser et al. (1996) compared the lipid compositions of beef, mutton, and pork, and found high amounts of saturated fat in all meat samples. In terms of PUFA, although low concentrations were observed for all samples, pork had a higher PUFA content than the other meats. The ω-6:ω-3 ratio in pork is, however, greater than those observed for the other red meats; thus the higher PUFA content in this meat is due to the presence of ω-6 fatty acid. Low ω-3 content in beef, mutton, pork, and even chicken meat has been reported, so it is very important to implement strategies to increase the levels of this fatty acid in these meats, given its importance in the human diet. By means of a metabolic pathway, meat's lipid profile may be modified by increasing the amount of certain fatty acids in the animals' diets, as well as changing the manufacturing processes (Rosenvold and Andersen, 2003). With regard to the lipid profile of meat products, reformulation strategies may be implemented mainly through the direct addition of fats and oils to the product of interest (Jiménez-Colmenero, 2007). In both cases, there are many factors contributing to the success of the operation, as well as factors that may produce a negative impact.

Despite the keen public interest in quinoa in recent years, its effects on human health have not been studied widely. One of the rare studies is that by De Carvalho et al. (2014) where 35 women consumed 25 g/day of either quinoa or corn flakes for 4 weeks. Both quinoa and corn flakes resulted in significant reductions in serum triglyceride and thiobarbituric acid reactive substances (TBARs), an indicator of lipid peroxidation induced oxidative stress. Nonsignificant decreases in total cholesterol and LDL and an increase in glutathione, a marker of antioxidant defense, were observed only in the quinoa flake group.

The main benefits associated with EPA and DHA fatty acids are their antiinflammatory properties, cardiovascular protective effects, reduction of the risk of depression and suicide, delaying of the onset of aging-associated neurological degeneration, and reduction of the risks for certain cancers. They also promote fetal development and improve infant cognitive functions. The ratio of ω-6 to ω-3 fatty acids in dietary intake has become of great significance in human nutrition because of those multiple health benefits, and a ratio equal to or less than 5:1 has been recommended. However, a higher dietary intake of ω-6 has been observed over recent decades, resulting in a ω-6:ω-3 ratio higher than 20:1 due to the low amounts of ω-3 found in food. This has led to a surge of interest in its enrichment in food products. As these compounds are highly susceptible to lipid oxidation, and particularly so in meat products, many parameters must be considered before ω-3 fatty acids may be added to foods in order to produce stable products that are acceptable to consumers.

The modification of dietary patterns over the last 100–150 years has led to a change in fatty acid consumption, with an increase in the consumption of ω-6 and a marked reduction in the consumption of ω-3. This in turn has given rise to an imbalance in the ω-6:ω-3 ratio, which is now very different from the original 1:1 ratio of humans in the past (Simopoulos, 2009). It can be stated that an adequate intake of both, ω-6 and ω-3, is essential for good health and for reducing the percentage of cardiovascular diseases—though it is not clear whether the ratio between them is of any use. During the 1960s–70s, an increase in ω-6 consumption proved central to dietetic counseling, in view of the effects observed in lowering LDL-cholesterol. The consumption of ω-6 doubled, and mortality due to coronary disease decreased 50%. Recently, the American Heart Association (AHA) published a review recommending the amount of ω-6 to represent between 5% and 10% of total energy consumed. The AHA indicates that the consumption of ω-6 from vegetable oils, nuts, and seeds is beneficial when forming part of a healthy diet plan in which saturated and trans-fats are replaced by PUFAs (Harris et al., 2009).

9.3 Polysaccharides

The beneficial aspects of dietary fiber depend on its physicochemical properties, which are grouped into four categories: (1) hydration properties (solubility, swelling, water-holding and absorption capacity, viscosity, and gelling); (2) cationic exchange capacity; (3) particle size, density, and surface characteristics (porosity and oil-holding capacity); and (4) organic molecule adsorption capacity (Lopez et al., 1997). The principal physiological effect of fiber is its ability to swell when absorbing water, which occurs due to the presence of carbohydrates with free polar groups, interaction with hydrophilic links, or retention within the matrix (Lopez et al., 1997). These lead to the formation of gel and a consequent increase in feces volume, which provokes more frequent peristaltic movements of the intestine. This in turn facilitates transit of the fecal bolus and intestinal distention, and thus aids in reducing the probability of intestinal tract disorders and constipation (Oliveira et al., 1991).

Fiber extracted from some grains and seeds exhibits physiological and functional properties that make it promising for use in food industry and health applications. This promise has led researchers to search for novel raw materials that meet needs in these areas, with a particular focus on the coproducts of protein extraction or other components from raw materials such as legumes (Betancur-Ancona et al., 2004; Goff et al., 2001). There is a parallel interest in new sources of dietary fiber that contain concentrations comparable to those in cereal and legume subproducts such as wheat, rice and oat bran, lupine, etc (Villarroel et al., 2003). Fiber source research has focused on tubers, cereals, vegetables, fruit, and algae, all of which are characterized by high dietary fiber content with low digestibility and low caloric content. The fiber fraction from chia seed has similar characteristics. A native of southern Mexico, chia has been under cultivation in the region for thousands of years, and was among the principal crops grown by ancient Mesoamerican cultures. Recent evaluation of chia properties and possible uses has shown that it has a high content of oil (32%) and 60% of this is linolenic acid, a fatty acid associated with various benefits to consumer health (Rosamond, 2002). However, after extracting the oil from the seeds, defatted chia has fiber (22 g/100 g) and protein (17 g/100 g) contents similar to those of other oil seeds currently used in the food industry (Ayerza and Coates, 1999).

Dietary fiber comprised more than 64% of carbohydrates in juçara pulp (*E. edulis*). Its consumption can provides a good intake of dietary fiber, since it has 27% (w/w) on a dry weight basis. A 100 g portion of juçara pulp (*E. edulis*) could supply approximately 20% of the recommended daily intake (Inada et al., 2015). The dietary fiber values are about 15% (w/w) on a dry weight basis for other fruits such as strawberry, blueberry, cherry (De Souza et al., 2014), banana, apple, and orange (Hui, 2006). Fruits of açai show fiber values from 20% to 30% (w/w) on a dry weight basis (Sanabria and Sangronis, 2007; Sangronis and Sanabria, 2011). Guergoletto et al. (2016) evaluated the potential prebiotic effect of juçara pulp and observed that this fruit can modulate the intestinal microbiota in vitro, resulting in a significant increase in numbers of bifidobacteria. The authors suggest that the fibers present in juçara fruit can be one of the compounds related to this beneficial effect.

The outer pericarp of basil (*Ocinum basilicum*) seeds, when soaked in water, swells into a gelatinous mass (Azuma and Sakamoto, 2003). The mucilaginous layer of the swollen seeds is a pectinous matrix, consisting of considerable amounts of unesterified galacturonic acid with a large capacity for hydration (Abraham and Werker, 1972). The optimum extraction condition of basil seed gum is temperature of 69°C, pH 8, and water/seed ratio 65:1. The extraction condition significantly alters the extraction yield, apparent viscosity, and protein content.

Larrauri (1999) described the "perfect fiber" as having the following characteristics:

- It must not contain any components that are nutritionally offensive.
- To maximize its use, it must be of high concentration in a small quantity.
- It should have no taste and no negative odor, color, or texture effects.
- It should contain a balance between soluble and insoluble fiber, with an acceptable presence of bioactive compounds.
- Its addition must not affect the food it is being added to, but it must also have a long shelf-life.
- It should work harmoniously with food processing.
- It should have a positive consumer image.
- It should contain the expected physiological effects.
- It should be of a reasonable price.

Typically, fibers such as wheat, corn, and rice have been used in food production in the past, both for their health attributes and technical functions. However, very recently, novel sources of fiber have been discovered and utilized. One of these sources is the byproduct fraction from different types of food processing. In particular, the byproducts obtained from fruit and vegetable processing (e.g., juices, drinks etc.) are gaining attention as novel and economic sources of a healthy functional ingredient (Ayala-Zavala et al., 2011).

Dietary fiber can also impart some functional properties to foods, e.g., increase water-holding capacity, oil-holding capacity, emulsification, and/or gel formation. Indeed, we shall illustrate that dietary fiber incorporated into food products (bakery products, dairy, jams, meats, soups) can modify textural properties, avoid synaeresis (the separation of liquid from a gel caused by contraction), stabilize high-fat food and emulsions, and improve shelf-life.

9.4 Bioactive Peptides

Proteins are important macronutrients of foods, serving as a source of energy and amino acids, that contribute to growth and maintenance of the body. Besides the nutritional role, proteins are responsible for various physicochemical and sensory properties of foods, and may act as functional and health-promoting ingredients (Shahidi and Zhong, 2008). The interest for health-promoting functional foods, dietary supplements, and pharmaceutical preparations containing bioactive peptides is markedly increasing (Coda et al., 2012). Numerous studies have been performed on bioactive peptides derived from animal proteins, especially from caseins, which appear to be proteins with high functional potential. More recently, the scientific community investigated the possibility of obtaining bioactive peptides from plants. In particular, the identification of bioactive peptides deriving from vegetable food proteins follows the growing interest of the scientific community and public opinion toward vegetable foods, due to their higher sustainability with respect to animal foods and the increased consumer requirements of healthy and balanced diets. Cereals (supplying half the world's protein needs) and legumes are the main target of this research, both being rich sources of proteins with a complementary spectrum of amino acids (García et al., 2013; Malaguti et al., 2014). Nevertheless bioactive peptides were found in many other vegetables (pseudocereals, algae, edible fungi, garlic, ginkgo biloba seeds, chia, amaranth, flaxseed, curcuma, sesame, peanut, alfalfa, spinach, sunflower, hempseeds, tubers, cocoa beans, and others) as the consequence of fermentation, enzymatic hydrolysis, but also not encrypted in any parent molecule (García et al., 2013).

Proteins from animals and plants are rich sources of bioactive peptides with specific physiological and biochemical activities besides providing basic nutritional benefits. Digestion of protein in vivo or in vitro produces free amino acids and peptides which enter in the circulatory system and exert a systemic effect. In living organisms, they function as hormones, neurotransmitters, control gland excretion, adjust blood pressure, impact body growth, affect sleep, learning, pain, sexual behavior, appetite, and stress via the central nervous system (Flores et al., 2011). They also play an essential role as neutraceuticals. Protein hydrolysates also serve as an alternative to intact protein and elemental formulas in the development of special formulations designed to provide nutritional support to patients with different needs. Thus, they forming an easy and cheap functional food suitable for patients with food allergy and chronic liver failure, in diets for the elderly, and in sports nutrition and weight control diets.

Bioactive peptides are small-protein fragments derived from digestive proteolysis during gastrointestinal digestion (in vivo), in vitro chemically by processing of foods through heat, acid or alkali conditions which hydrolyze proteins (Lule et al., 2015; Boschin et al., 2014), enzymatic hydrolysis of food proteins by digestive enzymes, fermentation with proteolytic starter cultures, and proteolysis by enzymes derived from microorganisms or plants (Korhonen and Pihlanto-Leppala, 2003). Such a mixture is composed of "hydrolysates". When the proteolysis of protein is induced by endogenous proteases then they are called as "autolysates." Bioactive peptides are several linked amino acids purified from these protein hydrolysates (Sarmadi and Ismail, 2010). Enzymatic methods are mild in nature and end products are more predictable. Generally endopeptidase enzymes (proteinases) such as pepsin, trypsin, subtilisin, chymotrypsin, papain, and commercial Alcalase and Flavourzyme are used. They are obtained from various sources such as plants, animals, and microbes and each one works in optimal conditions (temperature, pH, time course, enzyme/substrate ratio, etc.) for best performance and results in unique peptide sequences (Shahidi and Zhong, 2008). A study has shown that Flavourzyme hydrolysates have a low degree of hydrolysis, but show higher water absorption capacity and antioxidant activity than Alcalase hydrolysates. This might be due to the smaller-sized peptides produced (Cumby et al., 2008). Peptide concentration in the hydrolysates and its bioactivity are dependent on time, temperature, pH, enzyme concentration, and other factors. Each and every bioactive peptide shows different release kinetics during enzymatic hydrolysis of proteins. Mostly larger peptides appear in early stages of hydrolysis which cleave into smaller peptides with time showing different bioactivities (Zhao et al., 1996; Froidevaux et al., 2000).

Bioactive peptides are the focus of several investigations, mostly related to antioxidant, antihypertensive, and antimicrobial activities. In the last few decades, growing interest has emerged to identify and characterize bioactive peptides from vegetal and animal sources (Sarmadi and Ismail, 2010). Milk proteins contain within their sequences encrypted peptides, known as bioactive peptides, which may have potential to contribute to a wide range of bioactive properties in vitro (Fig. 9.1). The role of milk protein-derived peptides as antidiabetic agents has attracted a lot of interest. In particular, the inhibition of dipeptidyl peptidase IV (DPP-IV), a metabolic enzyme, with various food protein-derived peptides has been extensively studied (Lacroix and Li-Chan, 2016).

However, the enzyme to substrate ratio is an important factor to consider so as to obtain a good degree of hydrolysis. Peptide sequences and their biological activities may differ depending on the type of enzyme used (Mojica and de Mejía, 2016). Low-molecular-weight peptides (<10 kDa) have been found to be more effective antioxidants and antihypertensive peptides (García-Tejedor et al., 2014; Fernández-Musoles et al., 2013; Ruiz-Ruiz et al., 2013; Wattanasiritham et al., 2016) than high-molecular-weight peptides and hence proteases that yield low-molecular-weight peptides would be helpful for commercial production of antioxidant and antihypertensive peptides.

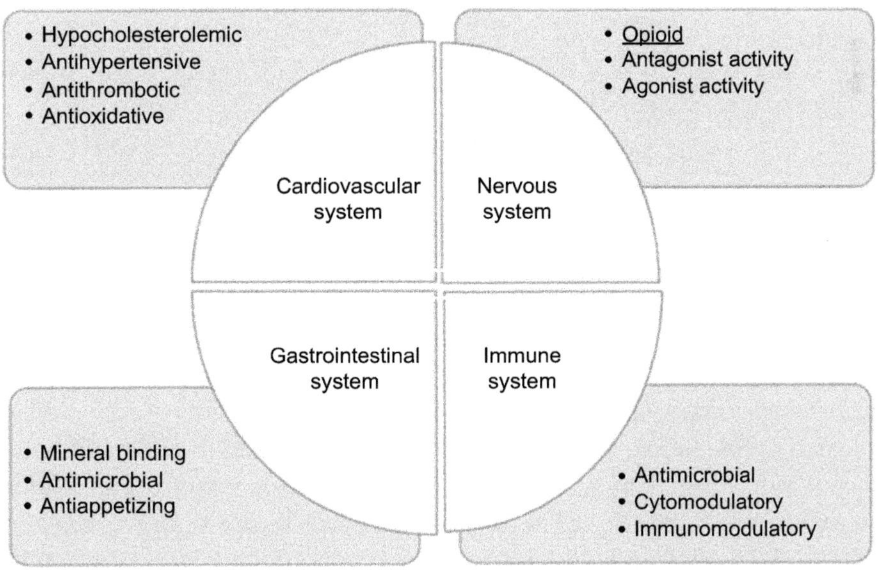

Figure 9.1
Bioactivity of enzymatically hydrolyzed milk peptides (Dhaval et al., 2016).

Fish frame resulting from filleting contains remarkable amounts of muscle protein and, due to the presence of essential amino acids, can be regarded as a complete protein source (Larsen et al., 2000). Protein quality is distinguished by the content of essential amino acids and bioavailability (WHO, 2007a,b). Nevertheless, in addition to utilizing fish protein directly, the use of fish protein hydrolysates has also been popular. One of the important uses of protein hydrolysate is the isolation of bioactive peptides (Kim and Mendis, 2006), and the peptides so procured have numerous bioactivities, such as antihypertensive or angiotensin-converting enzyme (ACE) inhibition, antiproliferative, anticoagulant, immunomodulatory, and chelating effects.

9.4.1 Antioxidant Capacities

Radicals are most commonly quenched by two mechanisms, transfer of either a hydrogen atom or an electron to convert the radical to a stable species:

HAT—hydrogen atom transfer (H atom transferred to target radical, possible secondary quenching by radical recombinations).

$$\mathrm{ROO^{\bullet} + AH/PheOH \rightarrow ROOH + A^{\bullet}/Phe\text{-}O^{\bullet}}$$

$$\mathrm{ROO^{\bullet} + A^{\bullet} \rightarrow ROOA}$$

SET—single electron transfer (one or more electrons transferred to reduce target compounds)

$$\mathrm{ROO^{\bullet} + AH/PheOH \rightarrow ROO^{-} + AH^{+}/PheOH^{+} \xrightarrow{+H_2O} A^{\bullet}/Phe\text{-}O^{\bullet} + H_3O^{+}}$$

$$\mathrm{ROO^{-} + H_3O^{+} \rightarrow ROOH + H_2O}$$

$$\left.\begin{array}{l} \text{Ascorbic acid } (\mathrm{AH_2}) \xrightarrow{O_2} \mathrm{O_2^{\bullet -}} \\ + \mathrm{AH^{\bullet}} \xrightarrow{O_2} \mathrm{A + O_2^{\bullet -}} \\ \mathrm{AH + O_2 \rightarrow O_2^{\bullet -}} \end{array}\right\} \rightarrow \mathrm{H_2O_2 \rightarrow 2\,HO^{\bullet}}$$

where AH = any antioxidant with donatable H, PheOH = phenol or polyphenol, M = redox-active metal.

Generally, antioxidant peptides are associated with low molecular mass (< 6.0 kDa), and can act through various antioxidant mechanisms: by acting as scavengers of free radicals, as inhibitors of lipid peroxidation, and as agents for chelating metal ions. Overall, the antioxidant activity is dependent on the amino acid sequence, conformation, and hydrophobicity. Antioxidant peptides have been produced from various plant and animal protein sources, especially from milk proteins. Nongonierma and Fitzgerald (2015) utilized papain and a microbial papain-like enzyme to hydrolyze quinoa protein isolates and demonstrated the antioxidant capacity of the hydrolysates with oxygen radical absorbance capacity (ORAC). The hydrolysates had antioxidant capacities twice as high as nonhydrolyzed quinoa protein isolates, indicating benefits of using exogenous proteases for releasing antioxidant peptides from quinoa proteins. At the same time, Rizzello et al. (2017) isolated lactic acid bacteria from quinoa flour and spontaneously fermented dough and tested the strains for the production of bioactive quinoa dough. The isolated lactic acid bacteria strains were selected according to their capacity to produce the desired techno-functional properties and the selected strains were studied further for the ability to produce antioxidant peptides from quinoa proteins. Three *L. plantarm* strains, T0A10, T1B6, and T6B4, were shown to produce high in vitro antioxidant activities in comparison with the synthetic antioxidant butylated hydroxytoluene at a concentration of 1 mg/mL. The antioxidant activity of the quinoa peptides was also studied in the human keratinocyte cell line, NCTC 2544. The cultured cells were subjected to oxidative stress and the purified quinoa peptide fractions exhibited a protective effect against intracellular production of reactive oxygen species (ROS) and detoxification effects similar to α-tocopherol.

The contribution of polyphenols to the antioxidant activity was also investigated and the results evidenced that peptides generated by the proteolytic effects of the lactic acid bacteria had an important role in antioxidant activity. Five putative antioxidant peptide sequences were identified from quinoa sourdough inoculated with *L. plantarum* T0A10. The identified peptides had structural characteristics known to correlate with antioxidant capacity: low molecular mass, high proportion of hydrophobic amino acid residues, and presence of cysteine and histidine residues. However, the antioxidative effects of the identified peptides were not assessed separately for each peptide and the total antioxidant capacity was a synergistic effect of the peptides. Also, more research is needed to elucidate the bioavailability of the antioxidant peptides and to investigate their human health potential using in vivo assays.

9.4.2 Antimicrobial Activity

Microbial contamination is one of the main problems that may affect the shelf-life of food and may also cause consumer illness. Therefore, many chemicals are used as preservatives to increase the safety and shelf-life of food products. As a result of the increased awareness of consumers about the deleterious effects of chemical preservatives and the increasing preference for natural components, researchers have focused on the generation of natural additives that demonstrate antimicrobial significance to be used in the food industry (Osman et al., 2013).

Antimicrobial peptides are usually amphipathic in nature, possess activity against Gram-negative and -positive bacteria, viruses, and fungi, and are generally not cytotoxic. Their amphipathicity allows them to interact with and disrupt membrane lipids. Antimicrobial peptides are major components of the innate immune systems of several living organisms, where they act as a first line of defense against invading pathogens. Usually they have less than 50 amino acids, of which about 50% are hydrophobic and have a molecular weight below 10 kDa. They have beneficial effects on native immunity by directly interacting with bacteria and killing them. Their mechanism of action is as follows: the positively charged amino acids of these peptides bind to negatively charged molecules and substances in the membranes of pathogens, forming pores which degrade the membranes of bacteria (Najafian and Babji, 2012; Rajanbabu and Chen, 2011). They interact with specific constituents of the bacterial cell envelope resulting in depolarization, destabilization, and/or disruption of the bacterial plasma membrane, leading to bacterial cell death within minutes (Pasupuleti et al., 2012).

Although hundreds of antimicrobial peptides have now been characterized as having widely diverse sequences, these peptides have been classified into relatively few conformational paradigms. Therefore, it may be argued that a high degree of degeneracy exists within the conformation code governing structure–activity relationships among antimicrobial peptides. Many of these molecules, within and beyond conformational classes, exhibit mechanisms of action that are highly complex and nonidentical. Moreover, new evidence points to targets that lie interior to the cytoplasmic membrane as being important in antimicrobial mechanisms of these peptides.

Recently, there have been more studies on the antimicrobial activity of plant and animal protein peptides. The antibacterial activity is enhanced by enzyme treatment (e.g., papain) and the resulting peptide hydrolysate exhibits significantly higher antibacterial activity than the native camel proteins (Abdel-Hamid et al., 2016). Among a total of 12 identified peptides of RuBisCO, three new pure antibacterial peptides, namely (Met-Asp-Asn), (Glu-Leu-Ala-Ala-Ala-Cys), and (Leu-Arg-Asp-Asp-Phe), were obtained under hydrolytic conditions. The latter peptides were highly active against the tested strains (Gram-negative [*Escherichia coli* and *Salmonella enteric*] and Gram-positive [*Listeria innocua*, *Staphylococcus aureus*, *Enterococcus faecalis*, *Micrococcus luteus*, and *Bacillus subtilis*] bacteria) (Kobbi et al., 2015). The fraction with a low molecular weight (<1 kDa) of peptides from flaxseed protein hydrolysates growth inhibition of *Pseudomonas aeruginosa* and *Escherichia coli* (Hwang et al., 2016).

9.4.3 Immunomodulatory Activity

In addition to direct antimicrobial activity, antimicrobial peptides display immunomodulatory activities. For example, they can prevent excessive activation of proinflammatory responses due to bacterial endotoxins such as lipopolysaccharide of Gram-negative bacteria, and peptidoglycan and lipoteichoic acid of Gram-positive bacteria. Antimicrobial peptides may improve clearance of bacterial biofilms by host defense systems (Mansour et al., 2014, 2015) as they may prevent derangement of immune responses after implantation of foreign bodies (Zaat et al., 2010; Heim et al., 2014, 2015). Immunomodulatory peptides can enhance immune cell functions, such as lymphocyte proliferation, natural killer (NK) cell activity, antibody synthesis, and cytokine regulation.

9.4.4 Antihypertensive Activity

Hypertension is the main cause of several risk factors such as heart failure, stroke, coronary heart disease, and myocardial infarction. Antihypertensive peptides are the most studied bioactive peptides in foods; they exhibit their activity by inhibiting angiotensin-converting enzyme. Angiotensin I-converting enzyme (peptidyl carboxy peptidase, EC 3.4.15.1, ACE) belongs to the class of zinc proteases that require zinc and chloride for activation. ACE plays an important role in blood pressure regulation via the renin–angiotensin system (RAS) and the kallikrein–kinin system (KKS). In the KKS, ACE inactivates the vasodilator bradykinin, while in the RAS, ACE acts as an exo peptidase cleaving His-Leu from the C-terminal of decapeptide angiotensin I and producing the potent vasoconstrictor octapeptide angiotensin II (Ko et al., 2012).

Interestingly, bioactive peptides are able to inhibit ACE, involved in the regulation of human blood pressure and fluid homeostasis via the renin–angiotensin system. In fact, ACE is responsible for converting angiotensin I into angiotensin II, which constricts the arteries and, as a consequence, increases the blood pressure. Further, it is involved in the inactivation of bradykinin, which is a known vasodilator. The identification of ACE-inhibitory bioactive peptide sequences requires sensible fractionation and isolation techniques, by using mass spectrometry analysis coupled to different separation techniques. A combination of chromatography analysis is usually done to simplify the complex mixture of peptides. Furthermore, a structure–activity relationship study was required, molecular docking simulation, to study the interaction between ACE and peptides in order to evaluate the binding interaction between the peptide (N-terminus or C-terminus) and ACE catalytic site. Short sequences (with 2–3 amino acids) are widely prevalent in the literature and the most popular commercial functional foods claiming antihypertensive effects contain the tripeptides VPP and IPP (García-Mora et al., 2017).

Food protein-derived hydrolysates with multibioactivities, such as antihypertensive and antioxidant properties, have recently received special attention since both activities can play significant roles in preventing cardiovascular diseases. The smooth-hound viscera hydrolysates ≤ 1 kDa and 1-3 kDa fractions showed the best ACE-inhibitory activities with IC_{50} values of 53.31 and 75.05 μg/mL, respectively (Abdelhedi et al., 2018). The resulting hydrolysate from Tilapia skin had an ACE-inhibitory activity (IC_{50}) of 1.2 mg/mL, which was slightly reduced by simulated gastrointestinal digestion (Thuanthong et al., 2017).

9.4.5 Antiobesity Activity

Obesity is a major health problem in many developed countries that has been associated with a higher incidence of CVD and related disorders. Hyperinsulinemia, insulin resistance, and abnormalities in lipid metabolism have all been associated with obesity. Obesity may lead to metabolic syndrome and metabolic syndrome increases the risk of cardiovascular disease and type II diabetes because it increases very low-density lipoprotein (VLDL) and low-density lipoprotein (LDL) cholesterol, decreases high-density lipoprotein (HDL) cholesterol, and elevates triglycerides, hypertension, and fatty liver. The lipoprotein profile obtained in obese subjects revealed a pattern of higher levels of triglycerides, elevated LDL cholesterol, and low HDL-cholesterol.

Complex health issues stemming from obesity and overweight pose a high risk to the lives of patients. In 2013, approximately 2.1 billion people worldwide were obese or overweight, with a body mass index of 25 or more. The adult population was about 28% larger and that of adolescents was about 47% larger than the respective populations reported in 1980. Obesity and overweight often complicate conditions such as diabetes, dyslipidemia, osteoarthritis, and menstrual abnormality, and ultimately worsen the prognosis. It has been confirmed that glucose metabolism, blood pressure, and blood lipid levels are improved if a patient with obesity achieves a 3% or more weight loss. For the treatment of obesity, nonmedication therapies such as diet and exercise are used; however, those who do not achieve sufficient body weight reduction receive medical and surgical therapies. Although several anorectic agents have been approved to date, their clinical efficacies are not sufficient. Therefore, there are unmet medical needs for new antiobesity drugs with a potent efficacy and favorable safety profile.

A recent study done by Iemolo et al. (2015) has shown a new dimension of peptide activity which reduces the desire to eat. This aspect of peptide activity may reduce binge eating disorder, an illness characterized by excessive uncontrolled consumption of food resulting in uncomfortable fullness and feelings of self-disgust. This fact has been very well proved in the experimental model of researchers using pituitary adenylate cyclase-activating-peptide (PACAP), which is a peptide and hormone produced in the "central amygdala" of the brain. The amygdala is outside the hypothalamus in the brain, and is involved in fear and the emotional component of eating. PACAP affects food intake and body weight effects in the hypothalamus. Researchers have found that upon injection of PACAP in the amygdala, food intake was reduced in two ways: consumption of smaller meals or fewer meals of normal size during the day. In addition, PACAP reduced the amount of food eaten within meals. This activity was also found to be dependent on brain-derived neurotrophic factor (BDNF), the growth hormone. Moreover, continuous administration of a 14-amino-acid peptide has previously shown an antiobesity effect in a 2-week diet-induced obesity (DIO) study in mice (Nishizawa et al., 2017).

9.4.6 Hypocholesterolemic Activity

Hypercholesterolemia is a significant risk factor in the development of heart disease, which is one of the major causes of death in Western countries. Dietary proteins can affect serum cholesterol concentrations.Maier et al. (2000) verified that fasting serum total cholesterol decreased 4.5%

after a 28-day dietary intervention in humans that consumed 50 g/day of amaranth. Mendonça et al. (2009) tested the effect of the whole seeds or the protein isolated from amaranth in hamsters. The study indicated that amaranth protein reduced the cholesterol concentration in plasma, being the main component responsible for the hypocholesterolemic effect.

The enzyme 3-hydroxy-3-methylglutaryl coenzyme A reductase (HMGCR) is a rate-determining step in hepatic cholesterol production. For this reason, this enzyme is the target of several drugs aimed at the control of high-cholesterol levels. As a subsidiary mechanism, the absorption of dietary cholesterol is an alternative for the body to maintain proper cholesterol levels. The nonpolar character of the cholesterol molecule affects the kinetics and the dynamics prior to micelle formation of bile salts for its absorption and internalization (Cuccioloni et al., 2011; Nes, 2011).

Peptide fragments from some vegetable proteins exhibit cholesterol-lowering ability by reducing the HMGCR activity, inhibiting intestinal absorption of dietary cholesterol, disrupting cholesterol micelle formation, and/or interfering with the cellular cholesterol carrier (Zhang et al., 2012). Recently, it has been observed that the in vitro hydrolysis of amaranth (*Amaranthus cruentus*) proteins with pepsin and trypsin produced the peptides GGV, IVG, and VGVL, that strongly inhibited the activity of 3-hydroxy-3-methyl-glutaryl-CoA reductase (HMG-CoA reductase), a key enzyme in cholesterol biosynthesis, suggesting a potential hypocholesterolemic effect (Soares et al., 2015).

References

Abdel-Hamid, M., Goda, H.A., De Gobba, C., Jenssen, H., Osman, A., 2016. Antibacterial activity of papain hydrolysed camel whey and its fractions. International Dairy Journal 61, 91–98.

Abdelhedi, O., Nasri, R., Mora, L., Jridi, M., Toldrá, F., Nasri, M., 2018. In silico analysis and molecular docking study of angiotensin I-converting enzyme inhibitory peptides from smooth-hound viscera protein hydrolysates fractionated by ultrafiltration. Food Chemistry 239, 453–463.

Abraham, F., Werker, E., 1972. Anatomical mechanisms of seed dispersal. In: Koylowski, T.T. (Ed.), Seed Biology. Academic Press, New York, p. 152e217.

Amato, M., Caruso, M.C., Guzzo, F., Commisso, M., Bochicchio, R., Galgano, F., Labella, R., Favati, F., 2015. Nutritional quality of seeds and leaf metabolites of Chia (*Salvia hispanica* L.) from Southern Italy. European Food Research and Technology 241, 615–625.

Andersen, Ø.M., Markham, K.R., 2006. Flavonoids chemistry, biochemistry and applications. In: Separation and Quantification of Flavonoids. Taylor & Francis, New York.

Andrzejewska, J., Ignaczak, S., Barzyk, P., 2016. Oil content and fatty acid profile in seeds of Polish breeding lines and cultivars of legumes. *Acta Scientiarum Polonorum*. Agricultura 15, 3–13.

Annicchiarico, P., Manunza, P., Arnoldi, A., Boschin, G., 2014. Quality of *Lupinus albus* L. (*White lupin*) seed: extent of genotypic and environmental effects. Journal of Agriculture and Food Chemistry 62, 6539–6545.

Ayala-Zavala, J.F., Vega-Vega, V., Rosas-Domínguez, C., Palafox-Carlos, H., Villa-Rodriguez, J.A., Siddiqui, M.W., 2011. Agro-industrial potential of exotic fruit byproducts as a source of food additives. Food Research International 44 (7), 1866–1874.

Ayerza, R., Coates, W., 1999. An omega-3 fatty acid enriched chia diet: its influence on egg fatty acid composition, cholesterol and oil content. Canadian Journal of Animal Science 79, 53–58.

Azuma, J., Sakamoto, M., 2003. Cellulosic hydrocolloid system present in seed of plants. Trends in Glycoscience and Glycotechnology 15, 1–14.

Baliga, M.S., Shivashankara, A.R., Haniadka, R., Dsouza, J., Bhat, H.P., 2011. Phytochemistry, nutritional and pharmacological properties of *Artocarpus heterophyllus Lam* (jackfruit): A review. Food Research International 44, 1800–1811.

Benherlal, P.S., Arumughan, C., 2007. Chemical composition and in vitro antioxidant studies on Syzygium cumini fruit. Journal of the Science of Food and Agriculture 87, 2560–2569.

Betancur-Ancona, D., Peraza-Mercado, G., Moguel-Ordoñez, Y., Fuertes-Blanco, S., 2004. Physicochemical characterization of Lima bean (*Phaseolus lunatus*) and Jack bean (*Canavalia ensiformis*) fibrous residues. Food Chemistry 84, 287–295.

Bicudo, M.O.P., Ribani, R.H., Beta, T., 2014. Anthocyanins, phenolic acids and antioxidant properties of juçara fruits (*Euterpe edulis* M.) along the on-tree ripening process. Plant Foods for Human Nutrition 69, 142–147.

Borges, G.S.C., Vieira, F.G.K., Copetti, C., Gonzaga, L.V., Zambiazi, R.C., Mancini Filho, J., Fett, R., 2011. Chemical characterization, bioactive compounds, and antioxidant capacity of Jussara (*Euterpe edulis*) fruit from the Atlantic Forest in southern Brazil. Food Research International 44 (7), 2128–2133.

Borges, G.D.S.C., Gonzaga, L.V., Jardini, F.A., Mancini Filho, J., Heller, M., Micke, G., Fett, R., 2013. Protective effect of *Euterpe edulis* M. on Vero cell culture and antioxidant evaluation based on phenolic composition using HPLC–ESI–MS/MS. Food Research International 51, 363–369.

Boschin, G., Scigliuolo, G.M., Resta, D., Arnoldi, A., 2014. ACE inhibitory activity of enzymatic protein hydrolysates from lupin and other legumes. Food Chemistry 145, 34–40.

Bunea, A., Rugina, D., Pintea, A., Andrei, S., Bunea, C., Pop, R., Bele, C., 2012. Carotenoid and fatty acid profiles of bilberries and cultivated blueberries from Romania. Chemical Papers 66 (10), 935–939.

Cardoso, A.L., Di Pietro, P.F., Vieira, F.G.K., Boaventura, B.C.B., De Liz, S., Borges, G.S.C., Da Silva, E.L., 2015a. Acute consumption of juçara juice (*Euterpe edulis*) and antioxidant activity in healthy individuals. Journal of Functional Foods 17, 152–162.

Cardoso, L.M., Novaes, R.D., De Castro, C.A., Azevedo, A., Gonçalves, R.V., Ricci-Silva, M.E., Leite, V., 2015b. Chemical composition, characterization of anthocyanins and antioxidant potential of *Euterpe edulis* fruits: applicability on genetic dyslipidemia and hepatic steatosis in mice. Nutricion Hospitalaria 32 (2), 702–709.

Carvalho, C.A., Cristina, P., Fonsêca, D.A., Priore, S.E., Castro, C., De Novaes, J.F., 2015. Food consumption and nutritional adequacy in Brazilian children: a systematic review. Revista Paulista de Pediatria 33, 211–221.

Carvalho, A.V., da Silveira, T.F.F., Mattietto, R.A., de Oliveira, M.S.P., Godoy, H.T., 2017. Chemical composition and antioxidant capacity of açaí (*Euterpe oleracea*) genotypes and commercial pulps. Journal of the Science of Food and Agriculture 97, 1467–1474.

Chakraborty, M., Mitra, A., 2008. The antioxidant and antimicrobial properties of the methanolic extract from *Cocos nucifera mesocarp*. Food Chemistry 107, 994–999.

Cheng, P., Wang, J., Shao, W., 2016. Monounsaturated fatty acid intake and stroke risk: a meta-analysis of prospective cohort studies. Journal of Stroke and Cerebrovascular Diseases 25 (6), 1326–1334.

Coda, R., Rizzello, C.G., Pinto, D., Gobbetti, M., 2012. Selected lactic acid bacteria synthesize antioxidant peptides during sourdough fermentation of cereal flours. Applied and Environmental Microbiology 78, 1087–1096.

Consolini, A.E., Sarubbio, M.G., 2002. Pharmacological effects of Eugenia uniflora (*Myrtaceae*) aqueous crude extract on rat's heart. Journal of Ethnopharmacology 81, 57–63.

Costa, A., Garcia-Diaz, D.F., Jimenez, P., Silva, P.I., 2013. Bioactive compounds and health benefits of exotic tropical red-black berries. Journal of Functional Food 5, 539–549.

Cuccioloni, M., Mozzicafreddo, M., Spina, M., Tran, C.N., Falconi, M., Eleuteri, A.M., Angeletti, M., 2011. Epigallocatechin-3-gallate potently inhibits the in vitro activity of hydroxy-3-methyl-glutaryl-CoA reductase. Journal of Lipid Research 52 (5), 897–907.

Cumby, N., Zhong, Y., Naczk, M., Shahidi, F., 2008. Antioxidant activity and water-holding capacity of canola protein hydrolysates. Food Chemistry 109, 144–148.

Cunha Júnior, L.C., Teixeira, G.H.A., Nardini, V., Walsh, K.B., 2016. Quality evaluation of intact açaí and juçara fruit by means of near infrared spectroscopy. Postharvest Biology and Technology 112, 64–74.

Cunha Junior, L.C., Nardini, V., Khatiwada, B.P., Teixeira, G.H.D.A., Walsh, K.B., 2015. Classification of intact açaí (*Euterpe oleracea* Mart.) and juçara (*Euterpe edulis* Mart.) fruits based on dry matter content by means of near infrared spectroscopy. Food Control 50, 630–636.

Dabrowski, G., Konopka, I., Czaplicki, S., Tanska, M., 2016. Composition and oxidative stability of oil from *Salvia hispanica* L. seeds in relation to extraction method. European Journal of Lipid Science and Technology 119, 1–9.

Da Silva, P.P.M., do Carmo, L.F., Silva, G.M., Silveira-Diniz, M.F., Casemiro, R.C., Spoto, M.H.F., 2013. Physical, chemical, and lipid composition of juçara (*Euterpe edulis* Mart.) pulp. Brazilian Journal of Food and Nutrition 24, 7–13.

Da Silva, N.A., Rodrigues, E., Mercadante, A.Z., de Rosso, V.V., 2014. Phenolic compounds and carotenoids from four fruits native from the Brazilian Atlantic forest. Journal of Agricultural and Food Chemistry 62, 5072–5084.

Dhaval, A., Yadav, N., Purwar, S., 2016. Potential applications of food derived bioactive peptides in management of health. International Journal of Peptide Research and Therapeutics 22, 377–398.

De Benoist, B., McLean, E., Egli, I., Cogswell, M., 2008. Worldwide Prevalence of Anaemia 1993–2005: WHO Global Database on Anaemia. World Health Organization, Geneva, Switzerland.

De Bona, K.S., Bellé, L.P., Bittencourt, P.E.R., Bonfanti, G., Car gnelluti, L.O., Pimentel, V.C., Ruviaro, A.R., Schetinger, M.R.C., Emanuelli, T., Moretto, M.B., 2011. Erythrocytic enzymes and antioxida nt stat us in people with type 2 diabetes: Beneficial effect of *Syzygium cumini* leaf extract in vitro. Diabetes Research and Clinical Practice 94, 84–90.

De Carvalho, F.G., Ovídio, P.P., Padovan, G.J., Jordao Junior, A.A., Marchini, J.S., Navarro, A.M., 2014. Metabolic parameters of postmenopausal women after quinoa or corn flakes intake–a prospective and double-blind study. International Journal of Food Science and Nutrition 65, 380–385.

De Falco, B., Amato, M., Lanzotti, V., 2017. Chia seeds products: an overview. Phytochemistry Reviews 16 (4), 745–760.

De Souza, M.O., Silva, M., Silva, M.E., de Oliveira, R.P., Pedrosa, M.L., 2010. Diet supplementation with acai (*Euterpe oleracea* Mart.) pulp improves biomarkers of oxidative stress and the serum lipid profile in rats. Nutrition 26, 804–810.

De Souza, V.R., Pereira, P.A.P., da Silva, T.L.t., de Lima, L.C.O., Pio, R., Queiroz, F., 2014. Determination of the bioactive compounds, antioxidant activity and chemical composition of Brazilian blackberry, red raspberry, strawberry, blueberry and sweet cherry fruits. Food Chemistry 156, 362–368.

Del Pozo-Insfran, D., Percival, S.S., Talcott, S.T., 2006. Acai (*Euterpe oleracea* Mart.) polyphenolics in their glycoside and aglycone forms induce apoptosis of HL-60 leukemia cells. Journal of Agricultural and Food Chemistry 54, 1222–1229.

Denardin, C.C., Hirsch, G.E., da Rocha, R.F., Vizzotto, M., Henriques, A.T., Moreira, J.C.F., Guma, F.T.C.R., Emanuelli, T., 2015. Antioxidant capacity and bioactive compounds of four Brazilian native fruits. Journal of Food and Drug Analysis 23 (3), 387–398.

Enser, M., Hallett, K., Hewett, B., Fursey, G.A.J., Wood, J.D., 1996. Fatty acid content and composition of English beef, lamb and pork at retail. Meat Science 44, 443–458.

FAO, 2010. Fats and Fatty Acids in Human Nutrition — Report of an Expert Consultation. FAO, Rome.

FAO, 2011. Trade and Markets – Current Situation and Medium-term Outlook for Tropical Fruits.

Feio, C.A., Izar, M.C., Ihara, S.S., Kasmas, S.H., Martins, C.M., Feio, M.N., Maués, L.A., Borges, N.C., Moreno, R.A., Póvoa, R.M., Fonseca, F.A., 2012. Euterpe oleracea (açaí) modifies sterol metabolism and attenuates experimentally-induced atherosclerosis. Journal of Atherosclerosis and Thrombosis 19 (3), 237–245.

Felzenszwalb, I., da Marques, M.R.C., Mazzei, J.L., Aiub, C.A.F., 2013. Toxicological evaluation of *Euterpe edulis*: a potential superfruit to be considered. Food and Chemical Toxicology 58, 536–544.

Fernández-Musoles, R., Manzanares, P., Burguete, M.C., Alborch, E., Salom, J.B., 2013. In vivo angiotensin I-converting enzyme inhibition by long-term intake of antihypertensive lactoferrin hydrolysate in spontaneously hypertensive rats. Food Research International 54, 627–632.

Flores, M.E.J., Lugo-Cervantes, E.C., Chel-Guerrero, L., 2011. Nutraceuticals and Functional Foods. Conventional and Non-conventional Sources. Studium Press, New Delhi.

Froidevaux, R., Lignot, B., Nedjar-Arroume, N., Guillochon, D., Coddeville, B., Ricart, G., 2000. Kinetics of appearance of hemorphins from bovine hemoglobin peptic hydrolysates by a direct coupling of reversed-phase high-performance liquid chromatography and electrospray ionization mass spectrometry. Journal of Chromatografy A 873, 185–194.

García, M.C., Puchalska, P., Esteve, C., Marina, M.L., 2013. Vegetable foods: a cheap source of proteins and peptides with antihypertensive, antioxidant, and other less occurrence bioactivities. Talanta 106, 328–349.

García-Mora, P., Martín-Martínez, M., Bonache, M.A., González-Múniz, R., Peñas, E., Frias, J., Martinez-Villaluenga, C., 2017. Identification, functional gastrointestinal stability and molecular docking studies of lentil peptides with dual antioxidant and angiotensin I converting enzyme inhibitory activities. Food Chemistry 221, 464–472.

García-Tejedor, A., Sánchez-Rivera, L., Castelló-Ruiz, M., Recio, I., Salom, J.B., Manzanares, P., 2014. Novel antihypertensive lactoferrin-derived peptides produced by *Kluyveromyces marxianus*: gastrointestinal stability profile and in vivo angiotensin I-converting enzyme (ACE) inhibition. Journal of Agricultural and Food Chemistry 62, 1609–1616.

Goff, A., Renard, C., Bonnin, E., Thibault, J., 2001. Extraction, purification and chemical characterization of xylogalacturonans from pea hulls. Carbohydrate Polymers 45, 325–334.

Gordon, A., Jungfer, E., da Silva, B.A., Maia, J.G.S., Marx, F., 2011. Phenolic constituent s and antioxidant capacity of four underutilized fruits from the Amazon region. Journal of Agricultural and Food Chemistry 59, 7688–7699.

Guergoletto, K.B., Costabile, A., Flores, G., Garcia, S., Gibson, G.R., 2016. In vitro fermentation of juçara pulp (*Euterpe edulis*) by human colonic microbiota. Food Chemistry 196, 251–258.

Harris, W.S., Mozaffarian, D., Rimm, E., Kris-Etherton, P., Rudel, L.L., Appel, L.J., Engler, M.M., Engler, M.B., Sacks, F., 2009. Omega-6 fatty acids and risk for cardiovascular disease: a science advisory from the American heart association nutrition subcommittee of the council on nutrition, physical activity, and metabolism; council on cardiovascular nursing; and council on epidemiology and prevention. Circulation 119 (6), 902–907.

Heidenreich, P.A., Trogdon, J.G., Khavjou, O.A., Butler, J., Dracup, K., Ezekowitz, M., Finkelstein, E., Hong, Y., Johnston, C., Khera, A., Lloyd-Jones, D., Nelson, S., Nichol, G., Orenstein, D., Wilson, P., Woo, J., 2011. Forecasting the future of cardiovascular disease in the United States a policy statement from the American Heart Association. Circulation 123, 933–944.

Heim, C.E., Vidlak, D., Scherr, T.D., Kozel, J.A., Holzapfel, M., Muirhead, D.E., Kielian, T., 2014. Myeloid-derived suppressor cells contribute to *Staphylococcus aureus* orthopedic biofilm infection. The Journal of Immunology 192, 3778–3792.

Heim, C.E., Vidlak, D., Scherr, T.D., Hartman, C.W., Garvin, K.L., Kielian, T., 2015. IL-12 Promotes myeloid-derived suppressor cell recruitment and bacterial persistence during *Staphylococcus aureus* orthopedic implant infection. The Journal of Immunology 194, 3861–3872.

Horiguchi, T., Ishiguro, N., Chihara, K., Ogi, K., Nakashima, K., Sada, K., Hori-Tamura, N., 2011. Inhibitory effect of acai (*Euterpe oleracea* Mart.) pulp on IgE-mediated mast cell activation. Journal of Agricultural and Food Chemistry 59, 5595–5601.

Hui, Y.H., 2006. Handbook of Fruit and Fruit Processing. Blackwell Publishing, Iowa.

Hwang, S., Shih, P.H., Yen, G., 2012. Neuroprotective effects of citrus flavonoids. Journal of Agricultural and Food Chemistry 60, 877–885.

Hwang, C.F., Chen, Y.A., Luo, C., Chiang, W.D., 2016. Antioxidant and antibacterial activities of peptide fractions from flaxseed protein hydrolysed by protease from *Bacillus altitudinis* HK02. International Journal of Food Science and Technology 51, 681–689.

Iemolo, A., Ferragud, A., Cottone, P., Sabino, V., 2015. Pituitary adenylate cyclase-activating peptide in the central amygdala causes anorexia and body weight loss via the melanocortin and the TrkB systems. International Journal of Neuropsychopahrmacology 40 (8), 1846–1855.

Inada, K.O.P., Oliveira, A.A., Revorêdo, T.B., Martins, A.B.N., Lacerda, E.C.Q., Freire, A.S., Braz, B.F., Santelli, R.E., Torres, A.G., Perrone, D., Monteiro, M.C., 2015. Screening of the chemical composition and occurring antioxidants in jabuticaba (*Myrciaria jaboticaba*) and jussara (*Euterpe edulis*) fruits and their fractions. Journal of Functional Foods 17, 422–433.

Inoue, T., Komoda, H., Uchida, T., Node, K., 2008. Tropical fruit camu-camu (*Myrciaria dubia*) has anti-oxidative and anti-inflammatory properties. Journal of Cardiology 52, 127–132.

Ixtaina, V.Y., Martinez, M.L., Spotorno, V., Mateo, C.M., Maestri, D.M., Diehl, B.W.K., Nolasco, S.M., Tomas, M.C., 2011. Characterization of chia seed oils obtained by pressing and solvent extraction. Journal of Food Composition and Analysis 24, 166–174.

Jiménez-Colmenero, F., 2007. Healthier lipid formulation approaches in meat-based functional foods: technological options for replacement of meat fats by non-meat fats. Trends in Food Science and Technology 18, 567–578.

Kang, J., Xie, C.H., Li, Z.M., Nagarajan, S., Schauss, A.G., Wu, X.L., Wu, T., 2011. Flavonoids from acai (*Euterpe oler acea* Mart.) pulp and their antioxidant and anti-inflammatory activities. Food Chemistry 128, 152–157.

Kim, S., Mendis, E., 2006. Bioactive compounds from marine processing byproducts – a review. Food Research International 39, 383–393.

Ko, S.C., Kang, N., Kim, E.A., Kanga, M.C., Leea, S.H., Kanga, S.M., Lee, J.B., Jeonc, B.T., Kimd, S.K., Park, S.J., Parke, P.J., Jungf, W.K., Kimg, D., Jeon, Y.J., 2012. A novel angiotensin Iconverting enzyme (ACE) inhibitory peptide from a marine *Chlorella ellipsoidea* and its antihypertensive effect in spontaneously hypertensive rats. Process Biochemistry 47, 2005–2011.

Kobbi, S., Balti, R., Bougatef, A., Flem, G.L., Firdaous, L., Bigan, M., Chataigné, G., Chaabouni, S., Dhulster, P., Nedjar, N., 2015. Antibacterial activity of novel peptides isolated from protein hydrolysates of RuBisCO purified from green juice alfalfa. Journal of Functional Foods 18, 703–713.

Korhonen, H., Pihlanto-Leppala, A., 2003. Food-derived bioactive peptides: opportunities for designing future foods. Current Pharmaceutical Design 9, 1297–1308.

Lacroix, I.M.E., Li-Chan, E.C.Y., 2016. Food-derived dipeptidyl-peptidase IV inhibitors as a potential approach for glycemic regulation – current knowledge and future research considerations. Trends in Food Science and Technology 54, 1–16.

Larrauri, J.A., 1999. New approaches in the preparation of high dietary fibre powders from fruit by-products. Trends in Food Science and Technology 10 (1), 38.

Larsen, T., Thilsted, S.H., Kongsbak, K., Hansen, M., 2000. Whole small fish as a rich calcium source. British Journal of Nutrition 83, 191–196.

Leite, A.V., Malta, L.G., Riccio, M.F., Eberlin, M.N., Pastore, G.M., Marostica, M.R., 2011. Antioxidant potential of rat plasma by administration of freeze-dried jaboticaba peel (*Myrciaria jaboticaba* Vell Ber g). Journal of Agricultural and Food Chemistry 59, 2277–2283.

Leite-Legatti, A.V., Batista Â.G., Dragano N.R.V., Marques, A.C., Malta, L.G., Riccio, M.F., Eberlin, M.N., Machado, A.R.T., de Carvalho-Silva, L.B., Ruiz, A.L.T.G., de Carvalho, J.E., Pastore, G.M., Maróstica Júnior, M.R.M., 2012. Jaboticaba peel: Antioxidant compounds, antiproliferative and antimutagenic activities. Food Research International 49, 596–603.

Lenquiste, S.A., Batista, A.G., Marineli, R.D.S., Dra gano, N.R.V., Maróstica, M.R.J., 2012. Freeze-dried jaboticaba peel added to high-fat diet increases HDL-cholesterol and improves insulin resistance in obese rats. Food Research International 49, 153–160.

Lima, Z.P., dos Santos, R.C., Torres, T.U., Sannomiya, M., Rodrigues, C.M., dos Santos, L.C., Pellizzon, C.H., Rocha, L.R., Vilegas, W., Souza Brito, A.R., Cardoso, C.R., Varanda, E.A., de Moraes, H.P., Bauab, T.M., Carli, C., Carlos, I.Z., Hiruma-Lima, C.A., 2008. *Byrsonima fagifolia*: an integrative study to validate the gastroprotective, healing, antidiarrheal, antimicrobial and mutagenic action. Journal of Ethnopharmacology 120, 149–160.

Lopez, G., Ros, G., RincÓn, F., Periago, M., Martínez, M., Ortuño, J., 1997. Propiedades funcionales de la fibra dietética. Mecanismo de acción en el tracto gastrointestinal. Archivos Latinoamericanos de Nutricion 47 (3), 203–207.

Lule, V.K., Garg, S., Pophaly, S.D., Tomar, S.D., 2015. Potential health benefits of Lunasin: a multifaceted Soy—derived bioactive peptide. Journal of Food Science 80, R485–R494.

Maier, S.M., Turner, N.D., Lupton, J.R., 2000. Serum lipids in hypercholesterolemic men and women consuming oat bran and amaranth products. Cereal Chemistry 77, 297–302.

Malaguti, M., Dinelli, G., Leoncini, E., Bregola, V., Bosi, S., Cicero, A.F., Hrelia, S., 2014. Bioactive peptides in cereals and legumes: agronomical, biochemical and clinical aspects. International Journal of Molecular Sciences 15, 21120–21135.

Mansour, S.C., Pena, O.M., Hancock, R.E.W., 2014. Host defense peptides: front-line immunomodulators. Trends in Immunology 35, 443–450.

Mansour, S.C., de la Fuente-Núñez, C., Hancock, R.E.W., 2015. Peptide IDR-1018: modulating the immune system and targeting bacterial biofilms to treat antibiotic-resistant bacterial infections. Journal of Peptide Science 21, 323–329.

Marineli, R.S., Moraes, E.A., Lenquiste, S.A., Godoy, A.T., Eberlin, M.N., Maróstica Jr., M.R., 2014. Chemical characterization and antioxidant potential of Chilean chia seeds and oil (*Salvia hispanica* L.). Lebensmittel-Wissenschaft und -Technologie- Food Science and Technology 59, 1304–1310.

Mendonça, S., Saldiva, P.H., Cruz, R.J., Arêas, J.A.G., 2009. Amaranth protein presents cholesterol lowering effect. Food Chemistry 116, 738–742.

Mertens-Talcott, S.U., Rios, J., Jilma-Stohlawetz, P., Pacheco-Palencia, L.A., Meibohm, B., Talcott, S.T., Derendorf, H., 2008. Pharmacokinetics of anthocyanins and antioxidant effects after the consumption of anthocyanin-rich acai juice and pulp (*Euterpe oleracea* Mart.) in human healthy volunteers. Journal of Agricultural and Food Chemistry 56, 7796–7802.

Michalska, M., Gluba, A., Mikhailidis, D.P., No wak, P., Bielecka-Dabrowa, A., Rysz, J., Banach, M., 2010. The role of polyphenols in cardiovascular disease. Medical Science Monitor 16, RA110–RA119.

Mojica, L., de Mejía, E.G., 2016. Optimization of enzymatic production of anti-diabetic peptides from black bean (*Phaseolus vulgaris* L.) proteins, their characterization and biological potential. Food and Function 7, 713–727.

Morais, D.R., Rotta, E.M., Sargi, S.C., Bonafe, E.G., Suzuki, R.M., Souza, N.E., Matsushita, M., Visentainer, J.V., 2016. Proximate composition, mineral contents and fatty acid composition of the different parts and dried peels of tropical fruits cultivated in Brazil. Journal of the Brazilian Chemical Society 1–11.

Najafian, L., Babji, A.S., 2012. A review of fish-derived antioxidant and antimicrobial peptides: Their production, assessment, and applications. Peptides 33, 178–185.

Nes, W.D., 2011. Biosynthesis of cholesterol and other sterols. Chemical Reviews 111 (10), 6423–6451.

Nishizawa, N., Niida, A., Adachi, Y., Kanematsu-Yamaki, Y., Masuda, Y., Kumano, S., Yokoyama, K., Noguchi, Y., Asakawa, T., Hirabayashi, H., Amano, N., Takekawa, S., Ohtaki, T., Asami, T., 2017. Highly potent antiobesity effect of a short-length peptide YY analog in mice. Bioorganic and Medicinal Chemistry 25, 5718–5725.

Nongonierma, A.B., FitzGerald, R.J., 2015. The scientific evidence for the role of milk protein-derived bioactive peptides in humans. A review. Journal of Functional Foods 17, 640–656.

Noratto, G.D., Angel-Morales, G., Talcott, S.T., Mertens, T.S.U., 2011. Polyphenolics from acai (*Euterpe oleracea* Mart.) and red muscadine grape (*Vitis rotundifolia*) protect human umbilical vascular Endothelial cells (HUVEC) from glucose- and lipo polysacc haride (LPS)-induced inflammation and target microRN A-126. Journal of Agricultural and Food Chemistry 59, 7999–8012.

Novello, A.A., da Conceição, L.L., dos Dias, M.M.S., Cardoso, L.M., de Castro, C.A., Ricci, M.E., Leite, P.V., do Peluzio, M.C.G., 2015. Chemical characterization, antioxidant and antiatherogenic activity of anthocyanin-rich extract from *Euterpe edulis* Mart. in mice. Journal of Food and Nutrition Research 54 (2), 101–112.

Oliveira, A.L., Lopes, R.B., Cabral, F.A., Eberlin, M.N., 2006. Volatile compounds from pitanga fruit (*Eugenia uniflora* L.). Food Chemistry 99, 1–5.

Oliveira, V.B., Yamada, L.T., Fagg, C.W., Brandão, M.G.L., 2012. Native foods from Brazilian biodiversity as a source of bioactive compounds. Food Research International 48, 170–179.

Oliveira, S., Reyes, F., Sgarbieri, V., Areas, M., Ramalho, A., 1991. Nutritional attributes of a sweet corn fibrous residue. Journal of Agricultural and Food Chemistry 39 (4), 740–743.

Ooi, E.M.M., Watts, G.F., Ng, T.W.K., Barrett, P.H.R., 2015. Effect of dietary fatty acids on human lipoprotein metabolism: a comprehensive update. Nutrients 7, 4416–4425.

Osman, A., Mahgoub, S., Sitohy, M., 2013. Preservative action of 11S (glycinin) and 7S (b-conglycinin) soy globulin on bovine raw milk stored either at 4 or 25 C. Journal of Dairy Research 80, 174–183.

Oyama, L.M., Silva, F.P., Carnier, J., De Miranda, D.A., Santamarina, A.B., Ribeiro, E.B., 2016. Juçara pulp supplementation improves glucose tolerance in mice. Diabetology and Metabolic Syndrome 1–8.

Pasupuleti, M., Schmidtchen, A., Malmsten, M., 2012. Antimicrobial peptides: key components of the innate immune system. Critical Reviews in Biotechnology 32, 143–171.

Patil, B.S., Jayaprakasha, G.K., Chidambara Murthy, K.N., Vikram, A., 2009. Bioactive compounds: historical perspectives, opportunities, and challenges. Journal of Agricultural and Food Chemistry 57, 8142–8160.

Paz, M., Gullon, P., Barroso, M.F., Carvalho, A., Domingues, V.F., Gomes, A.M., Becker, H., Longhinotti, E., Delerue-Matos, C., 2015. Brazilian fruit pulps as functional foods and additives: evaluation of bioactive compounds. Food Chemistry 172, 462–468.

Petruk, G., Illiano, A., Giudice, R.D., Raiola, A., Amoresano, A., Rigano, M.M., Piccoli, R., Monti, D.M., 2017. Malvidin and cyanidin derivatives from açai fruit (Euterpe oleracea Mart.) counteract UV-A-induced oxidative stress in immortalized fibroblasts. Journal of Photochemistry and Photobiology B: Biology 172, 42–51.

Pietta, P.G., 2000. Flavonoids as antioxidants. Journal of Natural Products 63, 1035–1042.

Proestos, C., Boziaris, I.S., Nychas, G.J.E., Komaitis, M., 2006. Analysis of flavonoids and phenolic acids in Greek aromatic plants: investigation of their antioxidant capacity and antimicrobial activity. Food Chemistry 95, 664–671.

Rajanbabu, V., Chen, J.C., 2011. Applications of antimicrobial peptides from fish and perspectives for the future, review. Peptides 32, 415–420.

Rauha, J.P., Remes, S., Heinonen, M., Hopia, A., Kahkonen, M., Kujala, T., Pihlaja, K., Vuorela, H., Vuorela, P., 2000. Antimicrobial effects of Finnish plant extracts containing flavonoids and other phenolic compounds. International Journal of Food Microbiology 56, 3–12.

Rendeiro, C., Rhodes, J.S., Spencer, J.P.E., 2015. The mechanisms of action of flavonoids in the brain: direct versus indirect effects. Neurochemistry International 89, 126–139.

Reynertson, K.A., Wallace, A.M., Adachi, S., Gil, R.R., Yang, H., Basile, M.J., D'Armiento, J., Weinstein, I.B., Kennelly, E.J., 2006. Bioactive depsides and anthocyanins from jaboticaba (*Myrciaria cauliflora*). Journal of Natural Products 69, 1228–1230.

Rizzello, C.G., Lorusso, A., Russo, V., Pinto, D., Marzani, B., Gobbetti, M., 2017. Improving the antioxidant properties of quinoa flour through fermentation with selected autochthonous lactic acid bacteria. International Journal of Food Microbiology 241, 252–261.

Robles-Sánchez, M., Astiazarán-García, H., Martín-Belloso, O., Gorinstein, S., Alvarez-Parrilla, E., de la Rosa, L.A., Yepiz-Plascencia, G., González-Aguilar, G.A., 2011. Influence of whole and fresh-cut mango intake on plasma lipids and antioxidant capacity of healthy adults. Food Research International 44, 1386–1391.

Rocha, A.P., Carvalho, L.C., Sousa, M.A., Madeira, S.V., Sousa, P.J., Tano, T., Schini-Kerth, V.B., Resende, A.C., Soares de Moura, R., 2007. Endothelium-dependent vasodilator effect of *Euterpe oleracea* Mart. (Acai) extracts in mesenteric vascular bed of the rat. Vascular Pharmacology 46, 97–104.

Rosamond, W.D., 2002. Dietary fiber and prevention of cardiovascular disease. Journal of the American College of Cardiology 39, 57–59.

Rosenvold, K., Andersen, H.J., 2003. Factors of significance for pork quality: a review. Meat Science 64, 219–237.

Rufino, M.D.S.M., Alves, R.E., de Brito, E.S., Pérez-Jiménez, J., Saura-Calixto, F., Mancini-Filho, J., 2010. Bioactive compounds and antioxidant capacities of 18 non-traditional tropical fruits from Brazil. Food Chemistry 121 (4), 996–1002.

Ruiz-Ruiz, J., Dávila-Ortíz, G., Chel-Guerrero, L., Betancur-Ancona, D., 2013. Angiotensin I-converting enzyme inhibitory and antioxidant peptide fractions from hard-to-cook bean enzymatic hydrolysates. Journal of Food Biochemistry 37, 26–35.

Sanabria, N., Sangronis, E., 2007. Caracterizacion del acai o manaca (*Euterpe oleracea* Mart.): Un fruto del Amazonas. Archivos Latinoamericanos de Nutricion 57, 94–98.

Sangronis, E., Sanabria, N., 2011. Impact of solar dehydration on composition and antioxidant properties of acai (*Euterpe oleracea* mart.). Archivos Latinoamericanos de Nutricion 61 (1), 74–80.

Santos, K.K.A., Matias, E.F.F., Tintino, S.R., Souza, C.E.S., Braga, M.F.B.M., Guedes, G.M.M., Rolón, M., Vega, C., de Arias, A.R., Costa, J.G.M., Menezes, I.R.A., Coutinho, H.D.M., 2012. Anti-trypanosoma cruzi and cytotoxic activities of *Eugenia uniflora* L. Experimental Parasitology 131, 130–132.

Sarmadi, B.H., Ismail, H., 2010. Antioxidative peptides from food proteins: a review. Peptides 31, 1949–1956.

Schreckinger, M.E., Lotton, J., Lila, M.A., de Mejia, E.G., 2010. Berries from South America: a comprehensive review on chemistry, health potential, and commercialization. Journal of Medicinal Food 13, 233–246.

Schulz, M., Borges, G.S.C., Gonzaga, L.V., Seraglio, S.K.T., Olivo, I.S., Azevedo, M.S., Nehring, P., Gois, J.S., de Almeida, T.S., Vitali, L., Spudeit, D.A., Micke, G.A., Borges, D.L.G., Fett, R., 2015. Chemical composition, bioactive compounds and antioxidant capacity of juçara fruit (*Euterpe edulis Martius*) during ripening. Food Research International 77, 125–131.

Shahidi, F., Zhong, Y., 2008. Bioactive peptides. Journal of AOAC International 91, 914–931.

Šiler, B., Zivkovic, S., Banjanac, T., Cvetkovic, J., Zivkovic, J.N., Ciric, A., Soković, M., Mišić, D., 2014. Centauries as underestimated food additives: antioxidant and antimicrobial potential. Food Chemistry 147, 367–376.

Simopoulos, A.P., 2003. Importance of the ratio of omega-6/omega-3 essential fatty acids: evolutionary aspects. World Review of Nutritional Diet 92, 1–22.

Simopoulos, A.P., 2009. Omega-6/omega-3 essential fatty acids: biological effects. World Review of Nutrition and Dietetics 99, 1–16.

Slavin, J.L., Lloyd, B., 2012. Health Benefits of Fruits and Vegetables. American Society for Nutrition: Advances in Nutrition 3, 506–516.

Soares, R.A.M., Mendonça, S., de Castro, L.Í.A., Menezes, A.C.C.C.C., Arêas, J.A.G., 2015. Major peptides from amaranth (*Amaranthus cruentus*) protein inhibit HMG-CoA reductase activity. International Journal of Molecular Sciences 16, 4150–4160.

Sujak, A., Kotlarz, A., Strobel, W., 2006. Compositional and nutritional evaluation of several lupin seeds. Food Chemistry 98, 711–719.

Teixeira, C.C., Pinto, L.P., Kessler, F.H.P., Knijnik, L., Pinto, C.P., Gastaldo, G.J., Fuchs, F.D., 1997. The effect of *Syzygium cumini* (L.) skeels on post-prandial blood glucose levels in non-diabetic rats and rats with streptozotocin-induced diabetes mellitus. Journal of Ethnopharmacology 56, 209–213.

Teixeira, C.C., Rava, C.A., Mallman da Silva, P., Melchior, R., Argenta, R., Anselmi, F., Almeida, C.R.C., Fuchs, F.D., 2000. Absence of antihyperglycemic effect of jambolan in experimental and clinical models. Journal of Ethnopharmacology 71, 343–347.

Thuanthong, M., De Gobba, C., Sirinupong, N., Youravong, W., Otte, J., 2017. Purification and characterization of angiotensin-converting enzyme inhibitory peptides from Nile tilapia (*Oreochromis niloticus*) skin gelatine produced by an enzymatic membrane reactor. Journal of Functional Foods 36, 243–254.

Torma, P.C.M.R., Brasil, A.V.S., Carvalho, A.V., Jablonski, A., Rabelo, T.K., Moreira, J.C.F., Gelain, D.P., Flôres, S.H., Augusti, P.R., Rios, A.O., 2017. Hydroethanolic extracts from different genotypes of açaí (*Euterpe oleracea*) presented antioxidant potential and protected human neuron-like cells (SH-SY5Y). Food Chemistry 222, 94–104.

Udani, J.K., Singh, B.B., Singh, V.J., Barrett, M.L., 2011. Effects of Acai (*Euterpe oleracea* Mart.) berry preparation on metabolic parameters in a healthy overweight population: a pilot study. Nutrition Journal 10, 45.

Villarroel, M., Acevedo, C., Yánez, E., Biolley, E., 2003. Propiedades funcionales de la fibra del musgo *Sphagnum magellanicum* y su utilización en la formulación de productos de panadería. Archivos Latinoamericanos de Nutricion 53, 400–407.

Victoria, F.N., Lenardão, E.J., Savegnago, L., Perin, G., Jacob, R.G., Alves, D., da Silva, W.P., da Motta de, A.S., Nascente Pda, S., 2012. Essential oil of the leaves of *Eugenia uniflora* L.: Antioxidant and antimicrobial properties. Food and Chemical Toxicology 50, 2668–2674.

Wattanasiritham, L., Theerakulkait, C., Wickramasekara, S., Maier, C.S., Stevens, J.F., 2016. Isolation and identification of antioxidant peptides from enzymatically hydrolyzed rice bran protein. Food Chemistry 192, 156–162.

Whelan, J., Rust, C., 2006. Innovative dietary sources of n-3 fatty acids. Annual Review of Nutrition 26, 75–103.

WHO, 2007a. Protein and Amino Acid Requirements in Human Nutrition: Report of the Joint WHO/FAO/UNU Expert Consultation. World Health Organization, Geneva.

WHO, 2007b. Assessing the Iron Status of Populations: Including Literature Reviews. World Health Organization, Geneva, Switzerland.

Wightman, J.D., Heuberger, R.A., 2015. Effect of grape and other berries on cardiovascular health. Journal of the Science of Food and Agriculture 95, 1584–1597.

Wong, D.Y.S., Musgrave, I.F., Harvey, B.S., Smid, S.D., 2013. Açaí (*Euterpe oleraceae* Mart.) berry extract exerts neuroprotective effects against B-amyloid exposure in vitro. Neuroscience Letters 556, 221–226.

Yazawa, K., Suga, K., Honma, A., Shirosaki, M., Koyama, T., 2011. Anti- inflammatory effects of seeds of the tropical fruit camu-camu (*Myrciaria dubia*). Journal of Nutritional Science and Vitaminology 57, 104–107.

Zaat, S.A.J., Kwakman, P.H.S., Drijfhour, J.W., 2010. Host tissue as a niche for biomaterial associated infection. Future Microbiology 5, 1149–1151.

Zhang, H., Yokoyama, W.H., Zhang, H., 2012. Concentration-dependent displacement of cholesterol in micelles by hydrophobic rice bran protein hydrolysates. Journal of the Science of Food and Agriculture 92 (7), 1395–1401.

Zhao, Q.Y., Sannier, F., Piot, J.M., 1996. Kinetics of appearance of four hemorphins from bovine hemoglobin peptic hydrolysates by HPLC coupled with photodiode array detection. Biochimica et Biophysica Acta 1295, 73–80.

CHAPTER 10

Health Benefits of Flavonoids

Cíntia Reis Ballard, Mário Roberto Maróstica Junior
Department of Food and Nutrition, Faculty of Food Engineering, University of Campinas, Campinas, Brazil

10.1 Introduction

In the last few decades there has been frequent interest regarding bioactive compounds and their health benefits. Bioactive compounds, in a broader definition, are essential (e.g., vitamins) and nonessential diet components found in plants, animals, or of synthetic origin that present probable effects (Biesalski et al., 2009; Guaadaoui et al., 2014).

A large proportion of these components remain unknown, but many have been isolated and identified. Among them, phenolic compounds are the group with the greatest amount of evidence about their therapeutic potential in health related to the reduction in the risk of developing chronic/degenerative diseases (Costa et al., 2017).

Phenolic compounds (PCs), nonessential human diet components, are products of secondary metabolism of plants (Quideau et al., 2011). Chemically, PCs have a variable structure, with one or more aromatic rings and hydroxyl groups. There are over 8000 PCs identified in plant foods, from simple molecules to those with a high degree of polymerization occurring in free or conjugated forms with sugars, acids, and other biomolecules (Tsao, 2010).

The most substantial group of PCs in the human diet is flavonoids (Marventano et al., 2017). Usually, flavonoids are classified into subgroups based on their chemical structure: flavones, flavonols, flavanols, flavanones, isoflavones, and anthocyanins. Its health benefits in preventing or treating diseases have been extensively studied (Kozłowska and Szostak-Wegierek, 2014).

Flavonoids may act by several mechanisms, some have been determined, while others remain unclear. These components, individually or in combination, showed important antioxidant, antiinflammatory, antidiabetic, anticancer, antiobesity, and cardioprotective effects in in vitro and in vivo models (Xiao et al., 2011). New evidence shows that not only flavonoids, but also their metabolites, can operate in multiple signaling cascades to exert their modulatory effects on different cell types (Rizza et al., 2011; Wang et al., 2010).

Although there are still a number of challenges associated with bioavailability, such as bioaccessibility, bioactivity, determination of safe doses of consumption and a better

Bioactive Compounds. https://doi.org/10.1016/B978-0-12-814774-0.00010-4

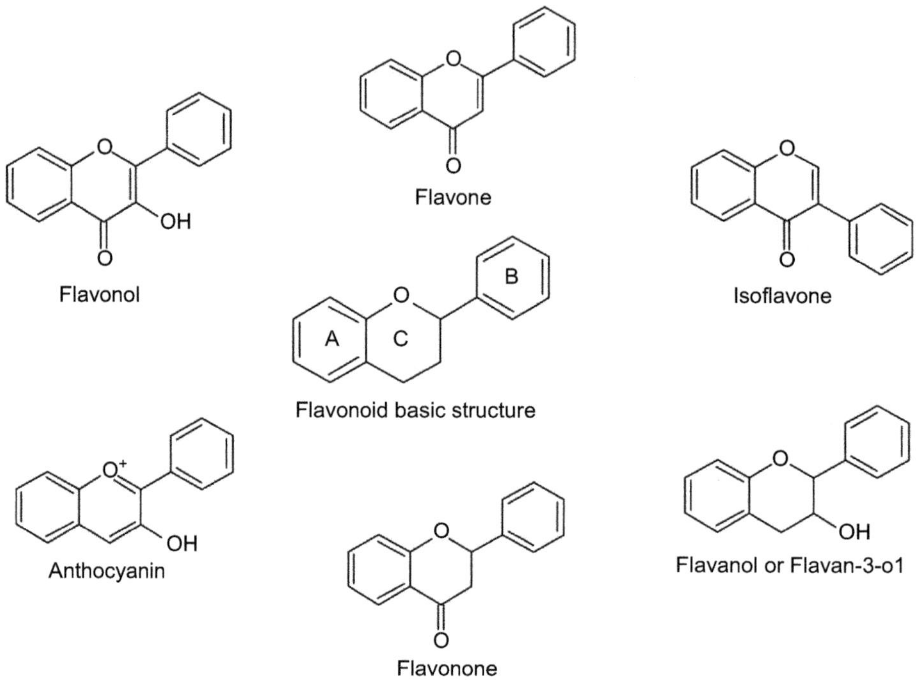

Figure 10.1
Chemical structure of flavonoids.

dose–response analysis, flavonoids show efficacy for not only the prevention and treatment of diseases but also to be used as complementary therapy (Wang and You, 2016).

In this context, it is interesting to know the potential of these compounds for the benefit of well-being to promote the consumption of flavonoid sources and reduce many types of damage to health. In order to have the size of the effect, data on flavonoids and their subgroups will be discussed either isolated or in combination.

10.2 Flavonoids: Chemistry, Food Sources, Estimated Food Intake

The basic structure of flavonoids is the flavan nucleus with 15 carbon atoms organized in three rings (Fig. 10.1). Generally, it is divided into six subclasses: flavones, flavonols, flavanols, flavanones, isoflavones, and anthocyanins (Table 10.1). This classification is according to the substitutions in the arrangements of hydroxyl, methoxy, prenyl, and glycosidic side groups and in the conjugation between the rings (Dai and Mumper, 2010).

The estimated intake of total flavonoids may range from 34.5 to 897 mg/day (Bai et al., 2014; Grosso et al., 2014; Jun et al., 2015; Kent et al., 2015; Miranda et al., 2016; Vogiatzoglou et al., 2015; Zamora-Ros et al., 2010). South America has the lowest consumption of

Table 10.1: Flavonoid Subclasses, Compounds, and Sources

Subclass	Examples of Compounds	Examples of Sources[a]
Flavones	Luteolin, apigenin, chrysin, baicalein	Chamomile, parsley, roots
Flavonols	Quercetin, kaempferol, myricetin, fisetin	Onion leaves, broccoli, apples, black tea, black grapes
Flavanols (flavan-3-ols)	Catechin, epicatechin, epigallocatechin-gallate, proanthocyanidins	Green tea, cocoa, legumes, red wine
Flavanones	Naringenin, hesperetin, naringin, hesperidin	Tomatoes, citrus fruits (peels)
Isoflavones	Genistein, daidzein, puerarin	Soy products, herbs
Anthocyanins	Cyanidin, delphinidin, pelargonidin, malvidin	Berry-type fruits, red wine

[a]Marzocchella, L., Fantini, M., Benvenuto, M., Masuelli, L., Tresoldi, I., Modesti, A., Bei, R., 2011. Dietary flavonoids: molecular mechanisms of action as anti- inflammatory agents. Recent Patents on Inflammation and Allergy Drug Discovery 5, 200–220. https://doi.org/10.2174/187221311797264937.

flavonoids, while European, Asian, and Oceanic countries have averages of 313.26–897 mg/day. There are no data, until the moment of this publication, for countries in Africa and Central America.

In relation to flavonoid subclasses, flavanols of the catechin family (catechin, epicatechin, and epigallocatechin-gallate), proanthocyanidins, flavanones, and anthocyanins are those with the highest consumption (Bai et al., 2014; Grosso et al., 2014).

Despite studies that have advanced details on the composition of foods, there are still many limitations in the literature on the ingestion of phenolic compounds. The studies do not use the same databases, many foods do not have their phenolic compounds identified and there are differences in analytical methodologies. Furthermore, a 24-h dietary recall is often used, which is a self-reported method, and as it is applied in a single moment it does not consider seasonality issues. These factors may lead to an underestimation of flavonoid consumption in the general population (Kent et al., 2018).

10.3 Flavonoid Intake and Prevention of Diseases

Prospective observational studies evaluating the association between total dietary flavonoid intake and risk of all causes, cardiovascular and cancer mortality in the general population, have supported favorable results for the intake of these phenolic compounds. The reduction in the risk of all-cause mortality had a significant variation from 18% to 40% for an intake of 200–350 mg/day of total flavonoids (Ivey et al., 2015; Liu et al., 2017).

High total flavonoid consumers may also present a risk reduction of 18%–50% (top of quintile) for mortality in cardiovascular disease and cancer compared to those whose intake was in the lowest quintile (Grosso et al., 2017; Ivey et al., 2015). An increase of 100 mg/day of total flavonoids in the diet may represent a reduction in risk of 6% and 4% for all-cause mortality and cardiovascular disease, respectively (Grosso et al., 2017).

Ischemic stroke, among cardiovascular diseases, is one of the leading causes of death around the world. An addition of 100 mg/day of food flavonoids can moderately reduce the risk of stroke (Tang et al., 2016). For type II diabetes mellitus (T2D), an intake of 500 mg/day of total flavonoids presented a reduction of the risk for development of disease in 5% (Liu et al., 2014).

Concerning the association between ingestion of subclasses of flavonoids, it has been observed that lower rates of all-cause mortality and cardiovascular disease are linked to the higher consumption of practically all classes of flavonoids for intakes of 1–10 mg/day (Grosso et al., 2017; Ivey et al., 2015). In relation to T2D, dietary intake of anthocyanin was associated with a 5% reduction in risk for a 7.5 mg/day increase in diet (Guo et al., 2016).

Regardless of the extent of the protective effect in prospective observational studies showing considerable variation, these findings state that a diet rich in flavonoids may reduce the risk of death from various diseases.

10.4 Flavonoids: General Mechanism of Action

Under normal physiological conditions, cellular metabolism produces reactive oxygen species (ROS) in equilibrium with the antioxidant defense system. When this homeostasis is interrupted by excess ROS, the result is oxidative stress. In this situation, ROS can cause lipid peroxidation (e.g., increased malondialdehyde [MDA]), and oxidative damage to DNA molecules (e.g., cancer), membrane phospholipids, and proteins, which constitutes the molecular basis of several diseases (Niemann et al., 2017; Fig. 10.2).

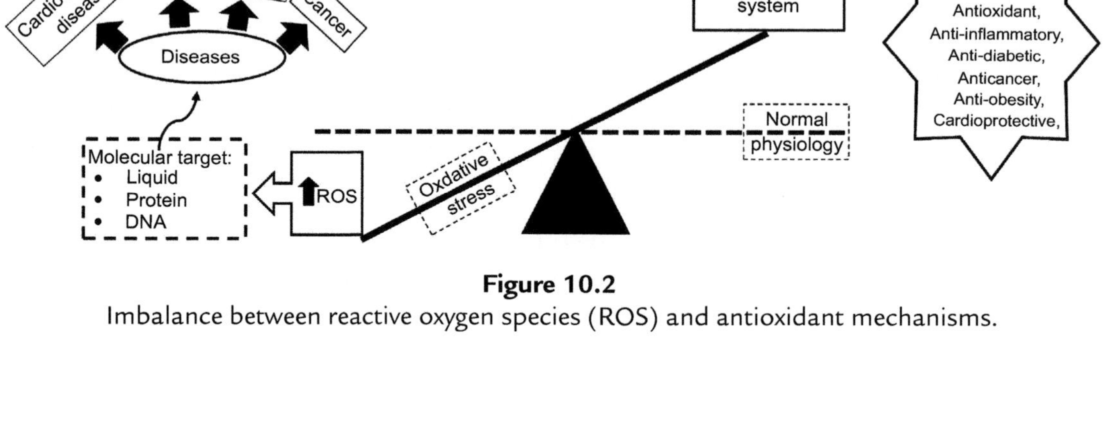

Figure 10.2
Imbalance between reactive oxygen species (ROS) and antioxidant mechanisms.

The endogenous antioxidant system of the cell includes superoxide dismutase (SOD), reduced glutathione (GSH), and glutathione peroxidase (GPx), in addition to exogenous antioxidants, such as vitamins, PCs, etc. (Khurana et al., 2013).

Flavonoids have an antioxidant action that attenuates ROS, by enhancing antioxidant enzymes or by inhibiting enzymes that employ a pro-oxidant effect on oxidative stress (e.g., NADPH oxidase, xanthine oxidase, catalase, myeloperoxidase [MPO]) (Kim et al., 2009; Xiao et al., 2017; Yan et al., 2013).

Furthermore, flavonoids are involved in many signaling pathways and the expression of inflammatory genes: they modulate the activities of enzymes that metabolize arachidonic acid (phospholipase A2, cyclooxygenase, lipoxygenase). They also inhibit the production of proinflammatory cytokines (IL-1β, IL-2, IL-6, IFN-γ, or TNF-α), nuclear factor kappa B (NF-κB), mitogen-activated protein kinases (ERK, p38, and JNK) and expression of nitric oxide synthase (iNOS) (Bhaskar et al., 2016; Song et al., 2014).

Reducing the activity of carbohydrate digestive enzymes, modulating carrier proteins (GLUT2 and 4) and glucose uptake are more actions attributed to flavonoids; and currently, on the release of gastrointestinal hormones, neuropeptides and beneficial changes in the intestinal microbiota (Habtemariam and Varghese, 2014; Porras et al., 2017; Williamson, 2013).

The most promising results on the health benefits of flavonoid intake are found in actions on the cardiovascular diseases, diabetes, obesity, and cancer.

10.5 Flavonoids and Cardiovascular Diseases

The three preeminent benefits of flavonoid intake on the cardiovascular system are those related to actions to protect against atherosclerosis, coronary artery disease, and coagulation disorders through its antioxidant, antiinflammatory, and chelating properties (Millar et al., 2017; Xiao et al., 2011).

Atherosclerosis is a multifactorial disease characterized by endothelial dysfunction involving ROS through the oxidation of LDL-cholesterol (ox-LDL) (Ho et al., 2013). This ox-LDL, in turn, increases the inflammatory process via expression of TLR4 and downstream activation of the transcription factor NF-κB.

Quercetins are capable of inactivating ROS by neutralizing ox-LDL and attenuating the TLR4/NF-κB signaling pathway in endothelial cells, thereby regulating the inflammatory process (Bhaskar et al., 2016). At the same time, proanthocyanidins, luteolin and apigenin, may inhibit LOX-1, which is an endothelial receptor for ox-LDL (Ishizuka et al., 2011; Jeong et al., 2007).

Daidzein can reduce the damage from myocardial infarct by inhibiting the release of TNF-α and IL-6 cytokines and by limiting nuclear translocation of NF-κB. Moreover, it may decrease MDA levels, MPO activity, catalase activity, and myocardial neutrophil infiltration (Kim et al., 2009).

Inhibition of enzymes such as NADPH oxidase, xanthine oxidase, and lipoxygenases, which increase ROS production, is also attributed to the antioxidant capacity of flavonoids. It was found that the consumption of quercetin regulated the expression of aortic NADPH and attenuated the formation of the atherosclerotic lesion in mice (Xiao et al., 2017). According to some studies, so far, quercetin is the flavonoid that has most evidence of its effect in the reduction of aortic atherosclerosis lesions by improving endothelial function and decreasing inflammation (Phie et al., 2017).

Luteolin inhibited xanthine oxidase in vitro (Yan et al., 2013). The 12/15-lipoxygenase, which plays a role in the development of atherosclerosis, hypertension, and heart failure, was inhibited by baicalein both in in vitro and in vivo studies (Song et al., 2014).

Protocatechuic acid, a metabolite of anthocyanins, may reduce atherosclerosis by inhibiting the adhesion of monocytes to tumor necrosis factor-alpha (TNF-α)-activated aortic endothelial cells by inhibiting the expression of vascular cell adhesion molecule 1 (sVCAM-1) and intercellular adhesion molecule 1 (sICAM-1) (Wang et al., 2010).

Flavonoids may suppress the activity of 3-hydroxy-3-methylglutaryl-coenzyme A reductase (HMG-CoA reductase), the enzyme involved in the synthesis of plasma cholesterol (Kwon et al., 2010). Epigallocatechin-gallate inhibited HMG-CoA reductase in vitro, revealing its potential in the treatment of hypercholesterolemic disorders (Cuccioloni et al., 2011).

Although most of the effects have been verified in cells, animal models, or prospective observational studies, there are promising results in clinical trials of intervention in the treatment of cardiovascular diseases (Dower et al., 2015; Rizza et al., 2011; Wang and You, 2016; Zhu et al., 2013).

Quercetin (160 mg/day) and epicatechin (100 mg/day) had cardioprotective effects in prehypertensive individuals (Dower et al., 2015). Both possibly contributed to improved endothelial function by reducing soluble endothelial selectin (sE-selectin) in plasma, and quercetin also reduced inflammation by promoting the decrease of IL-1β.

The ingestion of 500 mg/day of hesperidin, a hesperitin metabolite, for 3 weeks improved endothelial function in subjects with metabolic syndrome through vasculoprotective actions, increasing flow-mediated dilatation and reducing inflammatory biomarkers (Rizza et al., 2011).

Epigallocatechin-gallate (EGCG) has been shown to be a potential adjuvant in the therapy of patients with brain ischemic stroke coadministered with recombinant tissue plasminogen activator (rt-PA), extending the narrow treatment window of rt-PA (Wang and You, 2016). The results were related to an improvement in the NIHSS (National Institutes of Health Stroke Scale) scale; in addition to reduction of metalloproteinases 2 and 9 in plasma, both have oxidoreductase activity.

A supplement containing 320 mg of anthocyanins (delphinidin and cyanidin) for 24 weeks reduced biomarkers of inflammation (C-reactive, vascular cell adhesion molecule-1, IL1B) besides improving the lipid profile by increasing HDL and lowering LDL cholesterol in hypercholesterolemic participants (Zhu et al., 2013).

Regarding isoflavones, no conclusive evidence of their protective effect on the risk and treatment of cardiovascular disease was found in postmenopausal women (Dong et al., 2011). Other authors report a modest effect of isoflavones on vasomotor symptoms linked to the production of equol by intestinal bacteria in the metabolism of daidzein in menopausal women (Utian et al., 2015). Puerarin, an isoflavone not found in soy, may be effective in unstable angina when administered with conventional treatments, but more randomized controlled trials (RCTs) are needed to confirm its effects (Wang et al., 2006).

A recent meta-analysis of RCTs on the role of interindividual variability in the effect of flavonols on cardiometabolic biomarkers revealed that the country of origin and health status might influence the size of the effect (Menezes et al., 2017). The subgroup analysis showed a more pronounced effect in participants from Asian countries in patients with diagnosed disease or dyslipidemia compared to normal and healthy baseline values. Therefore, individual variability can influence effect size and should be controlled in clinical trials.

10.6 Flavonoids, Insulin Resistance, and Type II Diabetes Mellitus

Type II diabetes mellitus (T2D) and its complications are associated with various oxidative reactions, increased ROS, and subsequent oxidative stress. Chronic oxidative stress is a consequence of hyperglycemia, insulin resistance, inflammation, and dyslipidemia, and may result in defective expression of the insulin gene and insulin secretion, as well as destruction of β-cells by autoimmune cells (Singh et al., 2015).

In this manner, the benefit of flavonoid intake is in the sense of reducing the amount of ROS, with a direct effect on pro-oxidant enzymes or enzymes that have antidiabetic action (Habtemariam and Varghese, 2014). Another way is inhibition of the activity of the digestive enzymes for the production of glucose and of the carriers responsible for the transport (GLUT2 e 4) and absorption of glucose by the flavonoids (Williamson, 2013).

Therefore, flavonoids will have their actions in increasing insulin secretion, promoting the proliferation of pancreatic β cells, improving glycemia by regulating glucose metabolism, and thus reducing insulin resistance, inflammation, and oxidative stress (Babu et al., 2013).

EGCG protected β cells from cytokine-induced proinflammatory cytotoxicity by restoring insulin secretion in an in vitro study through the modulation of the antiapoptotic protein Bcl-2 expression (Zhang et al., 2011). Quercetin, apigenin, and luteolin protected against cytokine-induced pancreatic β cell damage (IL-1β and IFN-γ) by suppressing NF-κB activation (Kim et al., 2007).

Naringin and hesperidin (50 mg/kg), in a model of streptozotocin-induced diabetes (STZ), for 1 month improved hyperglycemia and oxidative stress (Mahmoud et al., 2012). There was an increase in hepatic antioxidant enzymes such as catalase, glutathione peroxidase, glutathione reductase, and superoxide dismutase.

Flavanones still have effects on the increase of the GLUT4 (glucose transporter type 4) protein expression in adipose tissue ameliorating insulin resistance (Ahmed et al., 2017). Similar results were found by trimeric and tetrameric proanthocyanidins, anthocyanins, and apigenin, promoting the translocation of GLUT4 to the muscle plasma membrane (Nizamutdinova et al., 2009; Yamashita et al., 2016).

Cyanidin, genistein, daidzein, and fisetin are able to reduce glucose 6-phosphatase in mice (Choi et al., 2008; Guo et al., 2012; Prasath and Subramanian, 2011). This enzyme hydrolyzes glucose 6-phosphate into phosphate and glucose, thereby controlling glycemia.

In a randomized, placebo-controlled, double-blind trial, hesperidin (500 mg/day) for 6 weeks relieved oxidative DNA damage and lipid peroxidation in T2D patients (Homayouni et al., 2017). Epicatechin supplementation (100 mg/day) for 4 weeks improved fasting plasma insulin and insulin resistance and had no effect on fasting plasma glucose in healthy subjects (Dower et al., 2015). Anthocyanin (320 mg/day) for 24 weeks significantly decreased fasting plasma glucose, in addition, it decreased LDL cholesterol and triglycerides, and increased HDL cholesterol in subjects with T2D (Li et al., 2015).

The clinical studies related to flavonoids and T2D, for the most part, analyze the effects of fruits and vegetables or their extracts, which are complex matrices that present not only flavonoids, but also other phenolic compounds and nutrients. Whether isolated, in combination, or present in foods, there is a benefit to flavonoid intake in the prevention or treatment of T2D. Further studies are required to determine safe doses and to clarify the mechanisms of action.

10.7 Flavonoids and Obesity

The antiobesity actions found in in vitro studies suggest that flavonoids may act on adipocytes (reducing viability and proliferation), suppressing triglyceride accumulation, stimulating lipolysis, β-oxidation of fatty acids, and reducing inflammation (Herranz-López et al., 2017; Lee et al., 2014; Leiherer et al., 2016).

Quercetin decreased intracellular ROS in a model of hypertrophied adipocytes (Herranz-López et al., 2017; Leiherer et al., 2016). Anthocyanins suppressed lipid accumulation in adipocytes due to the widespread inhibition of transcription factors that regulate lipogenesis, such as the liver X-receptor, the sterol regulatory element-binding protein-1c, proliferator-activated receptor-γ of peroxisome, and the binding protein of the CCAAT conjugator (Lee et al., 2014).

In animal model studies, flavonoids offer positive results for the prevention and treatment of obesity. Effects by increasing energy expenditure or by inhibiting food intake, through several processes that suppress the expression of oxidative stress and inflammatory markers; on the release of gastrointestinal hormones; and, recently, operating changes in the intestinal microbiota (Klaus et al., 2005; Panchal et al., 2012; Porras et al., 2017; You et al., 2017; Zhang et al., 2009).

Wistar rats receiving quercetin in the diet (0.8 g/kg) for 8 weeks had lower abdominal fat and systolic blood pressure, as well as lower NF-κB expression in the heart and liver than in the control group (Panchal et al., 2012). Recently, quercetin supplementation (0.5 g/kg diet) for 16 weeks promoted a reduction in body weight gain, epididymal fat accumulation, and liver weight, without differences in food intake, in addition, it repaired the intestinal microbiota imbalance and reversed the endotoxemia-mediated TLR-4 pathway induction (Porras et al., 2017).

The accumulation of body fat in mice was attenuated by EGCG for 4 weeks. The effect was dose-dependent (0.5% and 1% of the diet) and accompanied by a reduction in the digestibility of the diet verified in the energy content in the feces, without altering the food intake (Klaus et al., 2005). Isoflavones decreased body weight gain, total abdominal fat, and food intake. On food intake, there was a reduction in the levels of neuropeptide Y and ghrelin, which have an orexigenic action; and, at the same time, increase CCK and PYY, which can induce satiety by anorexic action (Zhang et al., 2009).

Most clinical studies focus on the effects of EGCG. In a recent meta-analysis, 300 mg/day of EGCG showed potential to increase the metabolic rate in adults, without presenting significant results on fat oxidation (Kapoor et al., 2017). Prospective trials are needed to confirm the findings and clarify the mechanisms.

It was examined in three prospective cohort studies that ingestion of flavonols, flavanols, and anthocyanins, even with adjustment for fiber intake, contribute to weight maintenance in healthy adults (Bertoia et al., 2016). In an RCT the intervention with 500 mg of hesperidin plus 50 or 75 mg/day of caffeine for 12 weeks was able to reduce body weight and BMI in overweight subjects (Ohara et al., 2015). There was a reduction of body fat only at the expense of subcutaneous adipose tissue.

Supplementation of EGCG (282 mg/day) plus resveratrol (280 mg/day) for 3 days in overweight individuals increased energy expenditure at rest and postprandial, and improved metabolic flexibility calculated as the postprandial increase for the highest respiratory quotient (Most et al., 2014). Studies with long-term supplementation are needed to evaluate whether EGCG plus resveratrol can be used in the treatment of obesity.

Unfortunately, to date, there is no RCT evidence on the effect of flavonoids isolated or in combination on intestinal microbiota. Health benefits are found in the consumption of fruits, seeds, vegetables, tea, cocoa products, and wine, which can modulate the levels of bacteria that bring health benefits, such as *Bacteroides* and those of the *Bifidobacterium* strain (Singh et al., 2017).

10.8 Flavonoids and Cancer

Flavonoids may have activity in the stages of initiation, promotion, and progression of carcinogenesis (Abdal Dayem et al., 2016). In the initiation and promotion stages, flavonoids can regulate phase 1 detoxification (e.g., CYP1A1) and phase 2 biotransformation (e.g., GSH) enzymes, on inhibition of cell proliferation (e.g., Ki-67) on DNA repair by reducing oxidative stress. Moreover, at the stage of progression, the flavonoids can inhibit proangiogenic factors (e.g., VEGF), regulate metastasis proteins (e.g., MMP2/9), and induce apoptosis (caspase-3) (Abd El-Rahman et al., 2017).

Exposure of human prostate epithelial cells (RWPE-1) to apigenin after H_2O_2 stress resulted in significant genoprotective effects by reduction in ROS levels and downregulation of NF-κB (Sharma et al., 2014). Fisetin significantly reduced lipid, DNA, and protein damage in vitro (V79-4 cells), thereby protecting against cell death, by eliminating ROS and increasing the antioxidant capacity of GSH (Kang et al., 2014). Similar results were observed for luteolin, EGCG, malvidin, and kaempferol in several cancer cell lines (Bestwick et al., 2005; Paixão et al., 2012; Ramos et al., 2010; Zhu et al., 2014).

Quercetin inhibited the development of hamster buccal carcinoma, impairing the production of ROS mediated by CYP by downregulation of CYP1A1 and CYP1B1 expression and positive regulation of antioxidant defenses (Priyadarsini and Nagini, 2012). EGCG decreased the size and number of tumors, improved markers of oxidative stress, and significantly inhibited the expression of CD44, VEGF, Ki-67, and MMP-2 associated with significantly increased expression of caspase-3 in breast cancer-induced rats (Abd El-Rahman et al., 2017).

The chemopreventive effect of flavonoids is described in meta-analyzes of observational studies (He et al., 2016; Hua et al., 2016; Hui et al., 2013). Concerning RCTs, many flavonoids are used to attenuate the effects of radiation therapy. Individuals in prostate cancer therapy benefited from the use of proanthocyanidins, resulting in a lower incidence of acute radiation cystitis (Hamilton et al., 2015). Topical EGCG may be effective treatment against radiation-induced dermatitis in breast cancer patients undergoing radiation therapy (Zhu et al., 2016). A mixture of flavonoids (20 mg of apigenin and 20 mg of EGCG) reduced the rate of recurrence of colon neoplasia in patients with resected colon cancer followed for 3–4 years (Hoensch et al., 2008).

As for cardiovascular diseases, diabetes, and obesity, RCTs rarely correlate cancer and flavonoids isolated or in mixture. Although flavonoids exert in vitro effects more strongly than in vivo, it is important to determine their mechanisms of action. Further studies are therefore required.

10.9 Conclusion

The consumption of food flavonoids, to date, is considered safe and shows benefits related to reducing risk for various diseases. Regarding treatment with supplementation of flavonoids, isolated or in combination, the results are promising and many mechanisms are established in vitro and in animal models (Fig. 10.3).

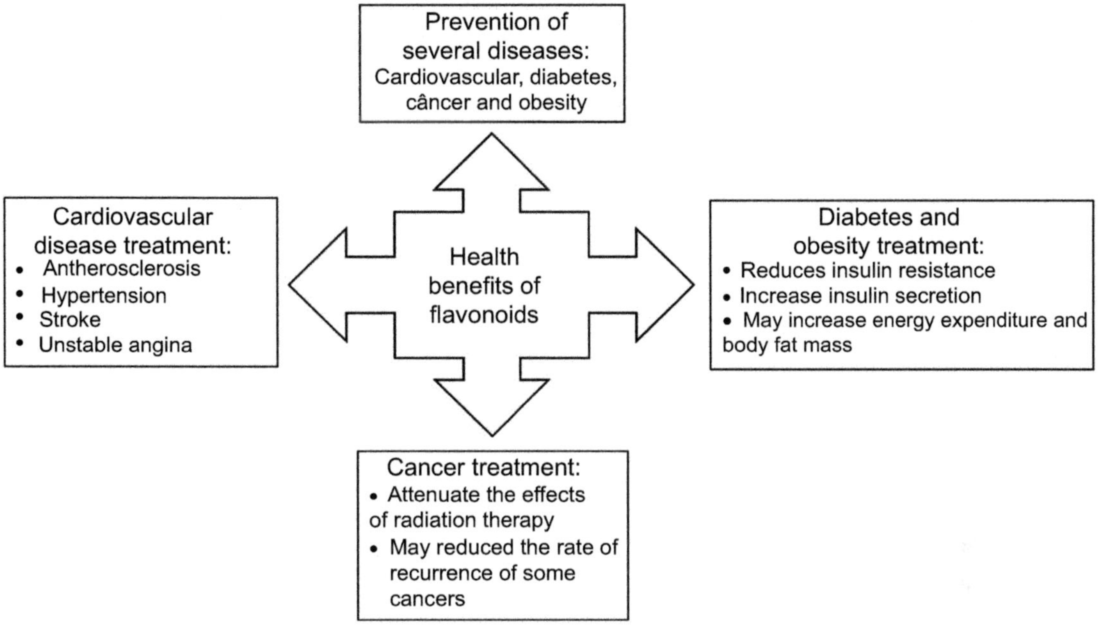

Figure 10.3
Summary of key health benefits of flavonoids.

In humans, flavonoid supplementation has the potential to be the main treatment for various diseases; or adjuvant primarily in heart disease and cancer. However, high-dose flavonoid supplements should be used sparingly and with health professional follow-up, because the toxicity of these concentrated compounds is unknown, as there are few studies of their interactions with drugs or other dietary components.

Furthermore, it is fundamental that more well-designed studies evaluate the bioavailability of the compounds, besides verifying how individual variability influences the effect of flavonoids, particularly, ethnicity and baseline.

References

Abd El-Rahman, S.S., Shehab, G., Nashaat, H., 2017. Epigallocatechin-3-Gallate: the prospective targeting of cancer stem cells and preventing metastasis of chemically-induced mammary cancer in rats. The American Journal of the Medical Sciences 354, 54–63. https://doi.org/10.1016/j.amjms.2017.03.001.

Abdal Dayem, A., Choi, H.Y., Yang, G.-M., Kim, K., Saha, S.K., Cho, S.-G., 2016. The anti-cancer effect of polyphenols against breast cancer and cancer stem cells: molecular mechanisms. Nutrients 8. https://doi.org/10.3390/nu8090581.

Ahmed, O.M., Hassan, M.A., Abdel-Twab, S.M., Abdel Azeem, M.N., 2017. Navel orange peel hydroethanolic extract, naringin and naringenin have anti-diabetic potentials in type 2 diabetic rats. Biomedicine and Pharmacotherapy 94, 197–205. https://doi.org/10.1016/j.biopha.2017.07.094.

Babu, P.V.A., Liu, D., Gilbert, E.R., 2013. Recent advances in understanding the anti-diabetic actions of dietary flavonoids. The Journal of Nutritional Biochemistry 24, 1777–1789. https://doi.org/10.1016/j.jnutbio.2013.06.003.

Bai, W., Wang, C., Ren, C., 2014. Intakes of total and individual flavonoids by US adults. International Journal of Food Sciences and Nutrition 65, 9–20. https://doi.org/10.3109/09637486.2013.832170.

Bertoia, M.L., Rimm, E.B., Mukamal, K.J., Hu, F.B., Willett, W.C., Cassidy, A., 2016. Dietary flavonoid intake and weight maintenance: three prospective cohorts of 124,086 US men and women followed for up to 24 years. BMJ 352, i17. https://doi.org/10.1136/BMJ.I17.

Bestwick, C.S., Milne, L., Pirie, L., Duthie, S.J., 2005. The effect of short-term kaempferol exposure on reactive oxygen levels and integrity of human (HL-60) leukaemic cells. Biochimica et Biophysica Acta - Molecular Basis of Disease 1740, 340–349. https://doi.org/10.1016/J.BBADIS.2004.10.005.

Bhaskar, S., Sudhakaran, P.R., Helen, A., 2016. Quercetin attenuates atherosclerotic inflammation and adhesion molecule expression by modulating TLR-NF- j B signaling pathway. Cellular Immunology 310, 131–140. https://doi.org/10.1016/j.cellimm.2016.08.011.

Biesalski, H.K., Dragsted, L.O., Elmadfa, I., Grossklaus, R., Müller, M., Schrenk, D., Walter, P., Weber, P., 2009. Bioactive compounds: definition and assessment of activity. Nutrition 25, 1202–1205. https://doi.org/10.1016/j.nut.2009.04.023.

Choi, M.S., Jung, U.J., Yeo, J., Kim, M.J., Lee, M.K., 2008. Genistein and daidzein prevent diabetes onset by elevating insulin level and altering hepatic gluconeogenic and lipogenic enzyme activities in non-obese diabetic (NOD) mice. Diabetes/Metabolism Research and Reviews 24, 74–81. https://doi.org/10.1002/dmrr.780.

Costa, C., Tsatsakis, A., Mamoulakis, C., Teodoro, M., Briguglio, G., Caruso, E., Tsoukalas, D., Margina, D., Dardiotis, E., Kouretas, D., Fenga, C., 2017. Current evidence on the effect of dietary polyphenols intake on chronic diseases. Food and Chemical Toxicology 110, 286–299. https://doi.org/10.1016/j.fct.2017.10.023.

Cuccioloni, M., Mozzicafreddo, M., Spina, M., Tran, C.N., Falconi, M., Eleuteri, A.M., Angeletti, M., 2011. Epigallocatechin-3-gallate potently inhibits the in vitro activity of hydroxy-3-methyl-glutaryl-CoA reductase. The Journal of Lipid Research 52, 897–907. https://doi.org/10.1194/jlr.M011817.

Dai, J., Mumper, R.J., 2010. Plant phenolics: extraction, analysis and their antioxidant and anticancer properties. Molecules 15, 7313–7352. https://doi.org/10.3390/molecules15107313.

Dong, J.-Y., Wang, P., He, K., Qin, L.-Q., 2011. Effect of soy isoflavones on circulating C-reactive protein in postmenopausal women. Menopause 18, 1256–1262. https://doi.org/10.1097/gme.0b013e31821bfa24.

Dower, J.I., Geleijnse, J.M., Gijsbers, L., Schalkwijk, C., Kromhout, D., Hollman, P.C., 2015. Supplementation of the pure flavonoids epicatechin and quercetin affects some biomarkers of endothelial dysfunction and inflammation in (Pre)Hypertensive adults: a randomized double-blind, placebo-controlled, crossover trial. Journal of Nutrition 145, 1459–1463. https://doi.org/10.3945/jn.115.211888.

Grosso, G., Stepaniak, U., Topor-Madry, R., Szafraniec, K., Pajak, A., 2014. Estimated dietary intake and major food sources of polyphenols in the Polish arm of the HAPIEE study. Nutrition 30, 1398–1403. https://doi.org/10.1016/j.nut.2014.04.012.

Grosso, G., Micek, A., Godos, J., Pajak, A., Sciacca, S., Galvano, F., Giovannucci, E.L., 2017. Dietary flavonoid and lignan intake and mortality in prospective cohort studies: systematic review and dose-response meta-analysis. American Journal of Epidemiology 185, 1304–1316. https://doi.org/10.1093/aje/kww207.

Guaadaoui, A., Benaicha, S., Elmajdoub, N., Bellaoui, M., Hamal, A., 2014. What is a bioactive compound ? A combined definition for a preliminary consensus. International Journal of Nutrition and Food Sciences 3, 174–179. https://doi.org/10.11648/j.ijnfs.20140303.16.

Guo, H., Xia, M., Zou, T., Ling, W., Zhong, R., Zhang, W., 2012. Cyanidin 3-glucoside attenuates obesity-associated insulin resistance and hepatic steatosis in high-fat diet-fed and db/db mice via the transcription factor FoxO1. The Journal of Nutritional Biochemistry 23, 349–360. https://doi.org/10.1016/j.jnutbio.2010.12.013.

Guo, X., Yang, B., Tan, J., Jiang, J., Li, D., 2016. Associations of dietary intakes of anthocyanins and berry fruits with risk of type 2 diabetes mellitus: a systematic review and meta-analysis of prospective cohort studies. European Journal of Clinical Nutrition 70, 1360–1367. https://doi.org/10.1038/ejcn.2016.142.

Habtemariam, S., Varghese, G., 2014. The antidiabetic therapeutic potential of dietary polyphenols. Current Pharmaceutical Biotechnology 15, 391–400. https://doi.org/10.2174/1389201015666140617104643.

Hamilton, K., Bennett, N.C., Purdie, G., Herst, P.M., 2015. Standardized cranberry capsules for radiation cystitis in prostate cancer patients in New Zealand: a randomized double blinded, placebo controlled pilot study. Supportive Care in Cancer 23, 95–102. https://doi.org/10.1007/s00520-014-2335-8.

He, X., Sun, L., He, X., Sun, L., 2016. Dietary intake of flavonoid subclasses and risk of colorectal cancer: evidence from population studies. Oncotarget 7, 26617–26627. https://doi.org/10.18632/oncotarget.8562.

Herranz-López, M., Borrás-Linares, I., Olivares-Vicente, M., Gálvez, J., Segura-Carretero, A., Micol, V., 2017. Correlation between the cellular metabolism of quercetin and its glucuronide metabolite and oxidative stress in hypertrophied 3T3-L1 adipocytes. Phytomedicine 25, 25–28. https://doi.org/10.1016/J.PHYMED.2016.12.008.

Ho, E., Karimi, K., Liu, C., Bhindi, R., Figtree, G.A., 2013. Redox Biology Biological markers of oxidative stress : applications to cardiovascular research and practice. Redox Biology 1, 483–491. https://doi.org/10.1016/j.redox.2013.07.006.

Hoensch, H., Groh, B., Edler, L., Kirch, W., 2008. Prospective cohort comparison of flavonoid treatment in patients with resected colorectal cancer to prevent recurrence. World Journal of Gastroenterology 14, 2187–2193. https://doi.org/10.3748/WJG.14.2187.

Homayouni, F., Haidari, F., Hedayati, M., Zakerkish, M., Ahmadi, K., 2017. Hesperidin supplementation alleviates oxidative DNA damage and lipid peroxidation in type 2 diabetes: a randomized double-blind placebo-controlled clinical trial. Phytotherapy Research 31, 1539–1545. https://doi.org/10.1002/ptr.5881.

Hua, X., Yu, L., You, R., Yang, Y., Liao, J., Chen, D., Yu, L., 2016. Association among dietary flavonoids, flavonoid subclasses and ovarian cancer risk: a meta-analysis. PLoS One 11, e0151134. https://doi.org/10.1371/journal.pone.0151134.

Hui, C., Qi, X., Qianyong, Z., Xiaoli, P., Jundong, Z., Mantian, M., 2013. Flavonoids, flavonoid subclasses and breast cancer risk: a meta-analysis of epidemiologic studies. PLoS One 8, e54318. https://doi.org/10.1371/journal.pone.0054318.

Ishizuka, B.T.N., Ujita, Y.F., Ato, Y.S., Akano, A.N., Akino, A.K., Hshima, S.O., Anda, T.K., Oshimoto, R.Y., Awamura, T.S., 2011. Procyanidins are potent inhibitors of LOX-1: a new player in the French paradox 87, 104–113.

Ivey, K.L., Hodgson, J.M., Croft, K.D., Lewis, J.R., Prince, R.L., 2015. Flavonoid intake and all-cause mortality. American Journal of Clinical Nutrition 101, 1012–1020. https://doi.org/10.3945/ajcn.113.073106.

Jeong, Y., Choi, Y., Choi, J., Kwon, H., Kang, S., Bae, J., Lee, S., Kang, J., Han, S.J., Kang, Y., 2007. Attenuation of monocyte adhesion and oxidised LDL uptake in luteolin-treated human endothelial cells exposed to oxidised LDL. The British Journal of Nutrition 447–457. https://doi.org/10.1017/S0007114507657894.

Jun, S., Shin, S., Joung, H., 2015. Estimation of dietary flavonoid intake and major food sources of Korean adults. British Journal of Nutrition 115, 480–489. https://doi.org/10.1017/S0007114515004006.

Kang, K.A., Piao, M.J., Kim, K.C., Cha, J.W., Zheng, J., Yao, C.W., Chae, S., Hyun, J.W., 2014. Fisetin attenuates hydrogen peroxide-induced cell damage by scavenging reactive oxygen species and activating protective functions of cellular glutathione system. In Vitro Cellular and Developmental Biology – Animal 50, 66–74. https://doi.org/10.1007/s11626-013-9681-6.

Kapoor, M.P., Sugita, M., Fukuzawa, Y., Okubo, T., 2017. Physiological effects of epigallocatechin-3-gallate (EGCG) on energy expenditure for prospective fat oxidation in humans: a systematic review and meta-analysis. The Journal of Nutritional Biochemistry 43, 1–10. https://doi.org/10.1016/J.JNUTBIO.2016.10.013.

Kent, K., Charlton, K.E., Russell, J., Mitchell, P., Flood, V.M., 2015. Estimation of flavonoid intake in older Australians: secondary data analysis of the blue mountains eye study. Journal of Nutrition in Gerontology and Geriatrics 34, 388–398. https://doi.org/10.1080/21551197.2015.1088917.

Kent, K., Charlton, K.E., Lee, S., Mond, J., Russell, J., Mitchell, P., Flood, V.M., 2018. Dietary flavonoid intake in older adults: how many days of dietary assessment are required and what is the impact of seasonality? Nutrition Journal 17, 1–9. https://doi.org/10.1186/s12937-017-0309-7.

Khurana, S., Piche, M., Hollingsworth, A., Venkataraman, K., Tai, T.C., 2013. Oxidative stress and cardiovascular health: therapeutic potential of polyphenols. Canadian Journal of Physiology and Pharmacology 91, 198–212. https://doi.org/10.1139/cjpp-2012-0252.

Kim, E.-K., Kwon, K.-B., Song, M.-Y., Han, M.-J., Lee, J.-H., Lee, Y.-R., Lee, J.-H., Ryu, D.-G., Park, B.-H., Park, J.-W., 2007. Flavonoids protect against cytokine-induced pancreatic β-cell damage through suppression of nuclear factor κB activation. Pancreas 35, e1–e9. https://doi.org/10.1097/mpa.0b013e31811ed0d2.

Kim, J.W., Jin, Y.C., Kim, Y.M., Rhie, S., Kim, H.J., Seo, H.G., Lee, J.H., Ha, Y.L., Chang, K.C., 2009. Daidzein administration in vivo reduces myocardial injury in a rat ischemia/reperfusion model by inhibiting NF-kB activation. Life Sciences 84, 227–234. https://doi.org/10.1016/J.LFS.2008.12.005.

Klaus, S., Pültz, S., Thöne-Reineke, C., Wolfram, S., 2005. Epigallocatechin gallate attenuates diet-induced obesity in mice by decreasing energy absorption and increasing fat oxidation. International Journal of Obesity 29, 615–623. https://doi.org/10.1038/sj.ijo.0802926.

Kozłowska, A., Szostak-Wegierek, D., 2014. Flavonoids- Food sources and health benefits. Roczniki Państwowego Zakładu Higieny 65, 79–85.

Kwon, E.K., Lee, D.Y., Hyungjae, L., Kim, D.O.K., Baek, N.I.N., Kim, Y.E., Kim, H.Y., 2010. Flavonoids from the buds of rosa damascena inhibit the activity of 3-Hydroxy-3-methylglutaryl-coenzyme a reductase and angiotensin I-converting enzyme. Journal of Agricultural and Food Chemistry 58, 882–886. https://doi.org/10.1021/jf903515f.

Lee, B., Lee, M., Lefevre, M., Kim, H.-R., 2014. Anthocyanins inhibit lipogenesis during adipocyte differentiation of 3T3-L1 preadipocytes. Plant Foods for Human Nutrition 69, 137–141. https://doi.org/10.1007/s11130-014-0407-z.

Leiherer, A., Stoemmer, K., Muendlein, A., Saely, C., Kinz, E., Brandtner, E., Fraunberger, P., Drexel, H., 2016. Quercetin impacts expression of metabolism- and obesity-associated genes in SGBS adipocytes. Nutrients 8, 1–12. https://doi.org/10.3390/nu8050282.

Li, D., Zhang, Y., Liu, Y., Sun, R., Xia, M., 2015. Purified anthocyanin supplementation reduces dyslipidemia, enhances antioxidant capacity, and prevents insulin resistance in diabetic patients. Journal of Nutrition 145, 742–748. https://doi.org/10.3945/jn.114.205674.

Liu, Y.J., Zhan, J., Liu, X.L., Wang, Y., Ji, J., He, Q.Q., 2014. Dietary flavonoids intake and risk of type 2 diabetes: a meta-analysis of prospective cohort studies. Clinical Nutrition 33, 59–63. https://doi.org/10.1016/j.clnu.2013.03.011.

Liu, X., Liu, Y., Huang, Y., Yu, H., Yuan, S., Tang, B., Wang, P., He, Q., 2017. Dietary total flavonoids intake and risk of mortality from all causes and cardiovascular disease in the general population: a systematic review and meta-analysis of cohort studies. Molecular Nutrition and Food Research 61. 1601003 https://doi.org/10.1002/mnfr.201601003.

Mahmoud, A.M., Ashour, M.B., Abdel-Moneim, A., Ahmed, O.M., 2012. Hesperidin and naringin attenuate hyperglycemia-mediated oxidative stress and proinflammatory cytokine production in high fat fed/streptozotocin-induced type 2 diabetic rats. Journal of Diabetes and Its Complications 26, 483–490. https://doi.org/10.1016/j.jdiacomp.2012.06.001.

Marventano, S., Godos, J., Platania, A., Galvano, F., Mistretta, A., Grosso, G., 2017. Mediterranean diet adherence in the Mediterranean healthy eating, aging and lifestyle (MEAL) study cohort. International Journal of Food Sciences and Nutrition 7486, 1–8. https://doi.org/10.1080/09637486.2017.1332170.

Menezes, R., Rodriguez-Mateos, A., Kaltsatou, A., González-Sarrías, A., Greyling, A., Giannaki, C., Andres-Lacueva, C., Milenkovic, D., Gibney, E.R., Dumont, J., Schär, M., Garcia-Aloy, M., Palma-Duran, S.A., Ruskovska, T., Maksimova, V., Combet, E., Pinto, P., 2017. Impact of flavonols on cardiometabolic biomarkers: a meta-analysis of randomized controlled human trials to explore the role of inter-individual variability. Nutrients 9. https://doi.org/10.3390/nu9020117.

Millar, C.L., Duclos, Q., Blesso, C.N., 2017. Effects of dietary flavonoids on reverse cholesterol transport, HDL metabolism, and HDL function. Advances in Nutrition An International Review of Journal 8, 226–239. https://doi.org/10.3945/an.116.014050.

Miranda, A.M., Steluti, J., Fisberg, R.M., Marchioni, D.M., 2016. Association between polyphenol intake and hypertension in adults and older adults: a population-based study in Brazil. PLoS One 11, 1–14. https://doi.org/10.1371/journal.pone.0165791.

Most, J., Goossens, G.H., Jocken, J.W.E., Blaak, E.E., 2014. Short-term supplementation with a specific combination of dietary polyphenols increases energy expenditure and alters substrate metabolism in overweight subjects. International Journal of Obesity 38, 698–706. https://doi.org/10.1038/ijo.2013.231.

Niemann, B., Rohrbach, S., Miller, M.R., Newby, D.E., Fuster, V., Kovacic, J.C., 2017. Oxidative stress and cardiovascular risk: obesity, diabetes, smoking, and pollution. Journal of the American College of Cardiology 70, 230–251. https://doi.org/10.1016/j.jacc.2017.05.043.

Nizamutdinova, I.T., Jin, Y.C., Chung, J.Il, Shin, S.C., Lee, S.J., Seo, H.G., Lee, J.H., Chang, K.C., Kim, H.J., 2009. The anti-diabetic effect of anthocyanins in streptozotocin-induced diabetic rats through glucose transporter 4 regulation and prevention of insulin resistance and pancreatic apoptosis. Molecular Nutrition and Food Research 53, 1419–1429. https://doi.org/10.1002/mnfr.200800526.

Ohara, T., Muroyama, K., Yamamoto, Y., Murosaki, S., 2015. Oral intake of a combination of glucosyl hesperidin and caffeine elicits an anti-obesity effect in healthy, moderately obese subjects: a randomized double-blind placebo-controlled trial. Nutrition Journal 15 (6). https://doi.org/10.1186/s12937-016-0123-7.

Paixão, J., Dinis, T.C.P., Almeida, L.M., 2012. Protective role of malvidin-3-glucoside on peroxynitrite-induced damage in endothelial cells by counteracting reactive species formation and apoptotic mitochondrial pathway. Oxidative Medicine and Cellular Longevity 2012, 1–12. https://doi.org/10.1155/2012/428538.

Panchal, S.K., Poudyal, H., Brown, L., 2012. Quercetin ameliorates cardiovascular, hepatic, and metabolic changes in diet-induced metabolic syndrome in rats. Journal of Nutrition 142, 1026–1032. https://doi.org/10.3945/jn.111.157263.

Phie, J., Krishna, S.M., Moxon, J.V., Omer, S.M., Kinobe, R., Golledge, J., 2017. Flavonols reduce aortic atherosclerosis lesion area in apolipoprotein E deficient mice: a systematic review and meta-analysis. PLoS One 12, 1–13. https://doi.org/10.1371/journal.pone.0181832.

Porras, D., Nistal, E., Martínez-Flórez, S., Pisonero-Vaquero, S., Olcoz, J.L., Jover, R., González-Gallego, J., García-Mediavilla, M.V., Sánchez-Campos, S., 2017. Protective effect of quercetin on high-fat diet-induced non-alcoholic fatty liver disease in mice is mediated by modulating intestinal microbiota imbalance and related gut-liver axis activation. Free Radical Biology and Medicine 102, 188–202. https://doi.org/10.1016/j.freeradbiomed.2016.11.037.

Prasath, G.S., Subramanian, S.P., 2011. Modulatory effects of fisetin, a bioflavonoid, on hyperglycemia by attenuating the key enzymes of carbohydrate metabolism in hepatic and renal tissues in streptozotocin-induced diabetic rats. European Journal of Pharmacology 668, 492–496. https://doi.org/10.1016/j.ejphar.2011.07.021.

Priyadarsini, R.V., Nagini, S., 2012. Quercetin suppresses cytochrome P450 mediated ROS generation and NFκB activation to inhibit the development of 7,12-dimethylbenz[a]anthracene (DMBA) induced hamster buccal pouch carcinomas. Free Radical Research 46, 41–49. https://doi.org/10.3109/10715762.2011.637204.

Quideau, S., Deffieux, D., Douat-Casassus, C., Pouységu, L., 2011. Plant polyphenols: chemical properties, biological activities, and synthesis. Angewandte Chemie International Edition 50, 586–621. https://doi.org/10.1002/anie.201000044.

Ramos, A.A., Pereira-Wilson, C., Collins, A.R., 2010. Protective effects of Ursolic acid and Luteolin against oxidative DNA damage include enhancement of DNA repair in Caco-2 cells. Mutation Research/Fundamental and Molecular Mechanisms of Mutagenesis 692, 6–11. https://doi.org/10.1016/J.MRFMMM.2010.07.004.

Rizza, S., Muniyappa, R., Iantorno, M., Kim, J.A., Chen, H., Pullikotil, P., Senese, N., Tesauro, M., Lauro, D., Cardillo, C., Quon, M.J., 2011. Citrus polyphenol hesperidin stimulates production of nitric oxide in endothelial cells while improving endothelial function and reducing inflammatory markers in patients with metabolic syndrome. The Journal of Clinical Endocrinology and Metabolism 96, 782–792. https://doi.org/10.1210/jc.2010-2879.

Sharma, H., Kanwal, R., Bhaskaran, N., Gupta, S., 2014. Plant flavone apigenin binds to nucleic acid bases and reduces oxidative DNA damage in prostate epithelial cells. PLoS One 9, e91588. https://doi.org/10.1371/journal.pone.0091588.

Singh, R., Devi, S., Gollen, R., 2015. Role of free radical in atherosclerosis, diabetes and dyslipidaemia: larger-than-life. Diabetes/Metabolism Research and Reviews 31, 113–126. https://doi.org/10.1002/dmrr.2558.

Singh, R.K., Chang, H.-W., Yan, D., Lee, K.M., Ucmak, D., Wong, K., Abrouk, M., Farahnik, B., Nakamura, M., Zhu, T.H., Bhutani, T., Liao, W., 2017. Influence of diet on the gut microbiome and implications for human health. Journal of Translational Medicine 15 (73). https://doi.org/10.1186/s12967-017-1175-y.

Song, L., Yang, H., Wang, H.X., Tian, C., Liu, Y., Zeng, X.J., Gao, E., Kang, Y.M., Du, J., Li, H.H., 2014. Inhibition of 12/15 lipoxygenase by baicalein reduces myocardial ischemia/reperfusion injury via modulation of multiple signaling pathways. Apoptosis 19, 567–580. https://doi.org/10.1007/s10495-013-0946-z.

Tang, Z., Li, M., Zhang, X., Hou, W., 2016. Dietary flavonoid intake and the risk of stroke: a dose-response meta-analysis of prospective cohort studies. BMJ Open 6. https://doi.org/10.1136/bmjopen-2015-008680.

Tsao, R., 2010. Chemistry and biochemistry of dietary polyphenols. Nutrients 2, 1231–1246. https://doi.org/10.3390/nu2121231.

Utian, W.H., Jones, M., Setchell, K.D.R., 2015. S-equol: a potential nonhormonal agent for menopause-related symptom relief. Journal of Women's Heal 24, 200–208. https://doi.org/10.1089/jwh.2014.5006.

Vogiatzoglou, A., Mulligan, A.A., Lentjes, M.A.H., Luben, R.N., Spencer, J.P.E., Schroeter, H., Khaw, K.T., Kuhnle, G.G.C., 2015. Flavonoid intake in European adults (18 to 64 Years). PLoS One 10. https://doi.org/10.1371/journal.pone.0128132.

Wang, X.-H., You, Y.-P., 2016. Epigallocatechin gallate extends therapeutic window of recombinant tissue plasminogen activator treatment for brain ischemic stroke. Clinical Neuropharmacology 40 (1). https://doi.org/10.1097/WNF.0000000000000197.

Wang, Q., Wu, T., Chen, X.Y., Duan, X., Zheng, J., Qiao, J., Zhou, L., Wei, J., Ni, J., 2006. Puerarin injection for unstable angina pectoris. In: Wu, T. (Ed.), Cochrane Database of Systematic Reviews. John Wiley & Sons, Ltd, Chichester, UK. https://doi.org/10.1002/14651858.CD004196.pub2.

Wang, D., Wei, X., Yan, X., Jin, T., Ling, W., 2010. Protocatechuic acid, a metabolite of anthocyanins, inhibits monocyte adhesion and reduces atherosclerosis in apolipoprotein E-deficient mice. Journal of Agricultural and Food Chemistry 58, 12722–12728. https://doi.org/10.1021/jf103427j.

Williamson, G., 2013. Possible effects of dietary polyphenols on sugar absorption and digestion. Molecular Nutrition and Food Research 57, 48–57. https://doi.org/10.1002/mnfr.201200511.

Xiao, Z.-P., Peng, Z.-Y., Peng, M.-J., Yan, W.-B., Ouyang, Y.-Z., Zhu, H.-L., 2011. Flavonoids health benefits and their molecular mechanism. Mini-Reviews in Medicinal Chemistry 11, 169–177. https://doi.org/10.2174/138955711794519546.

Xiao, L., Liu, L., Guo, X., Zhang, S., Wang, J., 2017. Quercetin attenuates high fat diet-induced atherosclerosis in apolipoprotein E knockout mice : a critical role of NADPH oxidase. Food and Chemical Toxicology 105, 22–33. https://doi.org/10.1016/j.fct.2017.03.048.

Yamashita, Y., Wang, L., Nanba, F., Ito, C., Toda, T., Ashida, H., 2016. Procyanidin promotes translocation of glucose transporter 4 in muscle of mice through activation of insulin and AMPK signaling pathways. PLoS One 11, 1–19. https://doi.org/10.1371/journal.pone.0161704.

Yan, J., Zhang, G., Hu, Y., Ma, Y., 2013. Effect of luteolin on xanthine oxidase: inhibition kinetics and interaction mechanism merging with docking simulation. Food Chemistry 141, 3766–3773. https://doi.org/10.1016/j.foodchem.2013.06.092.

You, Y., Yuan, X., Liu, X., Liang, C., Meng, M., Huang, Y., Han, X., Guo, J., Guo, Y., Ren, C., Zhang, Q., Sun, X., Ma, T., Liu, G., Jin, W., Huang, W., Zhan, J., 2017. Cyanidin-3-glucoside increases whole body energy metabolism by upregulating brown adipose tissue mitochondrial function. Molecular Nutrition and Food Research 61. 1700261 https://doi.org/10.1002/mnfr.201700261.

Zamora-Ros, R., Andres-Lacueva, C., Lamuela-Raventós, R.M., Berenguer, T., Jakszyn, P., Barricarte, A., Ardanaz, E., Amiano, P., Dorronsoro, M., Larrañaga, N., Martínez, C., Sánchez, M.J., Navarro, C., Chirlaque, M.D., Tormo, M.J., Quirós, J.R., González, C.A., 2010. Estimation of dietary sources and flavonoid intake in a Spanish adult population (EPIC-Spain). Journal of the American Dietetic Association 110, 390–398. https://doi.org/10.1016/j.jada.2009.11.024.

Zhang, Y., Na, X., Zhang, Y., Li, L., Zhao, X., Cui, H., 2009. Isoflavone reduces body weight by decreasing food intake in ovariectomized rats. Annals of Nutrition and Metabolism 54, 163–170. https://doi.org/10.1159/000217812.

Zhang, Z., Ding, Y., Dai, X., Wang, J., Li, Y., 2011. Epigallocatechin-3-gallate protects pro-inflammatory cytokine induced injuries in insulin-producing cells through the mitochondrial pathway. European Journal of Pharmacology 670, 311–316. https://doi.org/10.1016/j.ejphar.2011.08.033.

Zhu, Y., Ling, W., Guo, H., Song, F., Ye, Q., Zou, T., Li, D., Zhang, Y., Li, G., Xiao, Y., Liu, F., Li, Z., Shi, Z., Yang, Y., 2013. Anti-inflammatory effect of purified dietary anthocyanin in adults with hypercholesterolemia: a randomized controlled trial. Nutrition, Metabolism and Cardiovascular Diseases 23, 843–849. https://doi.org/10.1016/j.numecd.2012.06.005.

Zhu, W., Xu, J., Ge, Y., Cao, H., Ge, X., Luo, J., Xue, J., Yang, H., Zhang, S., Cao, J., 2014. Epigallocatechin-3-gallate (EGCG) protects skin cells from ionizing radiation via heme oxygenase-1 (HO-1) overexpression. Journal of Radiation Research 55, 1056–1065. https://doi.org/10.1093/jrr/rru047.

Zhu, W., Jia, L., Chen, G., Zhao, H., Sun, X., Meng, X., Zhao, X., Xing, L., Yu, J., Zheng, M., Zhu, W., Jia, L., Chen, G., Zhao, H., Sun, X., Meng, X., Zhao, X., Xing, L., Yu, J., Zheng, M., 2016. Epigallocatechin-3-gallate ameliorates radiation-induced acute skin damage in breast cancer patients undergoing adjuvant radiotherapy. Oncotarget 7, 48607–48613. https://doi.org/10.18632/oncotarget.9495.

PART 8

Health Benefits of Functional Foods

CHAPTER 11

Functional Food Consumption and Its Physiological Effects

Laís Marinho Aguiar[1], Marina Vilar Geraldi[1], Cínthia Baú Betim Cazarin[2], Mário Roberto Maróstica Junior[2]

[1]*University of Campinas, School of Food Engineering, Department of Food and Nutrition, Campinas, Brazil;* [2]*Department of Food and Nutrition, Faculty of Food Engineering, University of Campinas, Campinas, Brazil*

11.1 Introduction

Hippocrates, the Greek physician father of Western medicine, left a well-known phrase ~2500 years ago: "Let food be your medicine and medicine your food," starting the interest in the physiological active of specific food components. It is well known that besides the nutrients needed for body nutrition there are many non-nutrients in foods that exert important functions, especially related to the prevention of some diseases. In regards to this topic a new concept was created.

The concept of functional foods was firstly introduced in Japan in the 1980s. Research programs were founded by the Japanese government, a national effort to reduce the growing costs of health care. Foods for Specific Health Use (FOSHU) was established in 1991, referring to foods that demonstrated physiological benefits or reduced disease risks, in addition to performing their normal basic functions (Ashwell, 2002).

The functional foods category is not recognized by United States regulations and has no universally accepted definition. The Institute of Medicine's Food and Nutrition Board (IOM/FNB) defined functional foods as "any food or food ingredient that may provide a health benefit beyond the traditional nutrients it contains" (Committee on Opportunities in the Nutrition and Food Sciences and Nutrition Board, 1994). The American Dietetic Association (ADA) defined functional foods, including "whole foods and fortified, enriched, or enhanced foods" that have a "potentially beneficial effect on health when consumed as part of a varied diet on a regular basis, at effective levels" (Thomson et al., 1999). The International Life Sciences Institute defines them as "foods that, by virtue of the presence of physiologically-active components, provide a health benefit beyond basic nutrition" (International Life Sciences Institute, 1999).

During the first half of the 20th century, scientists established nutritional reference values, dietary guidelines, and food guides, with the objective of preventing deficiencies and

Bioactive Compounds. https://doi.org/10.1016/B978-0-12-814774-0.00011-6

promoting adequate growth. The US Food and Drug Administration noticed health benefits related to fruit, vegetable, and grain intakes, especially in decreasing the risk of development of some diseases. In addition, in the last few years, researchers have been identifying the physiological actions of some specific food components, recognized as phytochemicals.

The interest in functional foods is growing, and the 21st century faces a world in deep transformation with new challenges, longer life expectancy, rising healthcare costs, rapid advances in science and technology, changes in lifestyle, and concern over the quality of life. The scientific community continues to increase its understanding to improve the quality of diet, focusing on its contents of nutrients and non-nutrients that play a role in physiological and biochemical functions to achieve an "optimal nutrition" (Ashwell, 2002).

11.2 Potential Health Areas of Interest for Functional Food

Functional foods affect biological responses in the body, promoting health benefits in some important areas of human physiology. An explanation of each will be given for the current physiological areas of interest: cancer prevention, gastrointestinal health, cardiovascular health, cognition and neurodegenerative diseases, and cardiometabolic syndrome (Fig. 11.1).

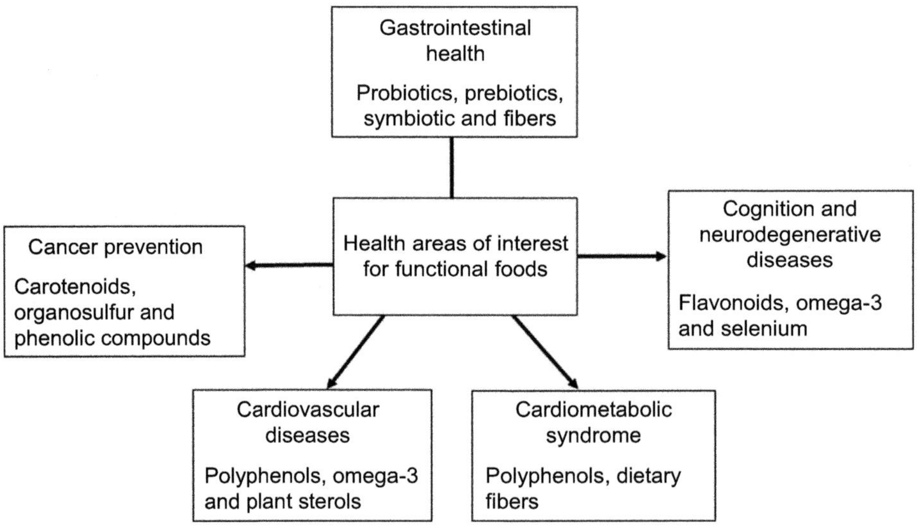

Figure 11.1
Health areas of interest for functional foods.

11.2.1 Functional Foods and Gut Health

The gut is considered to be the largest internal organ of the body, besides acting as a barrier against pathogens and intestinal lumen antigens (Gatt et al., 2007). In recent years, interest in research involving the human microbiota has increased, proving that it produces metabolites that play a key role in the host's immune system through a complex series of chemical interactions and signaling pathways (Sawicki et al., 2017).

The intestinal microbiota develops after birth; some factors, such as the mode of birth, infant nutrition, antibiotic use, diet, and age determine its colonization rate (Montalto et al., 2009). The distribution of different strains or species of bacteria within the gut will determine the metabolic profile of the microbiota, which could have potential physiological effects on health (Flint et al., 2015).

Eating habits, food consumption, and lifestyle have health impacts. In this way, some gut diseases result from an imbalance of intestinal microbiota, and are related to diet, therefore diet has implications on gut health (Cencic and Chingwaru, 2010). Some functional foods containing prebiotics, probiotics, synbiotics, and fibers have been used to promote healthier microbiota and better gut function (Tur and Bibiloni, 2016).

Probiotics are live microorganisms belonging to natural biota with low or no pathogenicity, but with functions of importance for the health and well-being of the host. Lactobacilli and bifidobacteria are the bacterial genera most often used as probiotics. The consumption of probiotics is more associated with the consumption of dairy products, such as yogurts and cheeses. This practice is already very popular in Japan and Europe (Roberfroid, 2000). Due to the benefits provided by these microorganisms, the food industry seeks to incorporate them into fermented products, other than just dairy products, such as fruits and vegetables, beverages, and breads (Di Cagno et al., 2016; Hittinger et al., 2018; Marco et al., 2017).

The use of probiotics can be positive for human health, modifying the intestinal microbiota, correcting the dysbiosis. Some probiotics have the ability to increase the production of lactate and short-chain fatty acids (SCFAs) and lower the pH of the intestinal lumen, which increases peristaltic movements and decreases the intestinal transit time, and other benefits as described below in the paragraphs about prebiotics (Dimidi et al., 2014).

The results from clinical studies have not been conclusive in that the effects of probiotics on the host are dependent on probiotic strain, type of infection, dose used, and duration of treatment but, the usual effective dosage in humans is 10^7–10^9 colony-forming units/mg per day (Minelli and Benini, 2008).

Prebiotics are fermented ingredients that result in specific changes in the composition and/or activity of the intestinal microbiota. The most recognized prebiotics as functional food ingredients are inulin-type fructans, which include native inulin, enzymatically hydrolyzed inulin or oligofructose, and synthetic fructooligosaccharides (Roberfroid and Delzenne, 1998). The most common natural sources are wheat, onion, banana, garlic, and leek (Van Loo et al., 1995).

Inulin and oligofructose exhibit functional attributes, including modulation of the gut microbiota, prevention of pathogen adhesion and colonization, induction of antiinflammatory effects, reduction of food intake, modulation of bowel habits, and regulation of alterations in lipid and glucose metabolism (Kleessen et al., 2001). The effects of these prebiotics on immune functions may be due to the induced changes in the gut microbiota and/or to the effects of the SCFAs generated and their binding to their receptors on leukocytes (Watzl et al., 2005). In addition, inulin and inulin-type fructans are considered dietary soluble fiber, and directly modulate bowel habits by slowing gastric emptying and intestine transit time, delaying absorption of glucose, and improving alterations in glucose metabolism (Laparra and Sanz, 2010).

Studies report a variable amount between 4 and 24 g of inulin alone or in combination. However, there is no consensus on the dose and the time of use to obtain benefits (Cani et al., 2009; Heap et al., 2015).

Probiotics and prebiotics share unique roles in human nutrition, largely centered on manipulation of populations or activities of the microbiota that colonize the human gastrointestinal tract (Douglas and Sanders, 2008). Regular consumption of probiotics or prebiotics has health implications that include enhanced immune function, improved colonic integrity, decreased incidence and duration of intestinal infections, downregulated allergic response, and improved digestion and elimination (Cencic and Chingwaru, 2010).

Furthermore, dietary fiber, including some nonstarch polysaccharides (cellulose, dextrins, chitins, pectins, beta-glucans, and lignin), can modulate the transit time through the gut providing similar beneficial effects to inulin-type fructans. These compounds are found in many foods such as cereals, nuts, oats, chia, etc. They are also partially susceptible to bacterial fermentation and may induce changes in bacterial populations, particularly in the number of bifidobacteria and lactobacilli. These dietary soluble fibers have been shown to exert additional beneficial effects, which could be partially a consequence of their effect on the microbiota composition (Laparra and Sanz, 2010).

Butyrate, acetate, and propionate, produced by the fermentation of dietary fibers, may play a role in energy homeostasis, immune function, and host–microbe signaling and prevention of diseases, such as bowel disease (Sawicki et al., 2017). Therefore, fiber-induced modulation of the gut microbiota has gained interest for its potential impact on health and disease (Flint et al., 2012).

11.2.2 Functional Foods in Cancer Prevention

Cancer is characterized by an interaction between some cell genes and their neighboring tissues, leading to the gradual conversion of healthy cells into cancerous cells (Mao et al., 2017). This is regarded as a preventable disease as 90%–95% have been linked to lifestyle factors and environment, including dietary habits (Aggarwal et al., 2009). Thus, rational dietary habits and behaviors, and consumption of sufficient amounts of antioxidants and bioactive plant-derived compounds, that have been demonstrated to have protective effects against carcinogenesis in preclinical and clinical studies may be the best way to prevent cancer (Willett, 1995).

An epidemiologic study showed that diet can modify carcinogenesis (Balsano and Alisi, 2009). The anticancer action of fruits and vegetables can be attributed to the presence of phytochemicals, which can act by increasing the activity of enzymes that detoxify carcinogens, inhibiting N-nitrosamine formation, altering estrogen metabolism, increasing apoptosis of cancer cells, and decreasing cell proliferation, as well as effects on cell differentiation (Gul et al., 2016; Roleira et al., 2015).

Some groups of phytochemicals are the most studied in cancer prevention, such as carotenoids, phenolics, and organosulfur compounds. Functional foods (e.g., garlic, tea, tomato, Brussels sprouts) have been associated with them (Gul et al., 2016).

Organosulfur compounds (OSCs) are phytochemicals of the *Allium genus*. In vitro and in vivo experimental studies demonstrated that OSCs have apoptotic effects which would protect against critical events that are involved in the cancer process (Arranz et al., 2007) and have been shown to inhibit tumorigenesis in several experimental models (Rafter, 2002).

Garlic and onion are rich in special and diverse OSCs, including diallyl sulfide (DAS), diallyl disulfide (DADS), and diallyl trisulfide (DATS), among which DATS has shown the highest biological activities. Studies have shown that dietary intake of garlic can decrease the risk of stomach and colorectal cancers (Jiang et al., 2017).

The anticancer action of *Allium* derivatives can be explained by the fact that these compounds inhibit the action of the cytochrome P4502E1, which is necessary for the activation of carcinogenic substances. In addition, *Allium* compounds improve the detoxification process by inducing phase II enzymes, such as glutathione S-transferases (GSTs), quinone reductase, and epoxide hydrolase. In addition, it stimulates the glutathione (GSH), which acts as an intracellular antioxidant (Herman-Antosiewicz and Singh, 2004).

Carotenoids are divided into two groups: carotenes and xanthophylls. These compounds are responsible for the color of some vegetables and fruits and receive considerable attention because of their unique physiological functions as provitamins and antioxidant effects, especially in scavenging singlet oxygen (Liu, 2013). For functional nutrition, beta-carotene and lycopene play a prominent role in relation to cancer prevention (Shami and Moreira, 2004).

Lycopene is a red pigment that occurs naturally in vegetable tissues. It is considered the most efficient antioxidant among all carotenoids, with twice the activity of beta-carotene. It is found in greater quantities in the peel of the food, increasing with maturation (Shami and Moreira, 2004).

The antioxidant property of lycopene is most likely the basis for its preventive role toward cancer, but others activities, including regulation of growth factor signaling, cell cycle arrest, and/or apoptosis induction, and changes in antioxidant and phase II detoxifying enzymes occur. Besides the antiinflammatory activity of lycopene, it is as an important determinant which suppresses the promotion and progression of carcinogenesis (Rafter, 2002; Trejo-Solis et al., 2013).

Studies show an inverse relationship between consumption of tomato and prostate cancer. However, there are inconsistent results on the real action of lycopene present in tomatoes and its protective action on cancer (Chen et al., 2015).

Catechins are found in red wine and chocolate; but green tea is the richest source. The catechins are the predominant and most significant of all tea polyphenols (Gul et al., 2016).

Epigallocatechin-3-gallate (EGCG), epicatechin gallate (ECG), epigallocatechin (EGC) and epicatechin (EC), gallocatechin gallate (GCG), catechin gallate (CG), and catechin (CT) are the main catechins found in green tea but EGCC is the major one (Hayat et al., 2015). This catechin can increase the junctional communication between cells, which could protect them from tumor development (Rashidi et al., 2017).

The literature has related that polyphenols present in tea are powerful antioxidants that induce phase 2 detoxification enzymes, which in turn reduce the risk of cancer by reducing damage to DNA in the cell and activation of cancer leading to malignancy (Hayat et al., 2015).

It is necessary to consider the preparation process of the tea, as well as the type of tea used, as the infusion time influences the final concentration of phenolic compounds (Cabrera et al., 2006).

High-dose oral green tea extract and EGCG could be related to hepatotoxic effects, however, a quantity of 800 mg per day of EGCG for up to 4 weeks could be safe and well-tolerated. Thus, green tea and its components, such as catechins, could be an option for cancer prevention (Rashidi et al., 2017).

Cruciferous vegetables (cauliflower, Brussels sprouts, and broccoli) are rich in glucosinolate which are degraded, releasing indoles and isothiocyanates, which show anticarcinogenic properties (Shapiro et al., 2001).

Indoles favor the production of enzymes that inhibit estrogen activity. Therefore, they reduce the risk of breast and uterine cancer, which are estrogen-dependent (Mills et al., 2003).

Isothiocyanate inhibits the metabolism and DNA attack of carcinogenic substances, such as nitrosamines (Mills et al., 2003). Sulforaphane, which is an isothiocyanate, has phytochemicals that induce phase 2 detoxification enzymes and bolster antioxidant activities in cells (Shapiro et al., 2001).

In general, the association between the consumption of brassica vegetables and the risk of cancer appears to be most consistent for lung, stomach, colon, and rectal cancers, and least consistent for prostatic, endometrial, and ovarian cancers (Verhoeven et al., 1996).

11.2.3 Functional Foods and Cardiometabolic Syndrome

Cardiometabolic syndrome is characterized by the presence of obesity, dyslipidemia, hypertension, and hyperglycemia. Inadequate eating habits, sedentary lifestyle, and weight gain strongly influence obesity and the development of other pathologies related to the cardiometabolic syndrome. Therefore, dietary components can prevent cardiometabolic syndrome or reduce symptoms (Mohamed, 2014). Three to four servings per day of fruits and vegetables was associated with a lower risk of disease (Miller et al., 2017).

The functional properties of the main food groups related to cardiometabolic syndrome are described in Table 11.1.

11.2.3.1 Obesity

Obesity is characterized by fat accumulation that results from complex interactions of genetic, behavioral, and environmental factors correlating with economic and social status and lifestyle (Ordovas and Shen, 2009).

A functional food that can act in the prevention of obesity should be able to regulate appetite and satiety, thus promoting adequate energy consumption (Myrie and Jones, 2011) and suppression of growth of adipose tissue by modulating adipocyte metabolism (Badimon et al., 2010).

The polyphenols present in many plants are related to the prevention of obesity (Mir et al., 2017). These compounds have the capacity to act as antioxidants and neutralizers of singlet oxygen, attenuating the deleterious effects of exacerbated ROS production, as well as obesity-associated inflammation (Wang et al., 2014a,b). They also act in induction of lipolysis, decrease lipid accumulation, and induce apoptosis in adipose tissue (Williams et al., 2013).

In an animal model investigation, chlorogenic and coumaric acid led to inhibition of cell growth and increased apoptosis and gallic acid, while enhancing the number of apoptotic cells, but did not affect the adipocyte cell cycle (Hsu and Yen, 2006). A study showed that resveratrol increased the level of GLP-1, a peptide involved with appetite, in the serum (Ordovas and Shen, 2009).

Table 11.1: Food Groups and Functional Properties in Cardiometabolic Syndrome

Characteristic of the Cardiometabolic Syndrome	Food Groups	Action	References
Obesity	Fruits and vegetables	• Promotes weight loss • Retards gastric emptying • Increases satiety	Hsu and Yen (2006), Lai et al. (2015) and Williams et al. (2013)
	Fish and fish oil	Increases lipid oxidation, reducing the accumulation of body fat	Chiu et al. (2017) and Pahlavani et al. (2017)
Insulin resistance	Olive oil and fish oil	• Increases insulin sensitivity	Kazeem and Davies (2016) and Rudkowska (2009)
	Vegetables	• Decreases postprandial blood glucose • Improves glycemic control and reduces the risk of diabetes	Beidokhti and Jäger (2017) and Bi et al. (2017)
Cardiovascular disease	Fish oil	Triglyceride-lowering benefits	Maehre et al. (2015)
	Plant sterols	Reduce levels of LDL	Calpe-Berdiel et al. (2009) and García-Llatas and Rodríguez-Estrada (2011)
	Polyphenols	• Lowering blood pressure • Improving blood vessel endothelial function • Improvement antioxidant and antiinflammatory activities	Tomé-Carneiro and Visioli (2016)

11.2.3.2 Diabetes

Diabetes is a modern epidemic, whose incidence is rapidly increasing, although it is one of the world's oldest diseases (Lakhtakia, 2010). The main clinical characteristic of diabetes is hyperglycemia but, when untreated, it can cause complications in organs such as the eyes and kidneys (Forbes and Cooper, 2013).

In general, foods that are related to diabetes can be divided into three groups, according to Table 11.2.

Dietary fiber intake is also associated with a reduced risk of diabetes. This is due to the action of fibers in the gastrointestinal tract, where they affect nutrient absorption and decrease the postprandial glucose response (American Diabetes Association, 2008).

Yacon, a tuberous root, presents many phytochemical compounds in its composition (caffeic acid, ferulic acid, chlorogenic acid), in addition to high amounts of water and Fructooligosaccharides (FOS) (Valentová et al., 2004). Rats supplemented with yacon showed improvement in insulin response after 5 weeks of root use (Satoh et al., 2013).

The antidiabetic activity of pumpkin has drawn attention in recent years. Studies show that pumpkin pulp, seeds, and oil may have a protective action on health, such as hypoglycemic properties, indicating it to be a beneficial food for diabetic patients (Williams et al., 2013).

Table 11.2: List of Functional Food Plants With Effects on Insulin

Functional Effects	Functional Food	References
Stimulation in the insulin secretion	Bitter melon	Keller et al. (2011)
	Chicory	Azay-Milhau et al. (2013)
	Yerba mate	Arçari et al. (2011)
	Spiral ginger	Ashwini et al. (2015)
Improvement in the response of insulin towards glucose	Flaxseed	Wang et al. (2015)
	Yacon	Satoh et al. (2013)
	Olive	Bock et al. (2013)
	Sweet potato	Chen et al. (2013)
Mimic the action of insulin	Turmeric	Mohankumar and McFarlane (2017)
	Pummelo	Rao et al. (2011)
	Pumpkin	Chang et al. (2014)

Cinnamon is associated with diabetes control. Men and women with metabolic syndrome supplemented with cinnamon showed a reduction in fasting glycemia and an improvement in body composition (Ziegenfuss et al., 2006). Experiments in vivo showed that cinnamon extract suppressed maltose- and sucrose-induced postprandial blood glucose in rats (Shihabudeen et al., 2011). Furthermore, it has been suggested that cinnamon has a potential role in the prevention of insulin resistance (Qin et al., 2010) and in increasing liver glycogen through regulating insulin signaling (Couturier et al., 2011). However, more studies are needed for a more specific recommendation.

In addition, green tea and black tea have significant activity in decreasing blood glucose levels, possessing preventive effects on diabetes in rats. Therefore, consumption of green tea or its main ingredient, catechin, is effective in lowering blood glucose levels in people and animals (Beidokhti and Jäger, 2017).

11.2.3.3 Cardiovascular Health

Cardiovascular disease (CVD) remains the number one cause of global mortality and morbidity, with an increasing prevalence. Major events are myocardial infarction, stroke, and atherosclerosis (Martínez-Augustin et al., 2012). The medical costs to the economy and health care need an alternative approach in treatment/prevention of the development of heart disease. Traditional lifestyle measures include smoking cessation, healthy body weight, regular exercise, and a balanced diet, rich in fruits and vegetables (Mente et al., 2009).

In addition to drug treatments, functional foods are been adding as an adjunct treatment to CVD or as a preventive in high-risk patients (Tomé-Carneiro and Visioli, 2016). Antioxidant-rich foods, mainly plant flavonoids, have potential cardiovascular protection. Epidemiological studies suggest the preventive potential of polyphenols present in cocoa, berries, grape, tea, coffee, and soy, in the incidence and risk factors to CVD (Arranz et al., 2013; Basu et al., 2010; Blumberg et al., 2015; Riso et al., 2013; Sarriá et al., 2015).

The production of reactive oxygen species (ROS) and oxidative stress are involved in endothelial damage, progression to atherosclerosis, myocardial infarction, and ischemia (Dhalla et al. 2000; Raedschelders et al., 2012). The effects of flavonoids are due to improvement of antioxidant defenses and antiinflammatory activities (Rodriguez-Mateos et al., 2013), lowering blood pressure, oxidation of low-density lipoproteins (LDL)-cholesterol, and improving blood vessel endothelial functions (Desch et al., 2010; Erlund et al., 2008; Hooper et al., 2008; Mathur et al., 2002; Widlansky et al., 2007). Results from recent meta-analyses and cohort studies have shown an inverse association between total flavonoid intake and incidence of CVD, coronary heart disease, and mortality in these diseases (Ivey et al., 2015; Jiang et al., 2015; Wang et al., 2014).

Studies have shown a correlation between high plasma triglycerides and cardiovascular risk (Hokanson and Austin, 1996). The triglyceride-lowering benefits of the very long-chain omega-3 fatty acids (O3FAs) include docosahexaenoic acid (DHA) and eicosapentaenoic acid (EPA), and are well documented.

The main dietary source of EPA and DHA is fatty fish, such as albacore tuna, salmon, mackerel, sardines, and herring, and the recommended consumption is 2–3 servings per week (Tur et al., 2012). After consumption, O3FAs are incorporated into cell membranes, where they modulate membrane protein function, cellular signaling, and gene expression. O3FA dietary intakes promote cardioprotective effects by reduction of triglyceride levels, attenuation of atherosclerotic plaques, the exertion of antidysrhythmic, antithrombotic, and antiinflammatory effects, lowering systolic and diastolic blood pressures and having an improvement in endothelial function (Bradberry and Hilleman, 2013). In addition, O3FAs play antiinflammatory and immunomodulatory roles via the attenuation of eicosanoids and leukotrienes, cytokines, oxidative stress, and altering endothelial and immune cell function, resulting in less inflammatory mediator (Calder, 2013).

However, recently scientific evidence has focused on fish and seafood instead of supplements as O3FAs sources. The benefits of supplementation have diminished, probably due to an increase in seafood consumption, uptake, and incorporation from seafood, which is better than from supplements and because food sources are rich in other nutrients that may provide synergetic effects (Maehre et al., 2015).

Plant sterols (PS) are natural components of plant cell membranes, that include phytosterols and phytostanols present in vegetable oil, nuts, seeds, and grains. Phytosterol has been proved to reduce levels of LDL cholesterol concentration and have a potential contribution to lowering the risk of CVD (Alemany et al., 2014), by reducing cholesterol absorption. The presence of a dietary intake competes with intestinal cholesterol for incorporation into micelles and chylomicrons to be absorbed into the bloodstream. Reduced absorption benefits feedback-regulation of enterohepatic cholesterol, resulting in a decrease in serum total and LDL-cholesterol levels. The hypothesis of the mechanisms has been revised, but it has not been completely elucidated. There is competition between cholesterol and PS for esterase activity in the intestine; since PS and cholesterol have similar structures, they may compete for the same transporter in the enterocyte; PS may inhibit the acyl-coenzyme A cholesterol acyltransferase activity inside the enterocyte, decreasing the esterification of cholesterol; and competition for the incorporation into chylomicrons, could be possible explanations (Calpe-Berdiel et al., 2009; García-Llatas and Rodríguez-Estrada, 2011; Marangoni and Poli, 2010; Rozner and Garti, 2006).

11.2.4 Cognition and Neurodegenerative Diseases

Neurodegenerative diseases (NDs) are a group of disorders in the nervous system, characterized by progressive loss of neurons that lead to memory impairment, locomotor dysfunction, cognitive defects, emotional and behavioral problems, as a consequence of environmental, hereditary, and brain aging factors. Alzheimer's disease (AD), Parkinson's disease, multiple sclerosis, Huntington's disease, and amyotrophic lateral sclerosis are the main age-related neurodegenerative diseases caused by neuronal degeneration (Amor et al., 2010). The nervous system has the highest amount of oxygen to produce energy, in this way it is especially vulnerable to the effects of ROS and reactive nitrogen species. The increase in oxidative stress, inflammatory response, activation of neuronal apoptosis, altered cell signaling, and gene expression, plays an important role in the pathogenesis of many disorders that involve neuronal degeneration (Jellinger, 2001).

Epidemiologic evidences have shown that a Mediterranean diet, rich in phenolic compounds, is effective in the prevention of age-related diseases such as AD (Sofi et al., 2010). Dietary flavonoid intake is associated with better preservation of cognitive performance with aging (Letenneur et al., 2007).

Dietary intervention studies in human and animals have shown benefits in the consumption of flavonoid-rich foods by protecting neurons and stimulating neuronal regeneration (Casadesus et al., 2004; Galli et al., 2002). Soy isoflavone supplementation (Casini et al., 2006; File et al., 2005), berries (Casadesus et al., 2004; Williams et al., 2008), green and white tea (Haque et al., 2006; Mandel and Youdim, 2004; Okello et al., 2012), and cocoa flavanols (Francis et al., 2006) have been observed to have positive effects on age-related deficits and cognitive function.

Studies have shown improvements in cognitive function by protecting vulnerable neurons, enhancing existing neuronal function, or by stimulating neuronal regeneration (Youdim and Joseph, 2001). Flavonoids can produce neuroprotective properties by different mechanisms. The neuroprotective actions of dietary flavonoids include their antioxidant capacity to protect neurons against oxidative stress by the inhibitory effect of nicotinamide adenine dinucleotide phosphate (NADPH) oxidase and reduction of ROS production, suppressing neuroinflammation via their inhibiting activity and attenuating the release of cytokines and downregulation of proinflammatory transcription factors, the potential to modulate cell signaling pathways and stimulate neuronal survival via induction of antiapoptotic genes (Solanki et al., 2016; Vauzour et al., 2008).

O3FAs, especially DHA, are associated with better performance and possibly prevention of age-related impairment. A lower plasma concentration of DHA is associated with a cognitive decline in health and in patients with AD (Beydoun et al., 2007; Heude et al., 2003).

Studies have shown an association between O3FA consumption with lower risks of dementia (Barberger-Gateau et al., 2011; Morris, 2016), better performance on neuropsychological tests (D'Ascoli et al., 2016), superior cognitive ability (van Duijn et al., 2016), and a reduction in the marker of neuroinflammation. Studies after the diagnosis of AD suggest no significant effect of O3FAs but support a preventive effect (Joffre et al., 2014; Quinn et al., 2010; Salem et al., 2015).

The essential trace element selenium (Se) has also been identified as playing a role in NDs. Selenoproteins, such as glutathione peroxidase, thioredoxin reductases, and selenoprotein P depend on sufficient availability of Se. Antioxidant selenoproteins protect neurons and astrocytes from oxidative damage, providing protection from ROS. An adequate intake is important for the maintenance of brain function, playing an important role in brain physiology and pathophysiology (Steinbrenner and Sies, 2013).

There is epidemiological evidence that lower Se levels in elderly people are associated with a faster decline in cognitive functions and lower performance in coordination and motor speed, suggesting an association between oxidative stress and Se deficiency (Berr et al., 2012; Gao et al., 2007; Rayman, 2012). However, studies of Se in ND patients are inconclusive, the available data on the Se status and the potential benefit of supplementation for prevention and/or treatment are inconclusive and insufficient (Cardoso et al., 2010; Loef et al., 2011). Future research is required to better assess the benefits of Se supplementation to prevent and delay the progression of NDs.

Dietary habits are one of the most promising modifiable risk factor for NDs. The literature shows that O3FAs, polyphenols, antioxidants, and Se have potential benefits in ND prevention. However, preventive approaches are necessary as when the symptoms are evident the therapeutic approach may be too late to intervene.

11.3 Conclusion

Functional foods contain bioactive components that can promote health, as long as they are combined with a balanced diet and a healthy lifestyle. Concern about health has been increasing among the population and, therefore, there has been a concomitant search for healthier foods. The field of study of functional foods is recent and solid scientific criteria are necessary to establish the real benefits obtained with the consumption of these foods. Therefore, standardized recommendations would be an alternative to evaluating the effects of these foods on health.

References

Aggarwal, B.B., Vijayalekshmi, R.V., Sung, B., 2009. Targeting inflammatory pathways for prevention and therapy of cancer: short-term friend, long-term foe. Clinical Cancer Research 15, 425–430. https://doi.org/10.1158/1078-0432.CCR-08-0149.

Alemany, L., Barbera, R., Alegría, A., Laparra, J.M., 2014. Plant sterols from foods in inflammation and risk of cardiovascular disease: a real threat? Food and Chemical Toxicology https://doi.org/10.1016/j.fct.2014.03.038.

American Diabetes Association, 2008. Nutrition recommendations and interventions for diabetes: a position statement of the American Diabetes Association. Diabetes Care 31. https://doi.org/10.2337/dc08-S061.

Amor, S., Puentes, F., Baker, D., Van Der Valk, P., 2010. Inflammation in neurodegenerative diseases. Immunology 129, 154–169. https://doi.org/10.1111/j.1365-2567.2009.03225.x.

Arçari, D.P., Bartchewsky, W., dos Santos, T.W., Oliveira, K.A., DeOliveira, C.C., Gotardo, É.M., Pedrazzoli, J., Gambero, A., Ferraz, L.F.C., de Carvalho, P.O., Ribeiro, M.L., 2011. Anti-inflammatory effects of yerba maté xtract (Ilex paraguariensis) ameliorate insulin resistance in mice with high fat diet-induced obesity. Molecular and Cellular Endocrinology 335, 110–115. https://doi.org/10.1016/J.MCE.2011.01.003.

Arranz, N., Haza, A.I., García, A., Delgado, E., Rafter, J., Morales, P., 2007. Effects of organosulfurs, isothiocyanates and vitamin C towards hydrogen peroxide-induced oxidative DNA damage (strand breaks and oxidized purines/pyrimidines) in human hepatoma cells. Chemico-Biological Interactions 169, 63–71. https://doi.org/10.1016/j.cbi.2007.05.006.

Arranz, S., Valderas-Martinez, P., Chiva-Blanch, G., Casas, R., Urpi-Sarda, M., Lamuela-Raventos, R.M., Estruch, R., 2013. Cardioprotective effects of cocoa: clinical evidence from randomized clinical intervention trials in humans. Molecular Nutrition and Food Research 57, 936–947. https://doi.org/10.1002/mnfr.201200595.

Ashwell, M., 2002. Concepts of Functional Foods, ILSI Europe Concise Monograph Series. https://doi.org/10.1108/nfs.2004.34.1.47.3.

Ashwini, S., Bobby, Z., Joseph, M., Jacob, S.E., Padmapriya, R., 2015. Insulin plant (Costus pictus) extract improves insulin sensitivity and ameliorates atherogenic dyslipidaemia in fructose induced insulin resistant rats: molecular mechanism. The Journal of Functional Foods 17, 749–760. https://doi.org/10.1016/J.JFF.2015.06.024.

Azay-Milhau, J., Ferrare, K., Leroy, J., Aubaterre, J., Tournier, M., Lajoix, A.-D., Tousch, D., 2013. Antihyperglycemic effect of a natural chicoric acid extract of chicory (Cichorium intybus L.): a comparative in vitro study with the effects of caffeic and ferulic acids. Journal of Ethnopharmacology 150, 755–760. https://doi.org/10.1016/J.JEP.2013.09.046.

Badimon, L., Vilahur, G., Padro, T., 2010. Nutraceuticals and atherosclerosis: human trials. Cardiovascular Therapy 28, 202–215. https://doi.org/10.1111/j.1755-5922.2010.00189.x.

Balsano, C., Alisi, A., 2009. Antioxidant effects of natural bioactive compounds. Current Pharmaceutical Design 15, 3063–3073. https://doi.org/10.2174/138161209789058084.

Barberger-Gateau, P., Samieri, C., Feart, C., Plourde, M., 2011. Dietary omega 3 polyunsaturated fatty acids and Alzheimers disease: interaction with apolipoprotein E genotype. Current Alzheimer Research 8, 479–491. https://doi.org/10.2174/156720511796391926.

Basu, A., Rhone, M., Lyons, T.J., 2010. Berries: emerging impact on cardiovascular health. Nutrition Reviews https://doi.org/10.1111/j.1753-4887.2010.00273.x.

Beidokhti, M.N., Jäger, A.K., 2017. Review of antidiabetic fruits, vegetables, beverages, oils and spices commonly consumed in the diet. Journal of Ethnopharmacology 201, 26–41. https://doi.org/10.1016/j.jep.2017.02.031.

Berr, C., Arnaud, J., Akbaraly, T.N., 2012. Selenium and cognitive impairment: a brief-review based on results from the EVA study. BioFactors 38, 139–144. https://doi.org/10.1002/biof.1003.

Beydoun, M.A., Kaufman, J.S., Satia, J.A., Rosamond, W., Folsom, A.R., 2007. Plasma n-3 fatty acids and the risk of cognitive decline in older adults: the Atherosclerosis Risk in Communities Study. American Journal of Clinical Nutrition 85, 1103–1111.

Bi, X., Lim, J., Henry, C.J., 2017. Spices in the management of diabetes mellitus. Food Chemistry 217, 281–293. https://doi.org/10.1016/j.foodchem.2016.08.111.

Blumberg, J.B., Vita, J.A., Oliver Chen, C.Y., 2015. Concord grape juice polyphenols and cardiovascular risk factors: dose-response relationships. Nutrients 7, 10032–10052. https://doi.org/10.3390/nu7125519.

Bock, M.de, Derraik, J.G.B., Brennan, C.M., Biggs, J.B., Morgan, P.E., Hodgkinson, S.C., Hofman, P.L., Cutfield, W.S., 2013. Olive (Olea europaea L.) leaf polyphenols improve insulin sensitivity in middle-aged overweight men: a randomized, placebo-controlled, crossover trial. PLoS One 8, e57622.

Bradberry, J.C., Hilleman, D.E., 2013. Overview of omega-3 fatty acid therapies. P and T 38, 681–691.

Cabrera, C., Artacho, R., Giménez, R., 2006. Beneficial effects of green tea—a review. Journal of the American College of Nutrition 25, 79–99. https://doi.org/10.1080/07315724.2006.10719518.

Calder, P.C., 2013. Omega-3 polyunsaturated fatty acids and inflammatory processes: nutrition or pharmacology? British Journal of Clinical Pharmacology 75, 645–662.

Calpe-Berdiel, L., Escolà-Gil, J.C., Blanco-Vaca, F., 2009. New insights into the molecular actions of plant sterols and stanols in cholesterol metabolism. Atherosclerosis 203, 18–31. https://doi.org/10.1016/j.atherosclerosis.2008.06.026.

Cani, P.D., Lecourt, E., Dewulf, E.M., Sohet, F.M., Pachikian, B.D., Naslain, D., De Backer, F., Neyrinck, A.M., Delzenne, N.M., 2009. Gut microbiota fermentation of prebiotics increases satietogenic and incretin gut peptide production with consequences for appetite sensation and glucose response after a meal. American Journal of Clinical Nutrition 90, 1236–1243. https://doi.org/10.3945/ajcn.2009.28095.

Cardoso, B.R., Ong, T.P., Jacob-Filho, W., Jaluul, O., Freitas, M.I.D.Á., Cozzolino, S.M.F., 2010. Nutritional status of selenium in Alzheimer's disease patients. British Journal of Nutrition 103, 803–806. https://doi.org/10.1017/S0007114509992832.

Casadesus, G., Shukitt-Hale, B., Stellwagen, H.M., Zhu, X., Lee, H.G., Smith, M.A., Joseph, J.A., 2004. Modulation of hippocampal plasticity and cognitive behavior by short-term blueberry supplementation in aged rats. Nutritional Neuroscience 7, 309–316. https://doi.org/10.1080/10284150400020482.

Casini, M.L., Marelli, G., Papaleo, E., Ferrari, A., D'Ambrosio, F., Unfer, V., 2006. Psychological assessment of the effects of treatment with phytoestrogens on postmenopausal women: a randomized, double-blind, crossover, placebo-controlled study. Fertility and Sterility 85, 972–978. https://doi.org/10.1016/j.fertnstert.2005.09.048.

Cencic, A., Chingwaru, W., 2010. The role of functional foods, nutraceuticals, and food supplements in intestinal health. Nutrients 2, 611–625. https://doi.org/10.3390/nu2060611.

Chang, C.-I., Hsu, C.-M., Li, T.-S., Huang, S.-D., Lin, C.-C., Yen, C.-H., Chou, C.-H., Cheng, H.-L., 2014. Constituents of the stem of *Cucurbita moschata* exhibit antidiabetic activities through multiple mechanisms. The Journal of Functional Foods 10, 260–273. https://doi.org/10.1016/J.JFF.2014.06.017.

Chen, P., Zhang, W., Wang, X., Zhao, K., Negi, D.S., Zhuo, L., Qi, M., Wang, X., Zhang, X., 2015. Lycopene and risk of prostate cancer: a systematic review and meta-analysis. Medicine (Baltimore) 94, e1260. https://doi.org/10.1097/MD.0000000000001260.

Chen, Y.-Y., Lai, M.-H., Hung, H.-Y., Liu, J.-F., 2013. Sweet potato [Ipomoea batatas (L.) lam. "Tainong 57"] starch improves insulin sensitivity in high-fructose diet-fed rats by ameliorating adipocytokine levels, pro-inflammatory status, and insulin signaling. Journal of Nutritional Science and Vitaminology 59, 272–280. https://doi.org/10.3177/jnsv.59.272.

Chiu, C.-Y., Chang, T.-C., Liu, S.-H., Chiang, M.-T., 2017. The regulatory effects of fish oil and chitosan on hepatic lipogenic signals in high-fat diet-induced obese rats. Journal of Food and Drug Analysis 25, 919–930. https://doi.org/10.1016/J.JFDA.2016.11.015.

Committee on Opportunities in the Nutrition and Food Sciences, Food and Nutrition Board of the Institute of Medicine, 1994. Enhancing the food supply. In: Thomas, P.R., Earl, R. (Eds.), Opportunities in the Nutrition and Food Sciences: Research Challenges and the Next Generation of Investigators. National Academy Press, Washington, DC.

Couturier, K., Qin, B., Batandier, C., Awada, M., Hininger-Favier, I., Canini, F., Leverve, X., Roussel, A.M., Anderson, R.A., 2011. Cinnamon increases liver glycogen in an animal model of insulin resistance. Metabolism 60, 1590–1597. https://doi.org/10.1016/j.metabol.2011.03.016.

D'Ascoli, T.A., Mursu, J., Voutilainen, S., Kauhanen, J., Tuomainen, T.P., Virtanen, J.K., 2016. Association between serum long-chain omega-3 polyunsaturated fatty acids and cognitive performance in elderly men and women: the Kuopio Ischaemic Heart Disease Risk Factor Study. European Journal of Clinical Nutrition 70, 970–975. https://doi.org/10.1038/ejcn.2016.59.

Desch, S., Schmidt, J., Kobler, D., Sonnabend, M., Eitel, I., Sareban, M., Rahimi, K., Schuler, G., Thiele, H., 2010. Effect of cocoa products on blood pressure: systematic review and meta-analysis. American Journal of Hypertension 23, 97–103. https://doi.org/10.1038/ajh.2009.213.

Dhalla, N.S., Temsah, R.M., Netticadan, T., 2000. Role of oxidative stress in cardiovascular diseases. Journal of Hypertension 18, 655–673.

Di Cagno, R., Filannino, P., Gobbetti, M., 2016. Fermented foods: fermented vegetables and other products. In: Encyclopedia of Food and Health. Elsevier, pp. 668–674. https://doi.org/10.1016/B978-0-12-384947-2.00284-1.

Dimidi, E., Christodoulides, K., Scott, M., Whelan, K., 2014. The effect of probiotics on functional constipation in adults : a systematic review and meta-analysis of randomized. American Journal of Clinical Nutrition 100, 1075–1084. https://doi.org/10.3945/ajcn.114.089151.1.

Douglas, L.C., Sanders, M.E., 2008. Probiotics and prebiotics in dietetics practice. Journal of the American Dietetic Association 108, 510–521. https://doi.org/10.1016/j.jada.2007.12.009.

Erlund, I., Koli, R., Alfthan, G., Marniemi, J., Puukka, P., Mustonen, P., Mattila, P., Jula, A., 2008. Favorable effects of berry consumption on platelet function, blood pressure, and HDL cholesterol. American Journal of Clinical Nutrition 87, 323–331. https://doi.org/87/2/323 (pii).

File, S.E., Hartley, D.E., Elsabagh, S., Duffy, R., Wiseman, H., 2005. Cognitive improvement after 6 weeks of soy supplements in postmenopausal women is limited to frontal lobe function. Menopause 12, 193–201. https://doi.org/10.1097/00042192-200512020-00014.

Flint, H.J., Duncan, S.H., Scott, K.P., Louis, P., 2015. Links between diet, gut microbiota composition and gut metabolism. Proceedings of the Nutrition Society 74, 13–22. https://doi.org/10.1017/S0029665114001463.

Flint, H.J., Scott, K.P., Louis, P., Duncan, S.H., 2012. The role of the gut microbiota in nutrition and health. Nature Reviews Gastroenterology and Hepatology 9, 577–589. https://doi.org/10.1038/nrgastro.2012.156.

Forbes, J.M., Cooper, M.E., 2013. Mechanisms of diabetic complications. Physiological Reviews 93, 137–188. https://doi.org/10.1152/physrev.00045.2011.

Francis, S.T., Head, K., Morris, P.G., Macdonald, I.A., 2006. The effect of flavanol-rich cocoa on the fMRI response to a cognitive task in healthy young people. Journal of Cardiovascular Pharmacology 47, 215–220. https://doi.org/10.1097/00005344-200606001-00018.

Galli, R.L., Sshukitt-Hale, B., Youdim, K.A., Joseph, J.A., 2002. Fruit polyphenolics and brain aging. Annals of the New York Academy of Sciences 959, 128–132. https://doi.org/10.1111/j.1749-6632.2002.tb02089.x.

Gao, S., Jin, Y., Hall, K.S., Liang, C., Unverzagt, F.W., Ji, R., Murrell, J.R., Cao, J., Shen, J., Ma, F., Matesan, J., Ying, B., Cheng, Y., Bian, J., Li, P., Hendrie, H.C., 2007. Selenium level and cognitive function in rural elderly Chinese. American Journal of Epidemiology 165, 955–965. https://doi.org/10.1093/aje/kwk073.

García-Llatas, G., Rodríguez-Estrada, M.T., 2011. Current and new insights on phytosterol oxides in plant sterol-enriched food. Chemistry and Physics of Lipids 164, 607–624. https://doi.org/10.1016/j.chemphyslip.2011.06.005.

Gatt, M., Reddy, B.S., MacFie, J., 2007. Review article: Bacterial translocation in the critically ill - Evidence and methods of prevention. Alimentary Pharmacology & Therapeutics 25, 741–757. https://doi.org/10.1111/j.1365-2036.2006.03174.

Gul, K., Singh, A.K., Jabeen, R., 2016. Nutraceuticals and functional foods: the foods for the future world. Critical Reviews in Food Science and Nutrition 56, 2617–2627. https://doi.org/10.1080/10408398.2014.903384.

Haque, A.M., Hashimoto, M., Katakura, M., Tanabe, Y., Hara, Y., Shido, O., 2006. Long-term administration of green tea catechins improves spatial cognition learning ability in rats. Journal of Nutrition 136, 1043–1047. https://doi.org/10.1093/jn/136.4.1043.

Hayat, K., Iqbal, H., Malik, U., Bilal, U., Mushtaq, S., 2015. Tea and its consumption: benefits and risks. Critical Reviews in Food Science and Nutrition 55, 939–954. https://doi.org/10.1080/10408398.2012.678949.

Heap, S., Ingram, J., Law, M., Tucker, A.J., Wright, A.J., 2015. Eight-day consumption of inulin added to a yogurt breakfast lowers postprandial appetite ratings but not energy intakes in young healthy females: a randomised controlled trial. British Journal of Nutrition 115, 262–270. https://doi.org/10.1017/S0007114515004432.

Herman-Antosiewicz, A., Singh, S.V., 2004. Signal transduction pathways leading to cell cycle arrest and apoptosis induction in cancer cells by Allium vegetable-derived organosulfur compounds: a review. Mutatation Research - Fundamental and Molecular Mechanism of Mutagenesis 555, 121–131. https://doi.org/10.1016/j.mrfmmm.2004.04.016.

Heude, B., Ducimetiere, P., Berr, C., 2003. Cognitive decline and fatty acid composition of erythrocyte membranes–the EVA Study. American Journal of Clinical Nutrition 77, 803–808. https://doi.org/10.1093/ajcn/77.4.803.

Hittinger, C.T., Steele, J.L., Ryder, D.S., 2018. Diverse yeasts for diverse fermented beverages and foods. Current Opinion in Biotechnology 49, 199–206. https://doi.org/10.1016/J.COPBIO.2017.10.004.

Hokanson, J.E., Austin, M.A., 1996. Plasma triglyceride level is a risk factor for cardiovascular disease independent of high-density lipoprotein cholesterol level: a metaanalysis of population-based prospective studies. European Journal of Cardiovascular Prevention and Rehabilitation 3, 213–219. https://doi.org/10.1177/174182679600300214.

Hooper, L., Kroon, P., Rimm, E., Cohn, J., Harvey, I., et al., 2008. Flavonoids, flavonoid-rich foods, and cardiovascular risk: a meta-analysis of randomized controlled trials. American Journal of Clinical Nutrition 88, 38–50. https://doi.org/10.1093/ajcn/88.1.38.

Hsu, C.L., Yen, G.C., 2006. Induction of cell apoptosis in 3T3-L1 pre-adipocytes by flavonoids is associated with their antioxidant activity. Molecular Nutrition and Food Research 50, 1072–1079. https://doi.org/10.1002/mnfr.200600040.

International Life Sciences Institute, 1999. Safety assessment and potential health benefits of food components based on selected scientific criteria. Critical Reviews in Food Science and Nutrition 39, 203–206. https://doi.org/10.1080/10408699991279169.

Ivey, K.L., Hodgson, J.M., Croft, K.D., Lewis, J.R., Prince, R.L., 2015. Flavonoid intake and all-cause mortality. American Journal of Clinical Nutrition 101, 1012–1020. https://doi.org/10.3945/ajcn.113.073106.

Jellinger, K.A., 2001. Cell death mechanisms in neurodegeneration. Journal of Cellular and Molecular Medicine 5, 1–17. https://doi.org/10.1111/j.1582-4934.2001.tb00134.x.

Jiang, W., Wei, H., He, B., 2015. Dietary flavonoids intake and the risk of coronary heart disease: a dose-response meta-analysis of 15 prospective studies. Thrombosis Research 135, 459–463. https://doi.org/10.1016/j.thromres.2014.12.016.

Jiang, X., Zhu, X., Huang, W., Xu, H., Zhao, Z., Li, S., Li, S., Cai, J., Cao, J., 2017. Garlic-derived organosulfur compound exerts antitumor efficacy via activation of MAPK pathway and modulation of cytokines in SGC-7901 tumor-bearing mice. International Immunopharmacology 48, 135–145. https://doi.org/10.1016/j.intimp.2017.05.004.

Joffre, C., Nadjar, A., Lebbadi, M., Calon, F., Laye, S., 2014. N-3 LCPUFA improves cognition: the young, the old and the sick. Prostaglandins, Leukotrienes and Essential Fatty Acids 91, 1–20. https://doi.org/10.1016/j.plefa.2014.05.001.

Kazeem, M.I., Davies, T.C., 2016. Anti-diabetic functional foods as sources of insulin secreting, insulin sensitizing and insulin mimetic agents. The Journal of Functional Foods 20, 122–138. https://doi.org/10.1016/j.jff.2015.10.013.

Keller, A.C., Ma, J., Kavalier, A., He, K., Brillantes, A.-M.B., Kennelly, E.J., 2011. Saponins from the traditional medicinal plant *Momordica charantia* stimulate insulin secretion in vitro. Phytomedicine 19, 32–37. https://doi.org/10.1016/J.PHYMED.2011.06.019.

Kleessen, B., Hartmann, L., Blaut, M., 2001. Oligofructose and long-chain inulin: influence on the gut microbial ecology of rats associated with a human faecal flora. British Journal of Nutrition 86 (291). https://doi.org/10.1079/BJN2001403.

Lai, C.S., Wu, J.C., Pan, M.H., 2015. Molecular mechanism on functional food bioactives for anti-obesity. Current Opinion in Food Science 2, 9–13. https://doi.org/10.1016/j.cofs.2014.11.008.

Lakhtakia, R., 2010. The history of diabetes mellitus. Textbook of Diabetes 13, 368–370. fourth ed. https://doi.org/10.1002/9781444324808.ch1.

Laparra, J.M., Sanz, Y., 2010. Interactions of gut microbiota with functional food components and nutraceuticals. Pharmacological Research 61, 219–225. https://doi.org/10.1016/J.PHRS.2009.11.001.

Letenneur, L., Proust-Lima, C., Le Gouge, A., Dartigues, J.F., Barberger-Gateau, P., 2007. Flavonoid intake and cognitive decline over a 10-year period. American Journal of Epidemiology 165, 1364–1371. https://doi.org/10.1093/aje/kwm036.

Liu, R.H., 2013. Health-promoting components of fruits and vegetables in the diet. Journal of Sensory Studies 52, 6503–6509. https://doi.org/10.3945/an.112.003517.convenient.

Loef, M., Schrauzer, G.N., Walach, H., 2011. Selenium and alzheimer's disease: a systematic review. Journal of Alzheimer's Disease https://doi.org/10.3233/JAD-2011-110414.

Maehre, H., Jensen, I.-J., Elvevoll, E., Eilertsen, K.-E., 2015. ω-3 fatty acids and cardiovascular diseases: effects, mechanisms and dietary relevance. International Journal of Molecular Sciences 16, 22636–22661. https://doi.org/10.3390/ijms160922636.

Mandel, S., Youdim, M.B., 2004. Catechin polyphenols: neurodegeneration and neuroprotection in neurodegenerative diseases. Free Radical Biology and Medicine 37, 304–317. https://doi.org/10.1016/j.freeradbiomed.2004.04.012.

Mao, Y., Hao, J., Jin, Z.-Q., Niu, Y.-Y., Yang, X., Liu, D., Cao, R., Wu, X.-Z., 2017. Network pharmacology-based and clinically relevant prediction of the active ingredients and potential targets of Chinese herbs in metastatic breast cancer patients. Oncotarget 8, 27007–27021. https://doi.org/10.18632/oncotarget.15351.

Marangoni, F., Poli, A., 2010. Phytosterols and cardiovascular health. Pharmacological Research 61, 193–199. https://doi.org/10.1016/J.PHRS.2010.01.001.

Marco, M.L., Heeney, D., Binda, S., Cifelli, C.J., Cotter, P.D., Foligné, B., Gänzle, M., Kort, R., Pasin, G., Pihlanto, A., Smid, E.J., Hutkins, R., 2017. Health benefits of fermented foods: microbiota and beyond. Current Opinion in Biotechnology 44, 94–102. https://doi.org/10.1016/J.COPBIO.2016.11.010.

Martínez-Augustin, O., Aguilera, C.M., Gil-Campos, M., de Medina, F.S., Gil, A., 2012. Bioactive anti-obesity food components. International Journal for Vitamin and Nutrition Research 82, 148–156. https://doi.org/10.1024/0300-9831/a000105.

Mathur, S., Devaraj, S., Grundy, S.M., Jialal, I., 2002. Cocoa products decrease low density lipoprotein oxidative susceptibility but do not affect biomarkers of inflammation in humans. Journal of Nutrition 132, 3663–3667. https://doi.org/10.1093/jn/132.12.3663.

Mente, A., De Koning, L., Shannon, H.S., Anand, S.S., 2009. A systematic review of the evidence supporting a causal link between dietary factors and coronary heart disease. Archives of Internal Medicine https://doi.org/10.1001/archinternmed.2009.38.

Miller, V., Mente, A., Dehghan, M., Rangarajan, S., Zhang, X., Swaminathan, S., Dagenais, G., Gupta, R., Mohan, V., Lear, S., Bangdiwala, S.I., Schutte, A.E., Wentzel-Viljoen, E., Avezum, A., Altuntas, Y., Yusoff, K., Ismail, N., Peer, N., Chifamba, J., Diaz, R., Rahman, O., Mohammadifard, N., Lana, F., Zatonska, K., Wielgosz, A., Yusufali, A., Iqbal, R., Lopez-Jaramillo, P., Khatib, R., Rosengren, A., Kutty, V.R., Li, W., Liu, J., Liu, X., Yin, L., Teo, K., Anand, S., Yusuf, S., Diaz, R., Orlandini, A., Linetsky, B., Toscanelli, S., Casaccia, G., Cuneo, J.M., Rahman, O., Yusuf, R., Azad, A., Rabbani, K., Cherry, H., Mannan, A., Hassan, I., Talukdar, A., Tooheen, R., Khan, M., Sintaha, M., Choudhury, T., Haque, R., Parvin, S., Avezum, A., Oliveira, G., Marcilio, C., Mattos, A., Teo, K., Yusuf, S., Dejesus, J., Agapay, D., Tongana, T., Solano, R., Kay, I., Trottier, S., Rimac, J., Elsheikh, W., Heldman, L., Ramezani, E., Dagenais, G., Poirier, P., Turbide, G., Auger, D., De Bluts, A.L., Proulx, M., Cayer, M., Bonneville, N., Lear, S., Gasevic, D., Corber, E., de Jong, V., Vukmirovich, I., Wielgosz, A., Fodor, G., Pipe, A., Shane, A., Lanas, F., Seron, P., Martinez, S., Valdebenito, A., Oliveros, M., Wei, L., Lisheng, L., Chunming, C., Xingyu, W., Wenhua, Z., Hongye, Z., Xuan, J., Bo, H., Yi, S., Jian, B., Xiuwen, Z., Xiaohong, C., Tao, C., Hui, C., Xiaohong, C., Qing, D., Xiaoru, C., Qing, D., Xinye, H., Bo, H., Xuan, J., Jian, L., Juan, L., Xu, L., Bing, R., Yi, S., Wei, W., Yang, W., Jun, Y., Yi, Z., Hongye, Z., Xiuwen, Z., Manlu, Z., Fanghong, L., Jianfang, W., Yindong, L., Yan, H., Liangqing, Z., Baoxia, G., Xiaoyang, L., Shiying, Z., Rongwen, B., Xiuzhen, T., Dong, L., Di, C., Jianguo, W., Yize, X., Tianlu, L., Peng, Z., Changlin, D., Ning, L., Xiaolan, M., Yuqing, Y., Rensheng, L., Minfan, F., Jing, H., Yu, L., Xiaojie, X., Qiang, Z., Lopez-Jaramillo, P., Lopez, P.C., Garcia, R., Jurado, L., Gómez-Arbeláez, D., Arguello, J., Dueñas, R., Silva, S., Pradilla, L., Ramirez, F., Molina, D., Cure-Cure, C., Perez, M., Hernandez, E., Arcos, E., Fernandez, S., Narvaez, C., Paez, J., Sotomayor, A., Garcia, H., Sanchez, G., David, T., Rico, A., Mony, P., Vaz, M., Bharathi, A.V., Swaminathan, S., Kurpad, K.S.A., Jayachitra, K., Kumar, N., Hospital, H., Mohan, V., Deepa, M., Parthiban, K., Anitha, M., Hemavathy, S., Rahulashankiruthiyayan, T., Anitha, D., Sridevi, K., Gupta, R., Panwar, R., Mohan, I., Rastogi, P., Rastogi, S., Bhargava, R., Kumar, R., Thakur, J.S., Patro, B., Lakshmi, P., Mahajan, R., Chaudary, P., Kutty, V.R., Vijayakumar, K., Ajayan, K., Rajasree, G., Renjini, A., Deepu, A., Sandhya, B., Asha, S., Soumya, H.,

Kelishadi, R., Bahonar, A., Mohammadifard, N., Heidari, H., Yusoff, K., Ismail, T., Ng, K., Devi, A., Nasir, N., Yasin, M., Miskan, M., Rahman, E., Arsad, M., Ariffin, F., Razak, S., Majid, F., Bakar, N., Yacob, M., Zainon, N., Salleh, R., Ramli, M., Halim, N., Norlizan, S., Ghazali, N., Arshad, M., Razali, R., Ali, S., Othman, H., Hafar, C., Pit, A., Danuri, N., Basir, F., Zahari, S., Abdullah, H., Arippin, M., Zakaria, N., Noorhassim, I., Hasni, M., Azmi, M., Zaleha, M., Hazdi, K., Rizam, A., Sazman, W., Azman, A., Khatib, R., Khammash, U., Khatib, A., Giacaman, R., Iqbal, R., Afridi, A., Khawaja, R., Raza, A., Kazmi, K., Zatonski, W., Szuba, A., Zatonska, K., Ilow, R., Ferus, M., Regulska-Ilow, B., Rózanska, D., Wolyniec, M., Alkamel, Ali, M., Kruger, M.A., Voster, H.H., Schutte, A.E., Wentzel-Viljoen, E., Eloff, F., de Ridder, H., Moss, H., Potgieter, J., Roux, A., Watson, M., de Wet, G., Olckers, A., Jerling, J., Pieters, M., Hoekstra, T., Puoane, T., Igumbor, E., Tsolekile, L., Sanders, D., Naidoo, P., Steyn, N., Peer, N., Mayosi, B., Rayner, B., Lambert, V., Levitt, N., Kolbe-Alexander, T., Ntyintyane, L., Hughes, G., Swart, R., Fourie, J., Muzigaba, M., Xapa, S., Gobile, N., Ndayi, K., Jwili, B., Ndibaza, K., Egbujie, B., Rosengren, A., Boström, K.B., Gustavsson, A., Andreasson, M., Snällman, M., Wirdemann, L., Oguz, A., Imeryuz, N., Altuntas, Y., Gulec, S., Temizhan, A., Karsidag, K., Calik, K., Akalin, A., Caklili, O., Keskinler, M., Erbakan, A., Yusufali, A., Almahmeed, W., Swidan, H., Darwish, E., Hashemi, A., Al-Khaja, N., Muscat-Baron, J., Ahmed, S., Mamdouh, T., Darwish, W., Abdelmotagali, M., Awed, S.O., Movahedi, G., Hussain, F., Al Shaibani, H., Gharabou, R., Youssef, D., Nawati, A., Salah, Z.A., Abdalla, R., Al Shuwaihi, S., Al Omairi, M., Cadigal, O., Alejandrino, R.S., Chifamba, J., Gwaunza, L., Terera, G., Mahachi, C., Murambiwa, P., Machiweni, T., Mapanga, R., 2017. Fruit, vegetable, and legume intake, and cardiovascular disease and deaths in 18 countries (PURE): a prospective cohort study. Lancet 390, 2037–2049. https://doi.org/10.1016/S0140-6736(17)32253-32255.

Mills, E., Ernst, E., Singh, R., Ross, C., Wilson, K., 2003. Health food store recommendations: implications for breast cancer patients. Breast Cancer Research: BCR 5, 8–12. https://doi.org/10.1186/bcr636.

Minelli, E.B., Benini, A., 2008. Relationship between number of bacteria and their probiotic effects. Microbial Ecology in Health and Disease 20, 180–183. https://doi.org/10.1080/08910600802408095.

Mir, S.A., Shah, M.A., Ganai, S.A., Ahmad, T., Gani, M., 2017. Understanding the role of active components from plant sources in obesity management. Journal of the Saudi Society of Agricultural Sciences https://doi.org/10.1016/J.JSSAS.2017.04.003.

Mohamed, S., 2014. Functional foods against metabolic syndrome (obesity, diabetes, hypertension and dyslipidemia) and cardiovasular disease. Trends in Food Science and Technology 35, 114–128. https://doi.org/10.1016/j.tifs.2013.11.001.

Mohankumar, S., McFarlane, J.R., 2017. An aqueous extract of *Curcuma longa* (turmeric) rhizomes stimulates insulin release and mimics insulin action on tissues involved in glucose homeostasis in vitro. An Aqueous Extract of *Curcuma longa* (Turmeric) Rhizomes Stimulates Insulin Release and Mimics Insulin Action on Tissues Involve 401, 396–401. https://doi.org/10.1002/ptr.3275.

Montalto, M., D'Onofrio, F., Gallo, A., Cazzato, A., Gasbarrini, G., 2009. Intestinal microbiota and its functions. Digestive and Liver Disease (Suppl. 3), 30–34. https://doi.org/10.1016/S1594-5804(09)60016-4.

Morris, M.C., 2016. Nutrition and risk of dementia: overview and methodological issues. Annals of the New York Academy of Sciences 1367, 31–37. https://doi.org/10.1111/nyas.13047.

Myrie, S.B., Jones, P.J.H., 2011. Functional foods and obesity. In: Functional Foods. Elsevier, pp. 234–260. https://doi.org/10.1533/9780857092557.2.234.

Okello, E.J., Leylabi, R., McDougall, G.J., 2012. Inhibition of acetylcholinesterase by green and white tea and their simulated intestinal metabolites. Food and Function 3 (651). https://doi.org/10.1039/c2fo10174b.

Ordovas, J.M., Shen, J., 2009. Gene–environment interactions and susceptibility to metabolic syndrome and other chronic diseases 79, 1508–1513. https://doi.org/10.1902/jop.2008.080232.Gene.

Pahlavani, M., Razafimanjato, F., Ramalingam, L., Kalupahana, N.S., Moussa, H., Scoggin, S., Moustaid-Moussa, N., 2017. Eicosapentaenoic acid regulates brown adipose tissue metabolism in high-fat-fed mice and in clonal brown adipocytes. The Journal of Nutritional Biochemistry 39, 101–109. https://doi.org/10.1016/J.JNUTBIO.2016.08.012.

Qin, B., Panickar, K.S., Anderson, R.A., 2010. Cinnamon: potential role in the prevention of insulin resistance, metabolic syndrome, and type 2 diabetes. Journal of Diabetes Science and Technology 4, 685–693. https://doi.org/10.1177/193229681000400324.

Quinn, J.F., Raman, R., Thomas, R.G., Yurko-Mauro, K., Nelson, E.B., Van Dyck, C., Galvin, J.E., Emond, J., Jack, C.R., Weiner, M., Shinto, L., Aisen, P.S., 2010. Docosahexaenoic acid supplementation and cognitive decline in Alzheimer disease: a randomized trial. The Journal of the American Medical Association 304, 1903–1911. https://doi.org/10.1001/jama.2010.1510.

Raedschelders, K., Ansley, D.M., Chen, D.D.Y., 2012. The cellular and molecular origin of reactive oxygen species generation during myocardial ischemia and reperfusion. Pharmacology and Therapeutics 133, 230–255. https://doi.org/10.1016/j.pharmthera.2011.11.004.

Rafter, J.J., 2002. Scientific basis of biomarkers and benefits of functional foods for reduction of disease risk: cancer. British Journal of Nutrition 88, S219. https://doi.org/10.1079/BJN2002686.

Rao, Y., Wang, H., Fan, L., Chen, G., 2011. Silencing MTA1 by RNAi reverses adhesion, migration and invasiveness of cervical cancer cells (SiHa) via altered expression of p53, and E-cadherin/β-catenin complex. Journal of Huazhong University of Science and Technology - Medical sciences 31, 1–9. https://doi.org/10.1007/s11596-011-0141-9.

Rashidi, B., Malekzadeh, M., Goodarzi, M., Masoudifar, A., Mirzaei, H., 2017. Green tea and its anti-angiogenesis effects. Biomedicine and Pharmacotherapy 89, 949–956. https://doi.org/10.1016/j.biopha.2017.01.161.

Rayman, M.P., 2012. Selenium and human health. Lancet 61452–61459. https://doi.org/10.1016/S0140-6736(11).

Riso, P., Klimis-Zacas, D., Del Bo', C., Martini, D., Campolo, J., Vendrame, S., Møller, P., Loft, S., De Maria, R., Porrini, M., 2013. Effect of a wild blueberry (Vaccinium angustifolium) drink intervention on markers of oxidative stress, inflammation and endothelial function in humans with cardiovascular risk factors. European Journal of Nutrition 52, 949–961. https://doi.org/10.1007/s00394-012-0402-9.

Roberfroid, M., 2000. Prebiotics and probiotics: are they functional foods? American Journal of Clinical Nutrition 71, 1682–1687.

Roberfroid, M.B., Delzenne, N.M., 1998. Dietary fructans. Annual Review of Nutrition 18, 117–143. https://doi.org/10.1146/annurev.nutr.18.1.117.

Rodriguez-Mateos, A., Rendeiro, C., Bergillos-Meca, T., Tabatabaee, S., George, T.W., Heiss, C., Spencer, J.P.E., 2013. Intake and time dependence of blueberry flavonoid-induced improvements in vascular function: a randomized, controlled, double-blind, crossover intervention study with mechanistic insights into biological activity. American Journal of Clinical Nutrition 98, 1179–1191. https://doi.org/10.3945/ajcn.113.066639.

Roleira, F.M.F., Tavares-da-Silva, E.J., Varela, C.L., Costa, S.C., Silva, T., Garrido, J., Borges, F., 2015. Plant derived and dietary phenolic antioxidants: anticancer properties. Food Chemistry 183, 235–258. https://doi.org/10.1016/J.FOODCHEM.2015.03.039.

Rozner, S., Garti, N., 2006. The activity and absorption relationship of cholesterol and phytosterols. Colloids and Surfaces A: Physicochemical and Engineering Aspects 282–283, 435–456. https://doi.org/10.1016/j.colsurfa.2005.12.032.

Rudkowska, I., 2009. Functional foods for health: focus on diabetes. Maturitas 62, 263–269. https://doi.org/10.1016/j.maturitas.2009.01.011.

Salem, N., Vandal, M., Calon, F., 2015. The benefit of docosahexaenoic acid for the adult brain in aging and dementia. Prostaglandins, Leukotrienes and Essential Fatty Acids 92, 15–22. https://doi.org/10.1016/j.plefa.2014.10.003.

Sarriá, B., Martínez-López, S., Sierra-Cinos, J.L., Garcia-Diz, L., Goya, L., Mateos, R., Bravo, L., 2015. Effects of bioactive constituents in functional cocoa products on cardiovascular health in humans. Food Chemistry 174, 214–218. https://doi.org/10.1016/j.foodchem.2014.11.004.

Satoh, H., Audrey Nguyen, M.T., Kudoh, A., Watanabe, T., 2013. Yacon diet (Smallanthus sonchifolius, Asteraceae) improves hepatic insulin resistance via reducing Trb3 expression in Zucker fa/fa rats. Nutrition and Diabetes 3, e70–e76. https://doi.org/10.1038/nutd.2013.11.

Sawicki, C.M., Livingston, K.A., Obin, M., Roberts, S.B., Chung, M., McKeown, N.M., 2017. Dietary fiber and the human gut microbiota: application of evidence mapping methodology. Nutrients 9, 1–21. https://doi.org/10.3390/nu9020125.

Shami, N.J.I.E., Moreira, E.A.M., 2004. Lycopene as an antioxidant agent. Revista de Nutrição 17, 227–236. https://doi.org/10.1590/S1415-52732004000200009.

Shapiro, T.A., Fahey, J.W., Wade, K.L., Stephenson, K.K., Talalay, P., 2001. Chemoprotective glucosinolates and isothiocyanates of broccoli sprouts: metabolism and excretion in humans. Cancer Epidemiology Biomarkers and Prevention 10, 501–508.

Shihabudeen, H.M.S., Priscilla, D.H., Kavitha, T., 2011. Cinnamon extract inhibits α-glucosidase activity and dampens postprandial glucose excursion in diabetic rats. Nutrition and Metabolism 8, 1–11. https://doi.org/10.1186/1743-7075-8-46.

Sofi, F., Macchi, C., Abbate, R., Gensini, G.F., Casini, A., 2010. Effectiveness of the mediterranean diet: can it help delay or prevent Alzheimer's disease? Journal of Alzheimer's Disease 20, 795–801. https://doi.org/10.3233/JAD-2010-1418.

Solanki, I., Parihar, P., Parihar, M.S., 2016. Neurodegenerative diseases: from available treatments to prospective herbal therapy. Neurochemistry International 95, 100–108. https://doi.org/10.1016/j.neuint.2015.11.001.

Steinbrenner, H., Sies, H., 2013. Selenium homeostasis and antioxidant selenoproteins in brain: implications for disorders in the central nervous system. Archives of Biochemistry and Biophysics 536, 152–157. https://doi.org/10.1016/j.abb.2013.02.021.

Thomson, C., Bloch, A.S., Hasler, C.M., Kubena, K., Earl, R., Heins, J., 1999. Position of the American dietetic association: functional foods. Journal of the American Dietetic Association 99, 1278–1285. https://doi.org/10.1016/S0002-8223(99)00314-4.

Tomé-Carneiro, J., Visioli, F., 2016. Polyphenol-based nutraceuticals for the prevention and treatment of cardiovascular disease: review of human evidence. Phytomedicine 23, 1145–1174. https://doi.org/10.1016/j.phymed.2015.10.018.

Trejo-Solis, C., Pedraza-Chaverri, J., Torres-Ramos, M., Jimenez-Farfan, D., Cruz Salgado, A., Serrano-Garcia, N., Osorio-Rico, L., Sotelo, J., 2013. Multiple molecular and cellular mechanisms of action of lycopene in cancer inhibition. Evidence-Based Complementary and Alternative Medicine 2013. 705121 https://doi.org/10.1155/2013/705121.

Tur, J.A., Bibiloni, M.M., 2016. Functional foods. In: Encyclopedia of Food and Health. first ed. Elsevier Ltd. https://doi.org/10.1016/B978-0-12-384947-2.00340-8.

Tur, J.A., Bibiloni, M.M., Sureda, A., Pons, A., 2012. Dietary sources of omega 3 fatty acids: public health risks and benefits. British Journal of Nutrition https://doi.org/10.1017/S0007114512001456.

Valentová, K., Moncion, A., De Waziers, I., Ulrichová, J., 2004. The effect of Smallanthus sonchifolius leaf extracts on rat hepatic metabolism. Cell Biology and Toxicology 20, 109–120. https://doi.org/10.1023/B:CBTO.0000027931.88957.80.

van Duijn, C.M., van der Lee, S.J., Ikram, M.A., Hofman, A., Hankemeier, T., Amin, N., Demirkan, A., 2016. Metabolites associated with cognitive function in the Rotterdam study and Erasmus Rucphen family study. Alzheimer's Dement 12, 165. https://doi.org/10.1016/j.jalz.2016.06.277.

Van Loo, J., Coussement, P., Leenheer, L., De Hoebregs, H., Smits, G., 1995. On the presence of inulin and oligofructose as natural ingredients in the western diet. Critical Reviews in Food Science and Nutrition 35 (6), 525–552. https://doi.org/10.1080/10408399509527714.

Vauzour, D., Vafeiadou, K., Rodriguez-Mateos, A., Rendeiro, C., Spencer, J.P.E., 2008. The neuroprotective potential of flavonoids: a multiplicity of effects. Genes and Nutrition 3, 115–126. https://doi.org/10.1007/s12263-008-0091-4.

Verhoeven, D.T.H., Goldbohm, R.A., Poppel, G., Van Verhagen, H., 1996. Epidemiological studies on brassica vegetables and cancer epidemiological studies on Brassica vegetables and cancer risk. Cancer Epidemiology Biomarkers and Prevention 5, 733–748. https://doi.org/10.1158/1055-9965.epi-08-0520.

Wang, S., Moustaid-Moussa, N., Chen, L., Mo, H., Shastri, A., Su, R., Bapat, P., Kwun, I., Shen, C.-L., 2014a. Novel insights of dietary polyphenols and obesity. The Journal of Nutritional Biochemistry 25, 1–18. https://doi.org/10.1016/J.JNUTBIO.2013.09.001.

Wang, X., Ouyang, Y.Y., Liu, J., Zhao, G., 2014b. Flavonoid intake and risk of CVD: a systematic review and meta-analysis of prospective cohort studies. British Journal of Nutrition https://doi.org/10.1017/S000711451300278X.

Wang, Y., Fofana, B., Roy, M., Ghose, K., Yao, X.-H., Nixon, M.-S., Nair, S., Nyomba, G.B.L., 2015. Flaxseed lignan secoisolariciresinol diglucoside improves insulin sensitivity through upregulation of GLUT4 expression in diet-induced obese mice. The Journal of Functional Foods 18, 1–9. https://doi.org/10.1016/J.JFF.2015.06.053.

Watzl, B., Girrbach, S., Roller, M., 2005. Inulin, oligofructose and immunomodulation. British Journal of Nutrition 93, S49. https://doi.org/10.1079/BJN20041357.

Widlansky, M.E., Hamburg, N.M., Anter, E., Holbrook, M., Kahn, D.F., Elliott, J.G., Keaney, J.F., Vita, J.A., 2007. Acute egcg supplementation reverses endothelial dysfunction in patients with coronary artery disease. Journal of the American College of Nutrition 26, 95–102. https://doi.org/10.1080/07315724.2007.10719590.

Willett, W.C., 1995. Diet, nutrition, and avoidable cancer. Environmental Health Perspectives 103, 165–170. https://doi.org/10.2307/3432305.

Williams, C.M., El Mohsen, M.A., Vauzour, D., Rendeiro, C., Butler, L.T., Ellis, J.A., Whiteman, M., Spencer, J.P.E., 2008. Blueberry-induced changes in spatial working memory correlate with changes in hippocampal CREB phosphorylation and brain-derived neurotrophic factor (BDNF) levels. Free Radical Biology and Medicine 45, 295–305. https://doi.org/10.1016/j.freeradbiomed.2008.04.008.

Williams, D.J., Edwards, D., Hamernig, I., Jian, L., James, A.P., Johnson, S.K., Tapsell, L.C., 2013. Vegetables containing phytochemicals with potential anti-obesity properties: a review. Food Research International 52, 323–333. https://doi.org/10.1016/j.foodres.2013.03.015.

Youdim, K.A., Joseph, J.A., 2001. A possible emerging role of phytochemicals in improving age-related neurological dysfunctions: a multiplicity of effects. Free Radical Biology and Medicine 30, 583–594. https://doi.org/10.1016/S0891-5849(00)00510-4.

Ziegenfuss, T.N., Hofheins, J.E., Mendel, R.W., Landis, J., Anderson, R.A., 2006. Effects of a water-soluble cinnamon extract on body composition and features of the metabolic syndrome in pre-diabetic men and women. Journal of the International Society of Sports Nutrition 3 (45). https://doi.org/10.1186/1550-2783-3-2-45.

Further Reading

Conquer, J.A., Tierney, M.C., Zecevic, J., Bettger, W.J., Fisher, R.H., 2000. Fatty acid analysis of blood plasma of patients with alzheimer's disease, other types of dementia, and cognitive impairment. Lipids 35, 1305–1312. https://doi.org/10.1007/s11745-000-0646-3.

Mink, P.J., Scrafford, C.G., Barraj, L.M., Harnack, L., Hong, C.-P., Nettleton Jr., J.A., 2007. Flavonoid intake and cardiovascular disease mortality: a prospective study in postmenopausal women 1 – 4. American Journal of Clinical Nutrition 85, 895–909. https://doi.org/10.1093/ajcn/85.3.895.

Nascimento, A.F., Sugizaki, M.M., Leopoldo, A.S., Lima-Leopoldo, A.P., Luvizotto, R.A.M., Nogueira, C.R., Cicogna, A.C., 2008. A hypercaloric pellet-diet cycle induces obesity and co-morbidities in Wistar rats. Arquivos Brasileiros de Endocrinologia e Metabologia 52, 968–974. https://doi.org/10.1590/S0004-27302008000600007.

Real, J.T., Martínez-Hervás, S., Tormos, M.C., Domenech, E., Pallardó, F.V., Sáez-Tormo, G., Redon, J., Carmena, R., Chaves, F.J., Ascaso, J.F., García-García, A.B., 2010. Increased oxidative stress levels and normal antioxidant enzyme activity in circulating mononuclear cells from patients of familial hypercholesterolemia. Metabolism 59, 293–298. https://doi.org/10.1016/j.metabol.2009.07.026.

CHAPTER 12

Bee Propolis: Properties, Chemical Composition, Applications, and Potential Health Effects

Roger Cauich-Kumul, Maira Rubi Segura Campos
Facultad de Ingeniería Química, Universidad Autónoma de Yucatán, Mérida, Mexico

12.1 Introduction

Propolis (from the Greek propolis "defense of the city") is a substance obtained by bees from the buds of plants and processed in the hive into a potent antibiotic to cover the walls of the hive (Saiz and Serrano, 2003). It is a resinous material collected by the worker bees, mainly of the species *Apis mellifera*, from the exudates of bark and diverse plant tissues. The collected materials are crushed, moistened with saliva and enzymatic secretions, mixed with the wax produced by the wax glands, and finally transported to the hive, where they fulfill various functions. The roles include keeping the hive free of pathogenic bacteria and fungi, acting as structural support to cover cracks and holes, regulating the temperature inside the hive, and preventing the entry of other organisms (Castaldo and Capasso, 2002; Pietta et al., 2002). There is a long history of the medicinal benefits of using propolis, but it has only been in the last century that scientists have been able to prove its activity as a therapeutic agent. The Egyptians recognized its antirotting properties and used it to embalm corpses. In Greece and Rome, it was used as a natural antiseptic and healing agent in the treatment of wounds and as an oral disinfectant. Likewise, the civilizations of the New World also appreciated its properties; it is known that the Incas used propolis as an antipyretic agent. In addition, the English pharmacopeia in the 17th century included propolis in the list of official medicines (Castaldo and Capasso, 2002). In recent decades, propolis has gained popularity as a potential health food in many parts of the world, including the United States, Japan, and the European Union, where it has become a priority food to improve human health and prevent pathologies, such as inflammation, heart disease, cancer, and diabetes. As a natural resource, propolis has attracted the attention of researchers for its wide range of biological properties, among which, the antimicrobial (Marcucci et al., 2001; Lu et al., 2005), antiinflammatory (Cardile et al., 2003), antiviral (Kujumgiev et al., 1999), anticancer (Banskota et al., 2000; Oršolić et al., 2004), and antiparasitic properties are highlighted (Dantas et al., 2006; De Carvalho et al.,

Bioactive Compounds. https://doi.org/10.1016/B978-0-12-814774-0.00012-8

2007; Wagh, 2013). Furthermore, some research has been directed to the in vitro study of antioxidant activity (Ahn et al., 2004; Wang et al., 2004; Trusheva et al., 2006) and biological and pharmacological activities of propolis (Saiz and Serrano, 2003).

12.2 Chemical Composition of Propolis

Propolis is a material of lipophilic nature, hard consistency, brittle, and when heated becomes flexible, gummy, and very sticky. It has a very characteristic and pleasant aromatic odor. Its coloration varies from yellow to red to dark brown, depending on its source and age. It has even been reported that there is transparent propolis (Marcucci et al., 1998; Bankova et al., 2000). Propolis is a complex mixture of the interaction made by the bee with the derivatives of plants. In general, crude propolis is composed of about 50% resins, 30% waxes, 10% essential oils, 5% pollen, and 5% various organic compounds (Table 12.1) (Park et al., 2002; Pietta et al., 2002). More than 300 constituents have been identified in a range of samples, and new components have been identified during the chemical characterization of new types of propolis (Marcucci, 1995; De Castro, 2011).

The proportions of the various substances present in the propolis depend on the place and time of collection. The compounds identified belong to the following groups of chemically similar compounds: polyphenols; benzoic acids and derivatives; cinnamic alcohol and cinnamic acid and their derivatives; sesquiterpenes and triterpene hydrocarbons; benzaldehyde derivatives; alcohols, ketones, and heteroaromatic compounds; terpenes and sesquiterpene alcohols and their derivatives; aliphatic hydrocarbons; minerals; sterols and steroidal hydrocarbons; sugars and amino acids (Wagh, 2013). Moreover, 32 amino acids (seven of them essential) are present, including vitamin B1 (thiamine), vitamin PP (nicotinic acid), and provitamin A. The following microelements have also been identified: calcium, potassium, sodium, magnesium, iron, aluminum, phosphorus, silicon, vanadium, and strontium.

The active ingredients of major scientific interest, of which about 50 compounds have been described, are phenolic compounds. These include, among others, benzoic acids, cinnamic acids, and their derivatives, as well as flavonoids: flavones (notably chrysin), flavonols (mainly quercetin and galangin), and flavanones, such as pinocembrin (Saiz and Serrano, 2003).

Table 12.1: Composition of Organic Propolis

Component	Quantity (%)
Resins and balsams	50
Waxes	25–30
Essential oils	10
Pollen	5
Organics and mineral substances	5

Volatile compounds (produced by the parent plants) are present in low amounts (De Castro, 2001, 2011). It is believed that the sugars are eitherintroduced accidentally during the preparation of the propolis or in an interaction of the bee with the resin, or both.

Among the main essential compounds responsible for the various biological activities are polyphenols, aromatic acids, and diterpene acids, but very few types of propolis differ in their major bioactive ingredients (Wagh, 2013). Table 12.2 lists the diverse range of solvents used for the extraction of bioactive compounds and other chemical compounds from propolis.

12.2.1 Phenols and Flavonoids Present in Propolis

Propolis is constituted mainly by flavonoids (Fig. 12.1), derivatives of esters and phenolic acids. Some of the first studies reported by Walker and Crane (1987) indicate the presence of 38 flavones, 12 benzoic acid derivatives, 14 cinnamic alcohol and cinnamic acid derivatives, 12 alcohols, ketones, and phenols, seven terpenes, 11 steroids, seven sugars, and two amino acids (Hegazi and El Hady, 2001; Murat et al., 2002; Park et al., 2002; Salamanca, 2002).

Propolis is a resinous substance, rich in polyphenols collected by honey bees from a variety of plant sources, such as trees and shrubs. In studies conducted in Chile, pinocembrin has

Table 12.2: Solvents Used for the Extraction of Active Components From Propolis (Wagh, 2013)

Water	Methanol	Ethanol	Chloroform	Dichloro-methane	Ether	Acetone
Anthocyanins, tannins, saponins, terpenoids, polypeptides, lectins	Anthocyanins, terpenoids, saponins, tannins, xanthoxylin, lactones, flavones, phenols, polyphe-nols, polypep-tides, lectins	Tannins, polyphenols, polyacety-lene, terpenoids, sterols, alkaloids	Terpenoids, flavonoids	Terpenoids, tannins, polyphenols, polyacetylene, sterols, alkaloids	Alkaloids, terpenoids, coumarins	Flavonols

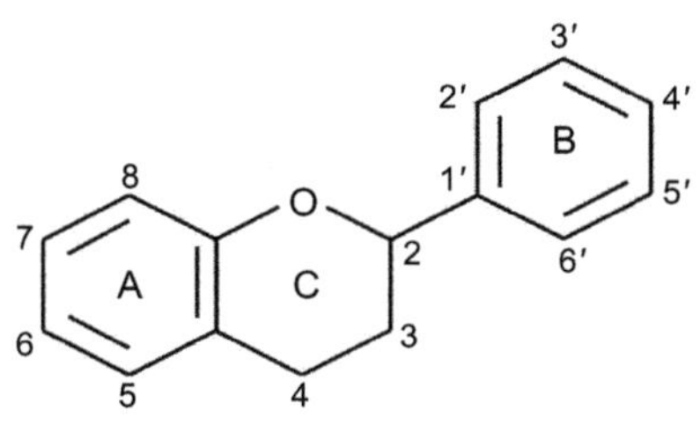

Figure 12.1
Basic structure of flavonoids (Russo and Speranza, 2006).

Table 12.3: Flavonoids Present in Propolis (Li et al., 2016)

Flavonoid	Quantity (%)
Pinocembrin	21.4
Galangin	5.0
Chrysin	4.8
Quercetin	2.2
Tectochrysin	1.1

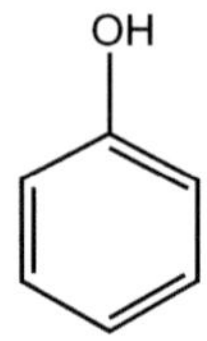

Figure 12.2
Chemical structure of phenol (Rodríguez, 2003).

been identified among its main components, which showed important biological activity as an isolated compound (Kumazawa et al., 2004; Rasul et al., 2013). Pinocembrin (5,7-dihydroxyflavone) is one of the main flavonoids of propolis. It can be extracted as a pure compound and has been incorporated into the pharmaceutical industry for its wide range of pharmacological effects (Rasul et al., 2013), including antimicrobial, antiinflammatory, antioxidant, and anticancer activities (Saad et al., 2015; Lan et al., 2016). Recently, a simple and efficient method was developed, using macroporous absorbent resin coupled with preparative high-performance liquid chromatography, for the separation of polyphenols from the aqueous extracts of Chinese propolis (Li et al., 2016). In that research, six phenolic acids and five flavonoids were detected (caffeic acid, ferulic acid, isoferulic acid, 3,4-dimethoxycinnamic acid, pinobanksin, caffeic acid benzyl ester, phenylethyl ester of caffeic acid, apigenin, pinocembrin, chrysin, and galangin) (Table 12.3) (Li et al., 2016).

12.2.2 Total Phenol Content

Phenolic compounds (Fig. 12.2) or polyphenols have been recognized for their broad spectrum of biological and pharmaceutical activities, for example, as anticancer, antiinflammatory, immunomodulatory, analgesic, and antioxidant agents (Gómez et al., 2006, 2016). It has been established and reported that most of the bioactive compounds found in propolis are phenolic compounds. However, the concentration of these can vary substantially, according to the origin of the samples, and consequently, their biological properties can also vary. For this reason, it is of great importance to carry out tests on propolis derived from different areas or types of natural vegetation that allow the detection and quantification of phenolic substances, to define the quality parameters for this bee product.

Currently, the content of total phenols in propolis is determined by the Folin–Ciocalteu method, in which the oxidation of the phenolic compounds by phosphoric molybdotungstate results in a colored product with a maximum absorbance at 760 nm wavelength (Singleton and Rossi, 1965). Several studies have shown a direct correlation between the content of total phenols and biological activities present in propolis, such as antioxidant and antimicrobial activities (Kumazawa et al., 2004; Choi et al., 2006; Moreira et al., 2008).

12.2.3 Total Flavonoid Content

As phenolic substances of scientific and therapeutic interest, flavonoids have been attributed various pharmaceutical properties since ancient times and have been established as the main functional component of formulations of plants and insects, for medical use (Havsteen, 2002). Some of the biological activities attributed to propolis are due to the high flavonoid content that is usually present in the samples.

The method for evaluating total flavonoid content is based on the use of visible, ultraviolet displacement reagents. Aluminum chloride ($AlCl_3$) is commonly used, and the absorbance is read at 425 nm wavelength (Kumazawa et al., 2004). This method has been reported in the evaluation of propolis by Marquele et al. (2005), Choi et al. (2006), Gómez et al. (2006), Ahn et al. (2007), Soleo de Funari et al. (2007), and Moreira et al. (2008). Some examples, highlighting the importance of the realization of this type of determination, are detailed as follows:

- Banskota et al. (2002) established that the antiproliferative activity of cancer cells presented by the analyzed propolis was due to the presence of cinnamic acid derivatives and flavonoids.
- Kumazawa et al. (2004) investigated the antioxidant activity of propolis from various regions of the world, using techniques, such as the oxidation of linoleic acid and the entrapment capacity of the 2,2-diphenyl-1-picrylhydrazyl (DPPH) radical. The authors found that propolis from Argentina, Australia, China, Hungary, and New Zealand had strong antioxidant activity, as a consequence of the total content of polyphenols and flavonoids present in the samples. Isla et al. (2005) determined the chemical composition and activity of different propolis samples against *Staphylococcus aureus*, stating that among the extracts, those with a higher flavonoid content, especially pinocembrin, showed high activity against the bacteria.

Isla et al. (2005) determined the chemical composition and activity of different propolis samples against *Staphylococcus aureus*, stating that among the extracts, those with a higher flavonoid content, especially pinocembrin, showed high activity against the bacteria.

12.3 Bioactivities

12.3.1 Biological and Pharmacological Activities of Propolis

Propolis is a natural product that has been used in traditional medicine since antiquity, due to its antitumor, antioxidant, antimicrobial, antiinflammatory, and immunological effects. These biological activities are attributed to the phenolic compounds present, mainly flavonoids. Flavonoids have been reported to exhibit a wide range of biological activities, including antibacterial, antiviral, antiinflammatory, antiallergic, and vasodilatory actions. Furthermore, these bioactive compounds inhibit lipid peroxidation, platelet aggregation, capillary permeability, fragility, and the activity of enzymatic systems, including cyclooxygenase (COX) and lipoxygenase (LOX) (Saiz and Serrano, 2003).

12.3.2 Propolis With Antioxidant Activity and Mechanism of Action

In recent decades, functional foods have attracted the attention of consumers, due to growing concerns about their health and the low toxicity of nonsynthetic additives, which has promoted research of these foods. In the context of health benefits, the antioxidant capacity of functional foods contributes to the prevention of certain diseases, including cardiovascular, cancer, and diabetes (Waris and Ahsen, 2006; Wagh, 2013). The importance of protecting cellular defense systems against damage caused by oxygen is well known. Free radicals and other oxidative agents are of great importance in the mechanism of action of many toxins (Nagai et al., 2003). Free radicals induce oxidative damage in biomolecules, such as carbohydrates, proteins, lipids, and nucleic acids, which can disrupt cells and cause death. Tissues of living organisms have endogenous defense mechanisms against oxidative damage, mainly enzymatic antioxidants, such as superoxide dismutase, catalase, peroxidase, and low-molecular-weight molecules, such as tocopherol, ascorbic acid, and polyphenols (Nagai et al., 2003).

Regarding the antioxidant activity of propolis, Yang et al. (2011) found that it possesses the capacity to trap free radicals, demonstrating that it is one of the mechanisms by which propolis exerts its antioxidant potential. Gülçin et al. (2010) have shown that propolis can chelate ferric (Fe^{3+}), cupric (Cu^{2+}), and ferrous (Fe^{2+}) ions, trap DPPH* and 2,2′-azino-bis(3-ethylbenzothiazoline-6-sulfonic acid) ($ABTS^{\bullet+}$) radicals and inhibit lipid peroxidation.

In this context, propolis can be used as a potential functional food because of its high antioxidant properties. The antioxidant activity of propolis is due to the phenolic compounds and the flavonoids they contain, where there is a high degree of correlation between these substances and antioxidant capacity. These bioactive compounds include amino acids, phenolic acids, flavonoids, terpenes, steroids, aldehydes, and ketones (Aljadi and Kamaruddin, 2004).

Free radicals and lipid peroxidation induce cellular damage and play an important role in the development of cardiovascular diseases, rheumatoid arthritis, cancer, inflammatory alterations, and aging processes. Despite the existence of antioxidant substances in the body, the search for new components that present these properties has become one of the most widespread research topics. In this sense, the importance of flavonoids present in many foods has been demonstrated. Research studies have verified that the antioxidant capacity of propolis is ascribed, in part, to its flavonoid content, which could also prevent or decrease the use of certain food preservatives. The antioxidant property of propolis can be attributed to its dose-dependent alkoxy radical scavenging effect and, to a lesser extent, against the superoxide radical.

Significant inhibition of the xanthine oxidase activity of the enzymes present in leukocytes (e.g., myeloperoxidase [MPO], nicotinamide adenine dinucleotide phosphate [NADPH] oxidase, and LOX) and an antilipoperoxidative capacity have also been observed (Saiz and Serrano, 2003; Farré et al., 2004).

12.3.3 Propolis With Antiinflammatory Activity and Mechanism of Action

Propolis is an excellent antiinflammatory. Its use as an extract in inhalers provides excellent results in upper respiratory, lung (bronchitis, tuberculosis) conditions and the treatment of pharyngitis and sinusitis. In an experimental model, in which a corneal lesion was chemically induced, Ozturk et al. (2000) indicated that the antiinflammatory effect of a propolis extract was comparable to that of dexamethasone, a potent inhibitor of angiogenesis and the enzymes, COX and LOX.

Antiinflammatory activity is the primary effect of host defense systems; various chemical or biological products, including proinflammatory enzymes and cytokines, low-molecular-weight compounds, such as eicosanoids or enzymatic degradation products of tissues, trigger the inflammatory process. Propolis has antiinflammatory effects on the activity of MPO, NADPH oxidase, and ornithine decarboxylase. This antiinflammatory activity is due to the presence of active flavonoids and cinnamic acid derivatives (Almeida and Menezes, 2002). Among these flavonoids, galangin is of particular interest, since this compound is capable of inhibiting COX and LOX activity, limiting the action of polygalacturonase and reducing the expression of the inducible COX-2 isoform (Mattace et al., 2001; Rossi et al., 2002). Phenethyl caffeic acid ester (CAPE) is another compound present in propolis that shows antiinflammatory activity. It acts by inhibiting the release of arachidonic acid from the cell membrane, leading to the suppression of COX-1 and -2 actions and inhibiting the activation of the gene expression of COX-2 (Mirzoeva and Calder, 1996; Lung-Ta et al., 2004).

Chrysin, another flavonoid present in propolis, also shows antiinflammatory activity and its mechanism of action is related to the suppression of the proinflammatory activity of COX-2 and inducible nitric oxide synthase (Cho et al., 2004). The first studies indicate that the antiinflammatory activity of propolis is due to the presence of caffeic acid, quercetin, and narigenin on the activity of MPO, NADPH oxidase, ornithine decarboxylase, and tyrosine protein kinases (Frenkel et al., 1993). However, the actions of CAPE (Mirzoeva and Calder, 1996) and galangin, kaempferol, and kaempferide (Krol et al., 1996) are considered to be the most notable. The potent free radical scavenging activity and in vitro antiinflammatory effects of ethanolic extracts of propolis (EEP) act by modulating key inflammatory mediators of mRNA transcription, as well as inhibition of the production of specific inflammatory cytokines and activation of the nuclear factor kappa B (Huang et al., 2014).

In vivo experiments in an arthritic rat model demonstrated the antiinflammatory effect was due to the inhibiting effect of prostaglandins (Park and Kahng, 1999). The antiinflammatory activity of phenolic acids and flavonoids present in propolis is a result of their antioxidant properties (Yao et al., 2004). In this sense, due to its high content of polyphenols, propolis shows an important antiinflammatory activity in both acute and chronic inflammatory processes (Borrelli et al., 2002).

12.3.4 Propolis With Activity on the Cardiovascular System

It has been documented that the polyphenols present in propolis, mainly flavonoids, stabilize and strengthen blood vessels. In this regard, and by the nature of their action, polyphenolic compounds can be used for the treatment and prevention of bleeding, ecchymosis, varicose veins, and atherosclerosis (Makowska-Was and Janeczko, 2004). Instead, the vasoprotective activity is related to the chelation of copper ions, which results in the inhibition of hyaluronidase, an enzyme that depolymerizes hyaluronic acid and hydrolyzes elastin, thereby reinforcing and sealing the endothelium of the vessels. The permeability of blood vessels is also decreased indirectly as a consequence of inhibiting the metabolism of catecholamines by the inactivation of catechol-*O*-methyltransferase (Olszewska, 2003).

In relation to the above, the polyphenols also show a beneficial influence on the coronary circulation and maintain a hypotensive effect, increasing the bioavailability of nitric oxide, which exerts a positive action on the activity of the endothelial nitric oxide synthase. Another property of polyphenols that has been reported since the last decade is the inhibition of angiotensin-converting enzyme (ACE) and exerts a vasodilatory effect, decreasing platelet aggregation and thromboxane concentration (Packer et al., 1999; Rohdewald, 2002).

In contrast, some flavonoids contained in propolis, such as quercetin, kaempferol, and rhamnetin, block the transport of calcium through the cell membranes in the cytoplasm, causing vasodilation and lowering of blood pressure. Cardioprotective effects were also achieved in the case of induced cardiomyopathy in rats that were treated by the intraperitoneal administration of a propolis extract (Konishi, 2005).

12.3.5 Propolis With Antihypertensive Activity

High blood pressure is the main risk factor for mortality and is considered one of the biggest public health issues, causing an estimated 7.5 million deaths per year worldwide (World Health Organization, 2013). In previous years, some antihypertensive effects of various foods and natural products have been reported for the treatment of hypertension. These effects include, for instance, ACE inhibitory activity, vasodilator action (Tokunaga et al., 2004), or inhibitory effects on the release of noradrenaline from sympathetic nerves (Suzuki et al., 2002).

According to previous research, hypertensive patients showed an increase in plasma superoxide and hydrogen peroxide levels (Lacy et al., 1998). Also, in spontaneously hypertensive rats, a valuable animal model of hypertension in humans, endothelial superoxide anion production and effects of impaired relaxation of the dependent endothelium were observed (Jameson et al., 1993; Kumar and Das, 1993). The treatment of hypertension consists of dietary measures, practices of physical exercise, and a reduction in the consumption of alcoholic beverages and cigarette smoking. However, when these measures are insufficient, pharmacological treatments involve the use of essential antihypertensives: diuretics, beta-blockers, calcium antagonists, ACE inhibitors, and sympatholytics; and nonessential antihypertensives: alpha-blockers and angiotensin-II (AG-II) receptor antagonists (Murray and Bakers, 1996).

According to Maruyama et al. (2009), EEP and its main constituents, among which the flavonoids stand out, also represent a benefit for the improvement of blood pressure. In this way, the extract of propolis of this investigation and its potential biological effect may represent an alternative for the inhibition of ACE-1. This protease hydrolyzes the vasodilator peptide bradykinin into vasoinactive peptides via the kallikrein–kinin system and, additionally, catalyzes the conversion of the biologically inactive decapeptide angiotensin-I (AG-I) to the vasoconstrictor octapeptide AG-II via the renin–angiotensin system (Scow et al., 2003).

In the kallikrein–kinin system, the liver secretes a protein substance (kininogen), which, in conjunction with kallikrein present in plasma and tissues, especially in the pancreas, forms the octapeptide bradykinin, and also a nonapeptide with the same actions, termed kallidin (lysyl-bradykinin). The duration of action of these local hormones is about 15 s, as they are disintegrated to fragments inactivated by kinase II or ACE. Tissue lesions, allergic reactions, viral infections, and other inflammatory processes, trigger a series of proteolytic events that produce bradykinin. This hormone causes pain, vasodilation, and increases capillary permeability, the release of nitric oxide from the vascular endothelium and increases the synthesis of prostaglandins.

The renin–angiotensin system is connected to a signal transduction system, wherein the renin separates the AG-I decapeptide from the N-terminal domain of the angiotensinogen.

The kidney is the only known site where prorenin is converted to renin and is the only source of plasma renin. Renin secretion may be due to three factors: a decrease in systemic blood pressure that is detected by receptors located in the afferent renal arterioles; hyponatremia detected by cells of the dense macula of the renal tubules, or stimulation of sympathetic activity of the cells of the juxtaglomerular apparatus, due to physical exercise or cardiovascular reflexes. The liver is the most important site of expression of the angiotensinogen gene, but angiotensinogen mRNA is expressed in several extrahepatic sites, including the brain, large arteries, kidney, adipose tissue, and heart. It has been estimated that more than 85% of AG-I is formed within tissues rather than in plasma (Scow et al., 2003).

Once AG-I is obtained from angiotensinogen by the action of renin, it is proteolytically converted to AG-II by ACE, mainly at the pulmonary level. AG-II is a potent vasoconstrictor because it acts directly on vascular smooth muscle cells and on the sympathetic nervous system, both peripherally and centrally, to increase vascular tone. It is the main active component of the renin–angiotensin–aldosterone system and inhibits the peptide vasodilator bradykinin and also causes volume expansion through sodium retention (via aldosterone and renal vasoconstriction) and fluid retention (via antidiuretic hormone). AG-II actions are mediated by AT1 receptors, which activate detrimental effects (causing vasoconstriction), and AT2 receptors, which lead to beneficial impacts (causing vasodilation) in the body (Scow et al., 2003).

12.3.6 Propolis With Antidiabetic Activity

According to the literature, perhaps the best-known propolis property, its antidiabetic activity, is the least studied. Some of the studies indicate that the aqueous extract of propolis shows a preventive effect on the destruction of pancreatic β-cells against streptozotocin toxicity that induces inhibition of interleukin-1β and generation of nitric oxide synthase (Matsushige et al., 1996; Sforcin and Bankova, 2011). Additionally, the antihyperglycemic effect of the water-soluble fraction of a sample of propolis has been studied, demonstrating a potent suppressive influence of postprandial elevation of blood glucose levels after oral administration of propolis extract in Sprague–Dawley rats (Matsui et al., 2004). This effect was attributed mainly to 3,4,5-tri-O-caffeoylquinic acid, a constituent derived from caffeic acid (Matsui et al., 2004). According to a study of rats with diabetes mellitus, Fuliang et al. (2005) demonstrated that propolis could positively modulate blood sugar and lipid levels, in addition to exerting a decrease in lipid peroxidation and a free radical scavenging activity.

In reference to the above, the effects caused by diabetes in the heart have been studied demonstrating that diabetes mellitus increases oxidative stress in cardiac tissue and that CAPE decreases oxidative stress through its antioxidant property by inhibiting the enzymes, superoxide dismutase and catalase (Okutan et al., 2005). Another study using propolis to examine the impact of fructose consumption in rats showed that it significantly decreases the plasma insulin level without affecting blood glucose or total cholesterol levels, indicating that propolis could prevent the development of insulin resistance (Zamami et al., 2007).

In 2010, an attempt was made to elucidate the mechanism of action of the antidiabetic effect of propolis, by investigating the expression and activity of the gluconeogenic glucose-6-phosphatase gene in HepG2 cells (Kang et al., 2010). The study demonstrated that propolis significantly reduced the expression and enzymatic activity of glucose-6-phosphatase, suggesting that this natural product may be used as a potential antidiabetic agent for the treatment of noninsulin-dependent diabetes (Kang et al., 2010). Besides, studies using propolis from countries, such as China and Brazil, indicate a significant inhibition of body weight loss and increased blood glucose in diabetic rats. In rats treated with propolis from China, a reduction in glycosylated hemoglobin levels was seen compared to untreated diabetic rats. Measurement of lipid metabolism in the blood showed signs of dyslipidemia in diabetic rats, and the effect of treatment with propolis reduced the level of total cholesterol (Ahn et al., 2007).

12.3.7 Potential Healthy and Biological Effects of Propolis of Apis mellifera *From Southeast Mexico*

In regards to human health, honeybee propolis has been used for the treatment of respiratory diseases, influenza, sinusitis, otitis, laryngitis, bronchitis, bronchial asthma, chronic pneumonia, and pulmonary tuberculosis. In dentistry, this natural product has been used for the treatment of mouth abscesses. In dermatology, it has been used to treat boils, cracks, warts, and abscesses. Besides the above effects, its therapeutic benefits have been exercised for over 300 years b.c., before finding its place in the modern pharmacopeia. Currently, propolis in any form (alone or combined) has analgesic, antibiotic, anesthetic antibacterial, antiinflammatory, antiviral, bactericidal, bacteriostatic, and fungicidal activities (Fokt et al., 2010). It has also shown protective action against photoinhibitory properties of sunlight, acts in the regenerating or healing of skin, and inhibits cancer cells, among other similar conditions, and its use in a healthy organism increases the natural immunity against various diseases (Fokt et al., 2010).

However, even though beneficial effects have been documented in other parts of the world, those of propolis samples from Yucatán State (Mexico) have only been partially elucidated. Research previously performed in tropical areas confirms the presence of prenylated benzophenones. Trusheva et al. (2004) isolated two new polyprenylated benzophenones; 18-ethyloxy-17-hydroxy-17,18-dihydroscrobiculatone A (Fig. 12.3, compound 52) and B (Fig. 12.3 compound 53). Also, scrobiculatones A (Fig. 12.3, compound 54) and B (Fig. 12.3, compound 55) were previously found by Porto et al. (2000) and isolated from resins exuded by the flowers of several *Clusia* species.

Several studies on propolis from different latitudes have shown that both, their composition and biological activity, are directly related to the plant species visited by pollinating bees

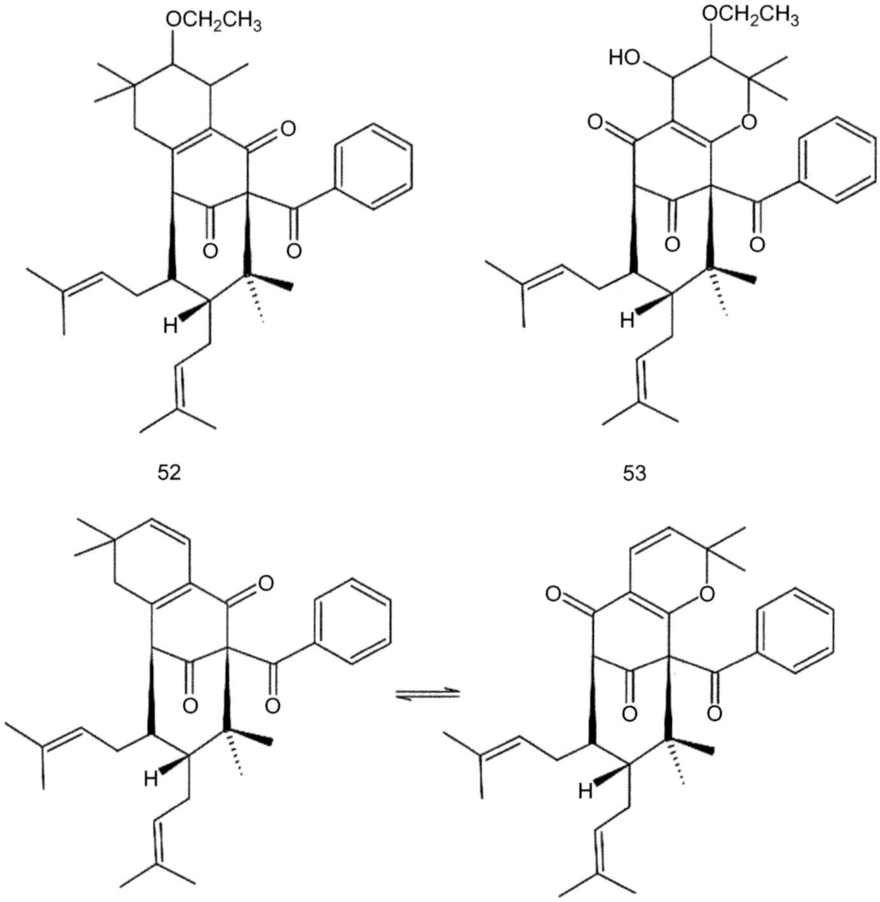

Figure 12.3
Prenylated benzophenones isolated from propolis (Trusheva et al., 2004).

(Bankova, 2005). For this reason, the botanical origin is crucial for the characterization of propolis, which is also affected by the geographical and climatic characteristics of the collection site (Bankova et al., 1998). An example of this variability is that in the temperate zones of the world, the predominant source of propolis is the exudate of *Populus* spp. (poplar), while in tropical regions bees need to look for different sources to make propolis (Gómez et al., 2006).

Analysis using chromatographic and magnetic resonance techniques has indicated that tropical propolis presents phytochemical differences, especially in phenol content and total flavonoids, due to differences in the vegetation surrounding the hives, compared to propolis from nontropical regions (Palomino et al., 2009). In a study carried out in the west of Yucatán, it was mentioned that the EEP obtained from the community of Santa Cruz, Maxcanú, Yucatán, had the highest total phenolic compounds content (94.29 ± 8.78 mg AG/g of EEP) and total flavonoids (42.6 ± 1.79 mg quercetin equivalents/g EEP) (Gutiérrez, 2016). Considering the above and that Yucatán, due to its geographical position, has a wide variety of vegetation types (low forest, coastal scrub, thorny jungle, medium forest), it is possible to assume that the total phenol content and flavonoids present in the propolis extracts differ remarkably among the three types of vegetation studied. Consequently, the biological effects of the EPP originating from the north, center, and west of Yucatán, where the presence of vegetation cover contrasts between one site and another, would also vary dramatically. The vegetation site represents an indicator of the content of bioactive compounds, representing a viable alternative for treatment and control for some of the diseases associated with the metabolic syndrome.

12.4 Conclusion

Several studies suggest the functionality of propolis. However, future researches are necessary to suggest it as a dietoterapeuctic alternative for the treatment and prevention of degenerative, chronic disease. Thus, propolis is a new prospect as a functional food. In this context, it is aim of study in food science and nutrition areas for its bioactive compounds and potential health benefits.

References

Ahn, M.R., Kumazawa, S., Hamasaka, T., Bang, K.S., Nakayama, T., 2004. Antioxidant activity and constituents of propolis collected in various areas of Korea. Journal of Agricultural and Food Chemistry 52, 7286–7292.

Ahn, M.R., Kumazawa, S.A., Usui, Y.A., Nakamura, J., Matzuca, M.B., Zhu, F.C., Tsutomu, N., 2007. Antioxidant activity and constituents of propolis collected in various areas of China. Food Chemistry 101 (4), 1383–1392.

Aljadi, A.M., Kamaruddin, M.Y., 2004. Evaluation of the phenolic contents and antioxidant capacities of two Malaysian floral honeys. Food Chemistry 85, 513–518.

Almeida, E.C.D., Menezes, H., 2002. Anti-inflammatory activity of propolis extracts: a review 2104. Journal of Venomous Animals and Toxins 8, 191–212.

Bankova, V., Boudourova-Krasteva, G., Popov, S., Sforcin, J.M., Funari, S.R.C., 1998. Seasonal variations of the chemical composition of Brazilian propolis. Apidologie 29 (4), 361–367.

Bankova, V.S., De Castro, S.L., Marcucci, M.C., 2000. Propolis: recent advances in chemistry and plant origin. Apidologie 31 (1), 3–15.

Bankova, V., 2005. Recent trends and important developments in propolis research. Evidence Based Complementary and Alternative Medicine 2, 29–32.

Banskota, A., Tezuka, Y., Midorikawa, K., Matsushige, K., Kadota, S., 2000. Two novel cytotoxic benzofuran derivatives from brazilian propolis. Journal of Natural Products 63, 1277–1279.

Banskota, A.H., Nagaoka, T., Sumioka, L.Y., Tezuka, Y., Awale, S., Midorikawa, K., Matsushige, K., Katoda, S., 2002. Antiproliferative activity of The Netherlands propolis and its active principles in cancer cell lines. Journal of Ethnopharmacology 80, 67–73.

Borrelli, F., Maffia, P., Pinto, L., Ianaro, A., Russo, A., Capasso, F., Ialenti, A., 2002. Phytochemical compounds involved in the antiinflammatory effect of propolis extract. Fitoterapia 73, 53–63.

Cardile, V., Panico, A., Gentile, B., Borrelli, F., Russo, A., 2003. Effect of propolis on human cartilage and chondrocytes. Life Sciences 73, 1027–1035.

Castaldo, S., Capasso, F., 2002. Propolis, an old remedy used in modern medicine. Fitoterapia 73 (1), 51–56.

Cho, M.J., Howard, L.R., Prior, R.L., Clark, J.R., 2004. Flavonoid glycosides and antioxidant capacity of various blackberry, blueberry and red grape genotypes determined by high-performance liquid chromatography/mass spectrometry. Journal of the Science of Food and Agriculture 84 (13), 1771–1782.

Choi, Y.M., Noh, D.O., Cho, S.Y., Suh, H.J., Kim, K.M., Kim, J.M., 2006. Antioxidant and antimicrobial activities of propolis from several regions of Korea. Lebensmittel-Wissenschaft and Technologie 39, 756–761.

Dantas, A.P., Olivieri, B.P., Gomes, F., De Castro, S.L., 2006. Treatment of Trypanosoma cruzi-infected mice with propolis promotes changes in the immune response. Journal of Ethnopharmacology 103, 187–193.

De Carvahlo, Machado.G.M., Leon, L.L., Lisboa, De Castro, S., 2007. Activity of Brazilian and Bulgarian propolis against different species of Leishmania. Memórias do Instituto Oswaldo Cruz, Rio de Janeiro 102 (1), 73–77.

De Castro, S.L., 2001. Propolis: biological and pharmacological activities. Therapeutic uses of this bee-product. Annual Review of Biomedical Sciences 3, 49–83.

De Castro, S.L., 2011. Propolis: biological and pharmacological activities. Therapeutic uses of this bee-product. Annual Review of Biomedical Sciences 3, 49–83.

Farré, R., Frasquet, I., Sánchez, A., 2004. Propolis and human health. Ars Pharmaceutica 45, 21–43.

Fokt, H., Pereira, A., Ferreira, A., Cunha, A., Aguiar, C., 2010. How do bees prevent hive infections? The antimicrobial properties of propolis. Current Research, Technology and Education Topics in Applied Microbiology and Microbial Biotechnology 1 (1), 481–493.

Frenkel, K., Wei, H., Bhimani, R., Ye, J., Zadunaisky, J.A., Huang, M.T., Ferrato, T., Conney, A.H., Gruenberger, D., 1993. Inhibition of tumor promoter-mediated processes in mouse skin and bovine lens by caffeic acid phenethyl ester. Cancer Research 53, 1255–1261.

Fuliang, H., Hepburn, H.R., Xuan, H., Chen, M., Daya, S., Radloff, S.E., 2005. Effects of propolis on blood glucose, blood lipid and free radicals in rats with diabetes mellitus. Pharmacological Research 51 (2), 147–152.

Gómez, C.A.M., Gómez, R.M., Arráez, R.D., Segura, C.A., Fernández, G.A., 2006. Advances in the analysis of phenolic compounds in products derived from bees. Journal of Pharmaceutical and Biomedical Analysis 41, 1220–1234.

Gómez, A.L., López, J.A., Rodríguez, A., Fortíz, A., Martínez, L.R., Apolinar, A., Enríquez, L.F., 2016. Producción de compuestos fenólicos por cuatro especies de microalgas marinas sometidas a diferentes condiciones de iluminación. Latin American Journal of Aquatic Research 44 (1), 137–143.

Gülçin, I., Bursal, E., Sehitoglu, M.H., Bilsel, M., Görden, A.C., 2010. Polyphenols contents and antioxidant activity of lyophilized aqueous extract of propolis from Erzurum, Turkey. Food and Chemical Toxicology 48, 2227–2238.

Gutiérrez, C.C.D., 2016. Evaluación de la calidad y actividad antioxidante del propóleo colectado en la región poniente de la península de yucatán, méxico. Tesis de maestría. Instituto Tecnológico de Mérida, Yucatán, México. 106 p.

Havsteen, B.H., 2002. The biochemistry and medical significance of the flavonoids. Pharmacology and Therapeutics 96, 67–202.

Hegazi, A., El Hady, F., 2001. Egyptian propolis: 1 Antimicrobial activity and chemical composition of upper Egypt propolis. Zeitschrift für Naturforschung 56, 82–88.

Huang, S., Zhang, C.P., Wang, K., Li, G., Hu, F.L., 2014. Recent advances in the chemical composition of propolis. Molecules 19, 19610–19632.

Isla, M.I., Paredes Guzman, J.F., Nieva Moreno, M.I., Koo, H., Park, Y.K., 2005. Some chemical composition and biological activity of northern Argentine propolis. Journal of Agricultural and Food Chemistry 53, 1166–1172.

Jameson, M., Dai, F.X., Luscher, T., Skopec, J., Diederich, A., Diederich, D., 1993. Hypertension. 21, 280–288.

Kang, L.J., Lee, H.B., Bae, H.J., Lee, S.G., 2010. Antidiabetic effect of propolis: reduction of expression of glucose-6-phosphatase through inhibition of Y279 and Y216 autophosphorylation of GSK-3α/β in HepG2 cells. Phytotherapy Research 24, 1554–1561.

Konishi, Y., 2005. Transepithelial transport of artepillin C in intestinal Caco-2 cell monolayers. Biochimica et Biophysica Acta 1713, 138–144.

Krol, W., Scheller, S., Czuba, Z., Matsuno, T., Zydowicz, G., Shani, J., Mos, M., 1996. Inhibition of neutrophils' chemiluminescence by ethanol extract of propolis (EEP) and its phenolic components. Journal of Ethnopharmacology 55, 19–25.

Kujumgiev, A., Tsvetkova, I., Serkedjieva, Y., Bankova, V., Christov, R., Popova, S., 1999. Antibacterial, antifungal and antiviral activity of propolis of different geographic origin. Journal of Ethnopharmacolog 64, 235–240.

Kumar, K.V., Das, U.N., 1993. Free radical Research Communication 19, 59–66.

Kumazawa, S., Hamasaka, T., Nakayama, T., 2004. Antioxidant activity of propolis of various geographic origins. Food Chemistry 84 (3), 329–339.

Lacy, F., O'Connor, D.T., Schmid-Schönbein, G.W., 1998. Journal of Hypertension 16, 291–303.

Lan, X., Wang, W., Li, Q., Wang, J., 2016. The natural flavonoid pinocembrin: molecular targets and potential therapeutic applications. Molecular Neurobiology 53 (3), 1794–1801.

Li, A., Xuan, H., Sun, A., Liu, R., Cui, J., 2016. Preparative separation of polyphenols from water-soluble fraction of Chinese propolis using macroporous absorptive resin coupled with preparative high performance liquid chromatography. Journal of Chromatography B 1012–1013, 42–49.

Lu, L.C., Chen, Y.W., Chou, C.C., 2005. Antibacterial activity of propolis against Staphylococcus aureus. International Journal of Food Microbiology 102, 213–220.

Lung-Ta, L., Ying-Tang, H., Jiuan-Jiuan, H., Amy, Y., Lee, L., Ferng-Chun, K., Chang-Jen, H., Chithan, K., Ping-Ping, H.L., Ming-Ting, L., 2004. Transinactivation of the epidermal growth factor receptor tyrosine kinase and focal adhesion kinase phosphorylation by dietary flavonoids: effect on invasive potential of human carcinoma cells. Biochemical Pharmacology 67 (11), 2103–2114.

Makowska-Was, J., Janeczko, Z., 2004. Bioavailability of plant polyphenols. Postepy Fitoterapii 3, 126–137.

Marcucci, M.C., Rodriguez, J., Ferreres, F., Bankova, V., 1998. Chemical composition of brazilian propolis from Sao Paulo state. Zeitschrift für Naturforschung C 53 (1–2), 117–119.

Marcucci, M.C., Ferreres, F., García-Viguera, C., Bankova, V. s., De Castro, S. l., Dantas, A.P., Valente, P.H.M., Paulino, N., 2001. Phenolic compounds from brazilian propolis with pharmacological activities. Journal of Ethnopharmacology 74,105–112.

Marcucci, M.C., 1995. Propolis: chemical composition, biological properties and therapeutic activity. Apidologie 26, 83–99.

Marquele, F.D., DI Mambro, V.M., Geogetti, S.R., Casagrande, R., Valim, Y., Fonseca, M.J., 2005. Assessment of the antioxidant activities of Brazilian extracts of propolis alone and in topical pharmaceutical formulations. Journal of Pharmaceutical and Biomedical Analysis 39, 455–462.

Maruyama, H., Sumitou, Y., Sakamoto, S., Araki, Y., Hara, H., 2009. Antihypertensive effects of flavonoids isolated from brazilian green propolis in spontaneously hypertensive rats. Biological and Pharmaceutical Bulletin 32 (7), 1244–1250.

Matsui, T., Ebuchi, S., Fujise, T., Abesundara, K.J., Doi, S., Yamada, H., Matsumoto, K., 2004. Strong antihyper-glycemic effects of watersoluble fraction of Brazilian propolis and its bioactive constituent, 3,4,5-triO-caffeoylquinic acid. Biological and Pharmaceutical Bulletin 27 (11), 1797–1803.

Matsushige, K., Basnet, P., Hase, K., Kadota, S., Tanaka, K., Namba, T., 1996. Propolis protects pancreatic beta-cells against the toxicity of streptozotocin (STZ). Phytomedicine 3, 203–209.

Mattace, R.G., Meli, R., Di Carlo, R., 2001. Inhibition of inducible nitric oxide synthase and cyclooxygenase-2 expression by flavonoids in macrophage J774A.1. Life Sciences 68 (8), 921–931.

Mirzoeva, O.K., Calder, P.C., 1996. The effect of propolis and its components on eicosanoid production during the inflammatory response. Prostaglandins Leukotrienes and Essential Fatty Acids 55 (6), 441–449.

Moreira, L., Dias, L.G., Pereira, J.A., Estevinho, L., 2008. Antioxidant properties, total phenols and pollen analysis of propolis samples from Portugal. Food and Chemical Toxicology 2008 (46), 3482–3485.

Murat, K., Serdar, K., Semra, K., 2002. GC/MS Analysis of propolis samples from two different regions of Turkey. eitschrift für Naturforschung 57, 905–909.

Murray, E., Bakers, G., 1996. New Approaches to antihipertensive therapy. Archives of Internal Medicine 156, 11–120.

Nagai, T., Inoue, R., Inoue, H., Ssuzuki, N., 2003. Preparation and antioxidant properties of water extract of propolis. Food Chemistry 80, 29–33.

Okutan, H., Ozcelik, N., Ramazan, Y.H., Uz, E., 2005. Effects of caffeic acid phenethyl ester on lipid peroxidation and antioxidant enzymes in diabetic rat heart. Clinical Biochemistry 38, 191–196.

Olszewska, M., 2003. Flavonoids and their use in therapy. Farmacja Polska 59, 391–402.

Oršolić, N., Knežević, A.H., Šver, L., Terzić, S., Bašić, I., 2004. Immunomodulatory and antimetastatic action of propolis and related polyphenolic compounds. Journal of Ethnopharmacology 94, 307–315.

Ozturk, F., Kurt, E., Cerci, M., Emiroglu, L., Inan, U.U., Turker, M., Ilker, S.S., 2000. Ophthalmic Research 32, 13–18.

Packer, L., Rimbach, G., Virgii, F., 1999. Antioxidant activity and biologic properties of a procyanidin — rich extract from pine (Pinus maritime) bark, pycnogenol. Free Radical Biology and Medicine 27, 704–724.

Palomino, L.R., García, P.C.M., Gil, G.J.H., Rojano, B.A., Durango, R.D., 2009. Determinación del contenido de fenoles y evaluación de la actividad antioxidante de propóleos recolectados en el departamento de Antioquia (Colombia). Vitae, Revista de la Facultad de Química Farmacéutica 16 (3), 2145–2660.

Park, E.H., Kahng, J.H., 1999. Suppressive effects of propolis in rat adjuvant arthritis. Archives of Pharmacal Research 22, 554–558.

Park, Y.K., Alencar, S.M., Aguiar, C.L., 2002. Botanical origin and chemical composition of Brazilian propolis. Journal of Agricultural and Food Chemistry 50 (9), 2502–2506.

Pietta, P.G., Gardana, C., Pietta, A.M., 2002. Analytical methods for quality control of propolis. Fitoterapia (73 Suppl. 1), S7–S20.

Porto, A.L., Machado, S.M., de Oliveira, C.M., Bittrich, V., Amaral, M.C., Marsaioli, A.J., 2000. Polyisoprenylated benzophenones from Clusia floral resins. Phytochemistry 55 (7), 755–768.

Rasul, A., Millimouno, F.M., Ali Eltayb, W., Ali, M., Li, J., Li, X., 2013. Pinocembrin: a novel natural compound with versatile pharmacological and biological activities. BioMed Research International 9 Article ID 379850.

Rodríguez, M., 2003. Fenton and UV-vis bases advanced oxidation processes in wastewater treatment: Degradation, mineralization and biodegradability enhancement. Tesis doctoral, Universitat de Barcelona, España 1–294.

Rohdewald, P., 2002. A review of the French maritime pine bark extract (Pycnogenol), a herbal medication with a diverse clinical pharmacology. International Journal of Clinical Pharmacology and Therapeutics 40, 158–168.

Rossi, M., Garavello, W., Talamini, E.N., Bosetti, C., Dal Mso, L., Lagiou, P., Tavani, A., Polesel, J., Barzan, L., Ramazzotti, V., Franceschi, S.Y., Vecchia, C.L., 2002. Flavonoids and the risk o oral and pharyngeal cáncer: a case-control study from Italy. Cancer Epidemiology, Biomarkers and Prevention 16 (8), 1621–1625.

Russo, R.O., Speranza, S.M., 2006. Los flavonoides en la terapia cardiovascular. Revisión. Revista Costarricense de Cardiología 8 (1), 37–42.

Saad, M.A., Abdel, S.S.M., Kenawy, S.A., Attia, A.S., 2015. Pinocembrin attenuates hippocampal inflammation, oxidative perturbations and apoptosis in a rat model of global cerebral ischemia reperfusion. Pharmacological Reports 67 (1), 115–122.

Saiz, C.M., Serrano, J., 2003. Propóleo: aplicaciones terapeúticas. Natura Medicatrix 21 (2), 94–104.

Salamanca, G.G., 2002. Origen naturaleza y características de los propóleos. In: Memorias XVI Seminario Americano de Apicultura. Secretaria de Agricultura Pesca y Alimentación, México, pp. 101–110.

Scow, D., Smith, E., Shaughnessy, P., 2003. Combination therapy ACE inhibitors and angiotensina-receptor blockers and hearth failure. American Family Physician 68 (9), 1795–1798.

Sforcin, J., Bankova, V., 2011. Propolis: is there a potential for the development of new drugs? Journal of Ethnopharmacology 133, 253–260.

Singleton, V.L., Rossi, J.A., 1965. Colorimetry of total phenolics with phosphomolybdic-phosphotungstic acid reagents. American Journal of Enology and Viticulture 16, 144–158.

Soleo De Funari, C., De Oliveira, F.V., Mathor, M.B., 2007. Analysis of propolis from Baccharis dracunculifolia DC. (Compositae) and its effects on mouse fibroblasts. Journal of Ethnopharmacology 111, 206–212.

Suzuki, A., Kagawa, D., Ochiai, R., Tokimitsu, I., Saito, I., 2002. Hypertension Research 25, 99–107.

Tokunaga, K., Yoshida, C., Suzuki, K., Maruyama, H., Futamura, Y., Araki, Y., Mishima, S., 2004. Antihypertensive effect of peptides from royal jelly in spontaneously hypertensive rats. Biological and Pharmaceutical Bulletin 27, 189–192.

Trusheva, B., Popova, M., Naydenski, H., Tsvetkova, I., Rodriguez, J.G., Bankova, V., 2004. New polyisoprenylated benzophenones from Venezuelan propolis. Fitoterapia 75, 683–689.

Trusheva, B., Popova, M., Bankova, V., Simova, S., Marcucci, M.C., Laguna, P., Da Rocha Pasin, F. Tsvetkova, I., 2006. Bioactive constituents of brazilian red propolis. Advance Acces Published, 1–6.

Wagh, V.D., 2013. Propolis: a wonder bees product and its pharmacological potentials. Advances in Pharmacological Sciences 11.

Walker, P., Crane, E., 1987. Constituents propolis. Apidologie 18, 327–334.

Wang, B.J., Lien, Y.H., Yu, Z.R., 2004. Supercritical fluid extractive fractionation- study of the antioxidant activities of propolis. Food Chemistry 86, 237–243.

Waris, G., Ahsen, H., 2006. Reactive oxigene species role in the development of cáncer and various chronic conditions. Journal of Carcinogenesis 5 (14), 1–8.

World Health Organization, 2013. Global Health Observatory (GHO). Raised blood pressure. Situation and trends Disponible en: http://www.who.int/gho/ncd/risk_factors/blood_pressure_prevalence/en/.

Yang, H., Dong, Y., Du, H., Shi, H., Peng, Y., Li, X., 2011. Antioxidant compounds from propolis collected in Anhui, China. Molecules 16, 3444–3455.

Yao, L.H., Jiang, Y.M., SHI, J., Tomas-Barberan, F.A., Datta, N., Singanusong, R., Chen, S.S., 2004. Flavonoids in food and their health benefits. Plant foods for Human Nutrition 59, 113–122.

Zamami, Y., Takatori, S., Koyama, T., Goda, M., Iwatani, Y., Doi, S., Kawasaki, H., 2007. Effect of propolis on insulin resistance in fructosedrinking rats. Yakugaku Zasshi 127 (12), 2065–2073.

PART 9

Bioactive Compounds and Functional Foods as Therapeutic Alternative

CHAPTER 13

Bioactive Compounds as Therapeutic Alternatives

Armando M. Martín Ortega, Maira Rubi Segura Campos
Facultad de Ingeniería Química, Universidad Autónoma de Yucatán, Mérida, Mexico

13.1 Introduction

The consumption of bioactive compounds started at the beginning of man's existence. Since then, foods of vegetable origin, including plants, fruits, and vegetables have been consumed, as evidenced by the discovery of remains of plants, legumes, and seeds in dental pieces of Neanderthal fossils (Henry et al., 2011). Likewise, there is evidence of the use of medicinal plants for the purpose of treating specific health conditions from ancient times, with more than 5000 years of antiquity coming from the Sumerian civilization. Followed by Chinese books dating from 2500 BC, mentioning herbs and medicinal roots, and describing their medicinal use for different conditions; many of the preparations include plants and roots still used today, such as: cinnamon bark, ephedra, and ginseng. Other civilizations, such as in India and Egypt, also have books over 1000 years old, which discuss the use of plants, spices, and foods for the treatment of different diseases (Petrovska, 2012).

The introduction of the terms functional foods and nutraceuticals illustrates the beneficial effect of the consumption of different foods and bioactive compounds in proper quantities and frequencies. Functional foods are considered to be: whole foods, enriched or fortified, the consumption of which, in adequate quantities, offers a health benefit beyond its nutritional value (Hasler, 2002). Meanwhile, the word nutraceutical is derived from the words "nutrient" and "pharmaceutical," referring to those supplements derived from bioactive compounds isolated from medicinal plants and/or nutrients, that besides offering a nutritional value can be used as a drug (Nasri et al., 2014).

Nowadays, thanks to technological and scientific advances, it is possible to extract, characterize, and evaluate bioactive compounds from foods and medicinal plants. Thus, there are a large number of in vitro, in vivo, randomized, and clinical studies evaluating the ability of bioactive compounds to provide health benefits. Therefore, this chapter is intended to describe the main beneficial health effects of bioactive compounds, starting from major diseases and health conditions that are in the first places of death worldwide, including cardiovascular disease, cancer, diabetes, neurodegenerative diseases, and aging.

Bioactive Compounds. https://doi.org/10.1016/B978-0-12-814774-0.00013-X

13.2 Bioactive Compounds

Bioactive compounds are essential or nonessential compounds that are found in nature or are created during the processing of foods or medicinal plants, and are capable of modulating different biological activities, benefiting health (Biesalski et al., 2009). Essential compounds are those that fulfill an essential biological function in the human body and the deficiency of which can lead to the development of diseases. Nonessential compounds are not necessary or do not fulfill an essential biological function in the human body and lacking them does not affect health (Biesalski et al., 2013).

Among the most studied essential compounds are vitamins, including vitamins A, B, C, D, and E. For example, vitamin C has been studied under different conditions, its therapeutic effect has been demonstrated in many diseases and health disorders, such as allergies, synthesis of neurotransmitters, conversion of cholesterol to bile acids, bone formation, wound healing, and maintenance of connective tissue, among others (Chambial et al., 2013).

Fatty acids have been well studied due to their participation in a great number of physiological and pathophysiological processes, finding great benefit besides the nutritional aspect. Also, there are different types of fatty acids, with monounsaturated and polyunsaturated being most beneficial. Among the polyunsaturated fatty acids are omega 3, which is considered antiinflammatory, neuroprotective, essential for adequate neurodevelopment, and protective in cardiovascular and metabolic diseases (Yan et al., 2013; Denis et al., 2015; Mori, 2014).

Phytochemicals are major bioactive compounds, and are considered nonessential (Biesalski et al., 2009). These, in turn, are very diverse and classified according to the nature of their chemical structures being: terpenoids, polyphenols, organosulfur compounds, phytosterols, and alkaloids (Somani et al., 2015). This classification of phytochemicals also has other classifications that take into account other characteristics, such as the number of carbons and number of rings, among other things (Tsao, 2010). This class of compounds is found mainly in fruits, vegetables, cereal grains, algae, and medicinal plants, as they are synthesized by the secondary metabolism of plant cells (Zhang et al., 2015). Among the most common fruits that are rich in phytochemicals are wild berries, blueberries, blackberries, pomegranates, and strawberries. The vegetables richest in phytochemicals are spinach, red peppers, broccoli, asparagus, and others that contain high antioxidant capacity and content of polyphenols (Liu, 2013). Among the cereals, red rice, black rice, and purple rice, are considered high in phytochemicals. Medicinal plants such as *Camellia sinensis, Dioscorea bulbifera*, and *Ephedra sinica*, and a great variety of plants and flowers have phytochemicals with potential beneficial effects on health (Zhang et al., 2015).

A third class of bioactive compounds is those generated by food processing. For example, biopeptides are peptide fractions of 3–20 amino acids that possess biological activity (Walther and Sieber, 2011). This class of bioactive compounds has become very important from the point of view of food science, since its use in the products offers a double benefit: protection of the food and transformation of the product into a functional food. A functional food is one that offers a health benefit beyond its nutritional qualities (Martínez Leo et al., 2016). The generation of these peptides is through controlled enzymatic hydrolysis, under in vitro conditions; bacterial fermentation, directly on the food; gastrointestinal digestion in vivo; and/or their generation during food preparation (Walther and Sieber, 2011). Among its main biological activities are antiinflammatory, antioxidant, antihypertensive, hypolipidemic, hypoglycemic, among others, offering health benefits such as cardiometabolic disease control, reduction of oxidation reactions (antiaging), blood pressure control, risk reduction of cardiovascular events, among the main ones (Nasri, 2016; Walther and Sieber, 2011).

13.3 Health Benefits

13.3.1 Antiaging

Aging is an inevitable process and has been defined as the time-dependent loss of functionality of the organism and is characterized by progressive dysfunction of organs and systems, with increasing vulnerability to death (López-Otín et al., 2013). This process is influenced by genetic, environmental, and lifestyle factors (including dietary habits). There is evidence that some phytochemicals have antiaging potential, among the most studied are resveratrol, epicatechin, and curcumin (Corrêa et al., 2016).

It has been considered that aging is due to different mechanisms that tend to happen simultaneously, including (López-Otín et al., 2013):

- Genomic instability (accumulation of DNA mutations, mitochondrial DNA damage, and changes in cell nucleus architecture);
- Wear of telomeres;
- Epigenetic alterations (histone modifications, DNA methylations, and chromatin remodeling);
- Loss of protein homeostasis (alterations in the synthesis and folding of proteins);
- Deregulated nutrient-sensing;
- Mitochondrial dysfunction;
- Cell senescence (loss of proliferative capacity);
- Stem cell exhaustion;
- Loss of intercellular communication.

Despite the aforementioned mechanisms, the aging process is not fully understood. However, evidence has shown that these mechanisms are related to oxidative processes mediated by free radicals (e.g., reactive oxygen species) and chronic inflammation (Si and Liu, 2014). Inflammation is the response to physical, chemical, or microorganism damage by vascularized living tissue. This process is characterized by an increase in different proinflammatory cytokines, such as tumor necrosis factor α (TNF-α), interleukin 1 (IL-1), and interleukin-6 (IL-6) and the recruitment of cells of the immune system to the tissue when the organ attacked. Proinflammatory cytokines modulate different intracellular pathways, including those mediated by nuclear factor kappa beta (NF-κβ) modulating the expression of various proinflammatory and antioxidant genes (Wardyn et al., 2015). Thus, the inflammatory process induces the generation of free radicals and reduces the antioxidant capacity of the cells, facilitating the oxidation of the fatty acids of the cell membranes, causing cell death. Free radicals can cause DNA damage, promote mutations, mitochondrial dysfunction, and alteration in transcription factors that regulate the expression of several genes (Fougère et al., 2016).

Several phytochemicals have been shown to have an antiaging potential by their ability to directly reduce oxidizing agents (e.g., reactive oxygen species), induce the expression of antioxidant enzymes, modulate nutrient- or energy-regulated intracellular pathways, regulate hormonal axes [(e.g., growth hormone GH)/insulin-like growth hormone-1 (IGF1)], antiinflammatory properties, and regulation of enzymes that produce epigenetic changes (e.g., nicotinamide adenine nucleotide-dependent protein deacetylases) (Si and Liu, 2014).

13.3.1.1 Resveratrol

Resveratrol (3,5,4′-trihydroxystilbene) is an antioxidant belonging to the family of polyphenols, found mainly in the skin of grapes, and also in red wine, peanuts, blueberries, pistachios, etc. (Mukherjee et al., 2010).

One of the main biological activities of this phytochemical is to reduce the inflammatory response by inhibiting NF-κβ, reducing free radicals, and increasing the expression of antioxidant enzymes and those involved in the process of cellular and hepatic detoxification (example, cytochromes P-450). Thus, they reduce cellular damage and prevent the generation of mutations that may lead to genomic instability and mitochondrial dysfunction (Lastra and Villegas, 2005). In addition, it also has the ability to increase the expression of nicotinamide adenine nucleotide (NAD)-dependent protein deacetylases. These enzymes are responsible for regulating various cellular processes, taking special importance in epigenetic changes, modulating gene silencing, DNA repair, and cell cycle regulation, reducing the aging process and lengthening life expectancy (Borra et al., 2005).

Resveratrol has been shown to have the ability to bind to estrogen receptors, there is evidence of having the ability to reduce symptoms related to the perimenopausal period and protect against the development of osteoporosis, which are both related to aging processes in women (Baxter, 2008).

13.3.1.2 Epigallo-Catechin-3-Gallate

Epigallo-catechin-3-gallate (EGCG) is a phytochemical found in the leaves of *Camellia sinensis* (green tea). Green tea is one of the most consumed and popular drinks in the world.

The consumption of such tea has been related to several health benefits, including prevention of cardiovascular diseases and cancer, antioxidant, antiinflammatory, lipid-lowering, antiviral, and neuroprotective effects among the main ones (Chacko et al., 2010).

EGCG has demonstrated a wide variety of biological effects, including some antiaging properties such as inhibition of epigenetic changes, induction of expression of antioxidant enzymes, and inhibition of NF-κβ (antiinflammatory effect) (Singh et al., 2011). Inhibition of epigenetic changes is regulated by the formation of EGCG hydrogen bonds with DNA methyltransferase (DNMT) enzymes, inhibiting their activity. The DNMT are responsible for attaching methyl groups to cytosine in cytosine-guanine dinucleotides (CpG) located in DNA and silencing gene expression (Singh et al., 2011). Inhibition of DNMT reduces the methylation of CpG in the DNA, inducing the expression of those genes that were silenced. Also, it has been found that EGCG increases the expression of the gene that encodes the enzyme glutathione-S-transferase, this enzyme catalyzes the conjugation of glutathione with a great variety of electrophilic compounds that have the potential to generate DNA damage. In addition, the reduction of methylation status is related to cancer prevention (Yang et al., 2010).

EGCG has direct and indirect antioxidant effects, counteracting free radicals and inducing the expression of antioxidant genes, respectively. However, several studies have shown that EGCG may have pro-oxidant effects. These pro-oxidant effects are related to the basal oxidative state of the person, with the finding that people with high basal levels of oxidative stress will have an antioxidant effect and vice versa. In spite of the above, EGCG regulates the expression of antioxidant enzymes, including glutathione peroxidase and reductase, catalase, quinone reductase, and superoxide dismutase, improving the oxidative state in in vitro and in vivo studies (Forester and Lambert, 2013). Moreover, consumption of green tea extract (18.6 mg/day) increases the antioxidant capacity of plasma by 15.6% in humans (Young et al., 2002). In addition, consumption of 400 mL of green tea increases the antioxidant capacity of human plasma by 4%, 40 min after consumption (Benzie et al., 2009)

13.3.1.3 Curcumin

Curcumin is the active compound present in *Curcuma longa*, traditionally used as a herbal remedy in traditional Chinese and Indian medicines and also as a cooking spice. This polyphenol has pleiotropic activities because of its ability to modulate different intracellular pathways, thus attributing antioxidant, antiinflammatory, and chemopreventive (cancer prevention) capabilities (Hatcher et al., 2008). In addition, clinical studies have shown that curcumin has the ability to combat various diseases including cancer, ulcerative colitis, gastric ulcer, diabetes, atherosclerosis, lupus, Crohn's disease, etc. (Gupta et al., 2013).

Meanwhile, curcumin, as with resveratrol and epigallocatechin, has antiaging properties, reducing inflammatory and oxidative processes through the regulation of NF-κβ, its antioxidant capacity, and by inducing antioxidant enzymes. Thus, curcumin reduces oxidative

DNA damage and inflammatory processes, reducing the aging process and preventing the development of age-related diseases such as cancer, and cardiovascular and neurodegenerative diseases (Sikora et al., 2010)

Also, other phytochemicals, such as those found in olive oil (Rigacci and Stefani, 2016), onion, garlic, and blueberries, present antiaging activities and prevent diseases related to aging (Si and Liu, 2014).

13.3.2 Prevention of Cardiovascular Diseases

Cardiovascular diseases (CVDs) represent the leading cause of death worldwide, being a relevant socioeconomic problem (Pagliaro et al., 2015). Among the main causes of CVD is atherosclerosis, a major risk factor for the development of acute myocardial infarction and cerebrovascular events (Golia et al., 2014). However, obesity and diabetes mellitus are pathological conditions considered as cardiovascular risk factors as they promote a hypercoagulant state, and chronic oxidative and inflammatory stress, doubling the risk of CVD (Bhupathiraju and Hu, 2016). Also, it is important to consider dyslipidemias, oxidative stress, and chronic inflammation as causal factors of atherosclerosis (Rizzo et al., 2009).

Dyslipidemia is the abnormal elevation of lipids in the blood, mainly linked to transport proteins called lipoproteins. Oxidation of low-density lipoproteins (LDLs), the main cholesterol containers, generates accumulation in the extracellular matrix rich in proteoglycans of endothelial cells. This accumulation of oxidized LDL in the endothelium stimulates the synthesis of macrophage chemotactic protein-1 (MCP-1), which promotes the accumulation and activation of monocytes (immune cells), which acquire characteristics of macrophages and phagocyte the oxidized LDL, accumulate in the arterial wall and become foam cells. This process is accompanied by the production of free radicals and the release of proinflammatory cytokines, thus promoting the accumulation of more macrophages forming atheromas that reduce the inner space of blood vessels. There are also lipoproteins that have a protective effect, which are called high-density lipoproteins (HDLs) (Sakakura et al., 2013).

Several active compounds found in foods and medicinal plants have protective properties against CVD. Among the main biological activities that prevent this kind of disease are lipid-lowering, antiinflammatory, and antithrombotic antioxidants, since they contribute to reducing the atherosclerotic process and reduce the risk of forming a clot produced by the rupture of an atheroma (Riccioni et al., 2015). Among the main active compounds studied in the prevention of CVD are phytochemicals, such as lycopene (Liu, 2013) and resveratrol (Riccioni et al., 2015), while among the essential compounds are vitamin C (Chambial et al., 2013), vitamin E (Saremi and Arora, 2010), and omega-3 fatty acids (Jain et al., 2015).

13.3.2.1 Resveratrol

As already mentioned in the previous section, resveratrol has antioxidant and antiinflammatory properties. Similarly, resveratrol in in vitro and in vivo studies has proven to be a potent inhibitor of lipid peroxidation, thus preventing LDL oxidation and exerting an antiatherogenic potential (Riccioni et al., 2015). Likewise, several studies have shown that resveratrol has lipid-lowering effects, reducing plasma levels of LDL-cholesterol and triglycerides. However, other studies do not show this effect (Riccioni et al., 2015). Clinical studies have shown that resveratrol improves endothelial function in patients with metabolic syndrome, reduces LDL oxidation by up to 20%, and improves the inflammatory and thrombotic state of patients (Tomé-Carneiro et al., 2013).

13.3.2.2 Lycopene

Lycopene is a phytochemical belonging to the family of carotenoids and the main pigment that gives the red color to tomato. Unlike the rest of the carotenoids, lycopene does not have provitamin A activity. It is found mainly in tomato and its derivatives such as ketchup, tomato juice, pizza sauce, etc. (Story et al., 2010). It is important to mention that lycopene is a fat-soluble phytochemical and its intestinal absorption is benefited by the consumption of fatty foods such as avocado, which increases its absorption by up to four times (Unlu et al., 2005).

Several preclinical studies have shown that consumption of tomato or lycopene reduces plasma LDL cholesterol levels, reduces oxidation, and elevates HDL levels significantly (Hadley et al., 2003; Bohn et al., 2013; Story et al., 2010) In addition, a study by Gajendragadkar et al. (2014) showed that supplementation with 7 mg/day of lycopene improves endothelial function in patients with a history of CVD disease and not in healthy volunteers.

13.3.2.3 Vitamin C

Vitamins are essential components required in various biochemical reactions in the human body and also participate in the regulation of various physiological functions. Most vitamins cannot be synthesized or at least not in sufficient quantity. It has been recognized that adequate doses of vitamins may have a pharmacological effect, so they have also been characterized as bioactive compounds (Chambial et al., 2013).

Meanwhile, vitamin C is also known as ascorbic acid and is found mainly in citrus fruits such as oranges and lemons, and in peppers, strawberries, tomatoes, broccoli, spinach, and other vegetables (Chambial et al., 2013).

Vitamin C is widely recognized for its antioxidant power and its involvement in oxidoreductive and hydroxylation reactions in the body; its deficiency has been linked to an increase in plasma levels of LDL-cholesterol and its oxidation (Chambial et al., 2013). Meanwhile, its adequate intake and supplementation with 500–1000 mg/day has been shown to reduce concentrations of oxidized LDL and lipid peroxidation. Thus it contributes to a reduction in the atherosclerotic process (Chambial et al., 2013). In addition, its adequate consumption improves vascular strength by stimulating the synthesis of collagen, a structural protein that maintains the integrity of the vascular endothelium and other tissues (Chambial et al., 2013).

13.3.2.4 Vitamin E

Vitamin E is a fat-soluble vitamin found in nature in eight different varieties. The most abundant and biologically active variety is alpha-tocopherol. This vitamin is found mainly in seeds such as peanuts, almonds, and sunflower seeds and also in avocado and olive oil (Saremi and Arora, 2010).

Several in vitro, animal and observational studies in humans have shown that vitamin E reduces lipid oxidation, reducing the atherosclerotic process (Saremi and Arora, 2010). The inhibition of lipid peroxidation also regulates the synthesis of proinflammatory prostaglandins through the inhibition of cyclooxygenase-2, reducing platelet aggregation and the inflammatory process. Thus reducing the risk of a CVD (Jiang, 2014). Despite the above, vitamin E supplementation is not recommended, but it is necessary to maintain a high consumption of foods that contain it (Saremi and Arora, 2010).

13.3.2.5 Omega 3 Fatty Acids

Omega-3 polyunsaturated fatty acids are dietary fats mainly found in marine foods such as algae and salmon. They are also found in seeds such as flaxseed and chia (Tur et al., 2012). This class of fatty acids is recognized for its ability to incorporate into the plasma membranes of different cells of the body, including those of the immune system and thus reducing inflammatory processes (Swanson et al., 2012). They have also been shown to have the ability to regulate different intracellular signaling pathways that regulate the expression of approximately 1040 genes, including those regulating the synthesis of triglycerides, eicosanoids, and adipogenesis (Swanson et al., 2012).

The most active omega 3 fatty acids are those of marine origin, mainly EPA and DHA. Consumption between 2 and 4 g/day of EPA and DHA has been shown to reduce plasma triglyceride and blood pressure levels, contributing to the control of dyslipidemias and arterial hypertension (Jain et al., 2015).

13.3.3 Prevention of Chronic Diseases

Chronic diseases are considered to be long-term diseases, which usually occur during a period of long latency before being clinically present; they have a multifactorial etiology and do not have a definitive cure. In this category are diseases such as diabetes, cancer, and neurodegenerative diseases (e.g., Alzheimer's disease) (Martin, 2007). This class of diseases is among the leading causes of death worldwide according to the World Health Organization

(Gostin et al., 2016). Chronic diseases are characterized by having a chronic inflammatory component and a pro-oxidant state, leading the body to a degenerative state and making it prone to the development of various diseases (Martínez-Leo et al., 2016).

13.3.3.1 Diabetes

Type II diabetes mellitus is a noncommunicable chronic disease. This disease is characterized by insulin resistance, reduced insulin secretion, progressive pancreatic β-cell dysfunction, chronic hyperglycemia, glucose intolerance, and chronic proinflammatory status (Al-Goblan et al., 2014). Obesity, excessive consumption of fats and sugars, lifestyle, and genetic factors are some of the major risk factors associated with its development (Wu et al., 2014).

Similarly, some dietary patterns have been linked to a reduction in the risk of developing diabetes mellitus, mainly the consumption of foods rich in active compounds such as fruits, vegetables, and whole grains. These foods have a wide variety of bioactive compounds, including a significant amount of polyphenols and carotenoids (Kiec-Wilk and Mykka, 2008).

An antioxidant reduction in diabetes has been documented and has been associated with elevated cardiovascular risk. Several studies have demonstrated a plasma reduction of antioxidants such as α and β carotene, lycopene, ascorbic acid, lutein, etc., being associated with complications such as atherosclerosis and endothelial dysfunction in patients with diabetes. Thus, consumption of polyphenols, carotenes, and vitamins C and E are important active compounds with the potential to reduce oxidative stress and cardiovascular risk (Kiec-Wilk and Mykka, 2008).

Several phytochemicals have been associated with benefits to people with diabetes mellitus, including cinnamaldehyde, epigallocatechin, and chlorogenic acid, showing hypoglycemic effects, α-amylase inhibitor and insulin sensitizer, respectively (Zhu et al., 2017; He et al., 2006; Meng et al., 2013).

Epidemiological studies have found an inverse relationship between serum levels of carotenoids (lycopene, β-carotene, zeaxanthin, α-carotene, and β-cryptoxanthin) and glucose intolerance, as well as type II diabetes mellitus. Thus, a diet rich in these active compounds may be an important factor in preventing the development of this disease (Coyne et al., 2005).

13.3.3.2 Cancer

Cancer is a worldwide public health problem and is among the leading causes of death. According to the World Health Organization's World Cancer Report (Stewart and Wild, 2014), it was estimated that in 2012 there were 14.1 million new cases diagnosed with cancer and 8.2 million cancer deaths. Cancer is a genetic disease with a multifactorial etiology, with exposure to cigarette smoke, drugs, synthetic carcinogenic substances, viruses, hormones, ultraviolet radiation, heavy metals, alcohol, and some culinary techniques being factors related to the appearance of this disease (Blackadar, 2016).

At present, a great number of researches have been carried out regarding cancer and various active compounds (phytochemicals and vitamins) in order to know its beneficial effects, as well as its deleterious effects on this disease. Thus, several epidemiological studies have shown that the consumption of foods rich in active compounds such as fruits, vegetables, whole grains, green tea, oilseeds, and legumes is a protective factor (Baena-Ruiz and Salinas-Hernández, 2014).

Consumption of polyphenols with estrogenic effects such as isoflavones and flavanone from soybeans and other beans has been associated with a 29% lower risk of relapse and death in patients with breast cancer in remission. Also, green tea consumption has been shown to reduce malignant cell counts by up to 30% in people with leukemia and to reduce tumor markers of cell proliferation such as prostate specific antigen (PSA) in men with prostate cancer (Thomas et al., 2015). Likewise, in people with skin cancer, the specific consumption of foods rich in lutein prevents the formation of new tumor cells (Heinen et al., 2007).

Some phytochemicals such as curcumin, lycopene, and resveratrol are found in clinical studies because of their anticancer potential and with adequate dosing have been shown to have a promising therapeutic effect (Dhillon et al., 2008; Kucuk et al., 2002). In addition, they are important adjuvant agents in the prevention of side effects of chemotherapy, radiotherapy, and comorbidities such as diabetes and atherosclerosis. However, it is important to mention that some phytochemicals may reduce the efficacy of various chemotherapeutic components, due to their ability to increase the metabolism of various drugs and/or reduce their mechanism of action.

13.3.3.3 Curcumin

Curcumin is a bioactive compound that in several clinical studies has been shown to have the potential to increase the sensitization of cancer cells to chemotherapeutic drugs. Doses between 6000 and 8000 mg have shown beneficial effects in patients with pancreatic cancer and breast cancer, reducing the inflammatory state (by reducing NF-κβ expression), reducing tumor size, and increasing the profile of antiinflammatory cytokines (Dhillon et al., 2008; Bayet-Robert et al., 2010).

In addition, in patients with colon cancer, supplementation with 360 mg three times a day increased expression of the p53 protein (an important protein in cell cycle regulation and programmed cell death) inducing more apoptosis. It also reduced TNF-α levels by improving the inflammatory status of patients (He et al., 2011).

13.3.3.4 Lycopene

Epidemiological studies have found an inverse association between dietary consumption of lycopene and the development of prostate cancer. Several mechanisms have been postulated to prevent cancer, including regulation of the expression of tumor suppressor genes, prevention of oxidative DNA damage, and inhibition of tumor cell growth (Kucuk et al., 2002).

Lycopene has been used in clinical trials with patients with prostate cancer, occasionally finding a reduction of the PSA tumor marker and levels of insulin-like growth factor-1 (a hormone that induces cell growth). In addition, an increase in proteins related to intercellular junctions has been found, which reduces the risk of metastasis (Kucuk et al., 2001).

Thus, bioactive compounds derived from foods and/or medicinal plants are important in the prevention, treatment, and remission of cancer.

13.3.3.5 Neurodegenerative Diseases

Neurodegenerative diseases are characterized by chronic and progressive loss of central nervous system (CNS) function, leading to partial loss of function and the appearance of mental disorders (Chen et al., 2016).

The causes associated with CNS degeneration are still poorly understood as multiple factors interact. However, it is well known that age is related to its onset, being more frequent in adults of older and middle age. This class of diseases encompasses Alzheimer's disease, Parkinson's disease, and multiple sclerosis. On the other hand, it has been identified that some viruses are capable of generating damage to neurons, their death or inducing them to undergo apoptosis. Thus leading to a neurodegenerative disease (Chen et al., 2016).

Other factors related to the appearance of neurodegenerative diseases are oxidative stress and neuroinflammation. Neuroinflammation has almost the same mechanisms as those mentioned in aging and in cardiovascular and/or metabolic diseases (Chen et al., 2016). In addition, the chronic systemic inflammatory state contributes to vascular alterations in the small blood vessels of the brain, reducing its blood perfusion and thus generating the chronic death of CNS cells, causing an increase in neuroinflammation and CNS degeneration (Chen et al., 2016). However, neuroinflammation is a necessary process for neuronal remodeling following damage caused by viruses, trauma, or ischemia, and thus allowing tissue regeneration. That said, chronic neuroinflammation is actually responsible for neurodegenerative diseases (Amor et al., 2010). In addition, the reduction of the expression of some neurotropic factors, related to the survival, maintenance, and regeneration of some neuronal populations, is associated with an increase in the rate of progression of neurodegenerative diseases (Weissmiller and Wu, 2012; Venkatesan et al., 2015).

The brain is a very delicate organ, which has different security measures to protect its integrity against any aggressor agent or stimulus. Among these measures is the blood–brain barrier (BBB), which is made up of vascular endothelium with highly effective tight junctions, transendothelial transport systems, enzymes, and a regulation of leukocyte infiltration, generating a physical barrier that regulates transport, immune system, and the enzymes that can pass into the brain (Abbott and Friedman, 2012). In this sense, the passage of nutrients and bioactive compounds is highly regulated. However, various phytochemicals have been found with CNS activities that have the ability to penetrate the BBB (Venkatesan et al., 2015).

Thus, several bioactive compounds have been shown to possess antioxidant, antiinflammatory, and neuroprotective activities, including resveratrol, theobromine, epigallocatechin gallate, curcumin, gallic acid, etc. (Venkatesan et al., 2015). The consumption of food and/or medicinal plants with bioactive compounds, together with a suitable proportion of macro- and micronutrient consumption, is able to induce neuronal regeneration and differentiation, neuronal survival, neuroprotection, improve memory, and also delay the appearance and reduce the rate of progression of various neurodegenerative diseases (Venkatesan et al., 2015).

Bioactive compounds achieve the task of preventing and delaying the appearance of neurodegenerative processes through various mechanisms of action, including induction of the expression of antioxidant enzymes (catalase and superoxide dismutase), direct inhibition of free radicals, reduction of the activation of NF-κβ (reducing the synthesis of TNF-α and IL-1β), and reduction of proinflammatory prostaglandin synthesis. Some components have the ability to interact with various neuronal receptors, thus inducing the synthesis of neurotropic factors that promote regeneration, growth, and neuronal survival, reducing the rate of progression of neurodegenerative diseases (Venkatesan et al., 2015).

13.3.3.6 6-Shogaol

6-Shogaol is a bioactive compound derived from ginger (*Zingiber officinale* Roscoe), a spice used in food preparation and in traditional Indian medicine. This compound has been shown in in vitro study to reduce the synthesis of IL-1β, prostaglandin E2, and TNF-α in cells of the brain's immune system. Also, this component has been shown to be effective in reducing the expression of cyclooxygenase 2, an enzyme that catalyzes the synthesis of proinflammatory prostaglandins (Ha et al., 2012). This compound has been shown to have similar effects to nerve growth factor (NGF) through the activation of intracellular signaling pathways, stimulating the neuritogenesis (Seow et al., 2017) process by neurites which are generated and later become axons and dendrites in mature neurons (Flynn, 2013). Thus, it is thought to have a preventive and therapeutic effect on neurodegenerative diseases, reducing inflammation and stimulating neuronal regeneration (Seow et al., 2017; Ha et al., 2012; Venkatesan et al., 2015).

13.3.3.7 Ginkgolide B

Ginkgolide is a phytochemical of the family of terpenoids, derived from the medicinal plant *Ginkgo biloba*. This medicinal plant has been used by traditional Chinese medicine for thousands of years to treat neurological diseases such as dementia and sensory disorders (Venkatesan et al., 2015). Studies in cerebral ischemic damage models in rats have shown antiinflammatory effects of this compound, reducing the activation of NF-κβ, and inducing the expression of Bcl2, a protein that stabilizes the mitochondria of cells, then inhibiting the induction of apoptosis (Gu et al., 2012). Furthermore, ginkgolide B, in in vitro studies with cell cultures of hippocampal neurons, demonstrated an increased expression of the brain-derived neurotropic factor, reducing the apoptosis of these cells and showing a neuroprotective effect (Xiao et al., 2010). Extracts of *Ginkgo biloba* also show a prophylactic effect against migraine (D'Andrea et al., 2009), and reduce cognitive, functional, and behavioral decline in patients with cognitive impairment and dementia, benefiting mainly patients with neuropsychiatric symptoms (Tan et al., 2015)

13.3.3.8 Quercetin

Quercetin is a phenolic compound found in various fruits and vegetables including onion, apple, berries, broccoli, tomato, tea, grapes, and also in medicinal plants such as *Ginkgo biloba, Hypericum perforatum*, and others. This bioactive compound has been widely studied for its antiinflammatory, anticancer, antiviral, antioxidant, and antiallergic properties, being a phytochemical with a great variety of beneficial effects (Li et al., 2016). Several studies have demonstrated the ability of quercetin to reduce free radicals and increase the expression of antioxidant enzymes, thus protecting neurons from oxidative stress (Costa et al., 2016). A study by Wang et al. (2014) in a model of Alzheimer's disease in mice, oral administration of quercetin for 16 weeks, reduced plaque buildup, mitochondrial dysfunction, reactive oxygen species, and increased ability to learning and memory.

13.4 Conclusions

Medicinal plants and various foods are important sources of bioactive compounds with multiple beneficial effects for health. Bioactive compounds are of different natures, being essential and nonessential, both generating beneficial effects in different health and disease conditions. Frequent consumption of foods or plants containing bioactive compounds, as well as direct supplementation of these compounds, offers multiple beneficial effects to health including antiaging effects, protection against cardiovascular diseases, control and prevention of metabolic diseases, prevention and treatment of cancer, and also protection against neurodegenerative diseases.

The beneficial effects of phytochemicals and vitamins are through different mechanisms of action, mainly regulating oxidative processes, increasing the expression of antioxidant enzymes, reducing the synthesis of proinflammatory cytokines, increasing the expression of neurotrophic factors, or interacting in different intracellular signaling pathways. The frequent consumption of foods and/or medicinal plants can prevent the development of a large number of diseases, as well as improve aspects of health, such as better memory, and delaying aging.

References

Abbott, N.J., Friedman, A., 2012. Overview and introduction: the blood-brain barriers in health and disease. Epilepsia 53 (6), 1–6. https://doi.org/10.1111/j.1528-1167.2012.03696.x.Overview.

Al-Goblan, A.S., Al-Alfi, M.A., Khan, M.Z., 2014. Mechanism linking diabetes mellitus and obesity. Diabetes, Metabolic Syndrome and Obesity: Targets and Therapy 7, 587–591.

Amor, S., Puentes, F., Baker, D., Van Der, P., 2010. Inflammation in neurodegenerative diseases. Immunology 129, 154–169. https://doi.org/10.1111/j.1365-2567.2009.03225.x.

Baxter, R.A., 2008. Anti-aging properties of resveratrol: review and report of a potent new antioxidant skin care formulation. Journal of Cosmetic Dermatology 7 (1), 2–7. https://doi.org/10.1111/j.1473-2165.2008.00354.x.

Bayet-Robert, M., Kwiatowski, F., Leheurteur, M., Gachon, F., Planchat, E., Abrial, C., Chollet, P., 2010. Phase I dose escalation trial of docetaxel plus curcumin in patients with advanced and metastatic breast cancer. Cancer Biology and Therapy 9 (1), 8–14. https://doi.org/10.4161/cbt.9.1.10392.

Benzie, I.F.F., Szeto, Y.T., Strain, J.J., Tomlinson, B., 2009. Consumption of green tea causes rapid increase in plasma antioxidant power in humans. Nutrition and Cancer 34 (1), 83–87. https://doi.org/10.1207/S15327914NC340112.

Bhupathiraju, S.N., Hu, F.B., 2016. Epidemiology of obesity and diabetes and their cardiovascular complications. Circulation Research 118 (11), 1723–1735. https://doi.org/10.1161/CIRCRESAHA.115.306825.

Biesalski, H.K., Dragsted, L.O., Elmadfa, I., Grossklaus, R., Müller, M., Schrenk, D., Weber, P., 2009. Bioactive compounds: definition and assessment of activity. Nutrition 25, 1202–1205. https://doi.org/10.1016/j.nut.2009.04.023.

Biesalski, H.K., Erdman Jr., J., Hathcock, J., Ellwood, K., Beatty, S., Johnson, E., Griggiths, J.C., 2013. Nutrient reference values for bioactives: new approaches needed? A conference report. European Journal of Clinical Nutrition 52, 1–9. https://doi.org/10.1007/s00394-013-0503-0.

Blackadar, C.B., 2016. Historical review of the causes of cancer. World Journal of Clinical Oncology 7 (1), 54–86. https://doi.org/10.5306/wjco.v7.i1.54.

Bohn, T., Blackwood, M., Francis, D., Tian, Q., Schwartz, S.J., Clinton, S.K., 2013. Bioavailability of phytochemical constituents from a novel soy fortified lycopene rich tomato juice developed for targeted cancer prevention trials torsten. Nutrition and Cancer 65 (6). https://doi.org/10.1080/01635581.2011.630156.

Borra, M.T., Smith, B.C., Denu, J.M., 2005. Mechanism of human SIRT1 activation by resveratrol. The Journal of Biological Chemistry 280 (17), 17187–17195. https://doi.org/10.1074/jbc.M501250200.

Chacko, S.M., Thambi, P.T., Kuttan, R., Nishigaki, I., 2010. Beneficial effects of green tea: a literature review. Chinese Medicine 5 (13), 1–9. https://doi.org/10.1186/1742-4933-7-1.

Chambial, S., Dwivedi, S., Shukla, K.K., John, P.J., Sharma, P., 2013. Vitamin C in disease prevention and cure: an overview. Indian Journal of Clinical Biochemistry 28 (4), 314–328. https://doi.org/10.1007/s12291-013-0375-3.

Chen, W.-W., Zhang, X., Huang, W.-J., 2016. Role of neuroinflammation in neurodegenerative diseases (review). Molecular Medicine Reports 13, 3391–3396. https://doi.org/10.3892/mmr.2016.4948.

Corrêa, R.C.G., Peralta, R.M., Haminiuk, C.W.I., Maciel, G.M., Bracht, A., Ferreira, I.C.F.R., 2016. New phytochemicals as potential human anti-aging compounds: reality, promise, and challenges. Critical Reviews in Food Sciencie and Nutrition 13. https://doi.org/10.1080/10408398.2016.1233860.

Costa, L.G., Garrick, J.M., Roquè, P.J., Pellacani, C., 2016. Mechanisms of neuroprotection by Quercetin: counteracting oxidative stress and more. Oxidative Medicine and Cellular Longevity 2016, 1–10. https://doi.org/10.1155/2016/2986796.

Coyne, T., Ibiebele, T.I., Baade, P.D., Dobson, A., Mcclintock, C., Dunn, S., Shaw, J., 2005. Diabetes mellitus and serum carotenoids: findings of a population-based study in Queensland, Australia. American Journal of Clinical Nutrition 82, 685–693.

D'Andrea, G., Bussone, G., Allais, G., Aguggia, M., Onofrio, F.D., Maggio, M., Benedetto, C., 2009. Efficacy of Ginkgolide B in the prophylaxis of migraine with aura. Neurological Sciences 30, 121–124. https://doi.org/10.1007/s10072-009-0074-2.

Denis, I., Potier, B., Heberden, C., Vancassel, S., 2015. Omega-3 polyunsaturated fatty acids and brain aging. Current Opinion in Clinical Nutrition and Metabolic Care 18 (2), 139–146. https://doi.org/10.1097/MCO.0000000000000141.

Dhillon, N., Aggarwal, B.B., Newman, R.A., Wolff, R.A., Kunnumakkara, A.B., Abbruzzese, J.L., Kurzrock, R., 2008. Phase II trial of curcumin in patients with advanced pancreatic cancer. Clinical Cancer Research 14 (14), 4491–4499. https://doi.org/10.1158/1078-0432.CCR-08-0024.

Flynn, K.C., 2013. The cytoskeleton and neurite initiation. BioArchitecture 3 (4), 86–109.

Forester, S.C., Lambert, J.D., 2013. Antioxidant effects of green tea. Molecular Nutrition and Food Research 55 (6), 844–854. https://doi.org/10.1002/mnfr.201000641.

Fougère, B., Boulanger, E., Nourhashémi, F., Guyonnet, S., Cesari, M., 2016. Chronic inflammation: accelerator of biological aging. Journals of Gerontology: Medical Sciences 0 (0), 1–8. https://doi.org/10.1093/gerona/glw240.

Gajendragadkar, P.R., Hubsch, A., Mäki-Petäjä, K.M., Serg, M., Wilkinson, I.B., Cheriyan, J., 2014. Effects of oral lycopene supplementation on vascular function in patients with cardiovascular disease and healthy volunteers: a randomised controlled trial. PLoS One 9 (6). https://doi.org/10.1371/journal.pone.0
099070.

Golia, E., Limongelli, G., Natale, F., Fimiani, F., Maddaloni, V., Pariggiano, I., et al., 2014. Inflammation and cardiovascular disease: from pathogenesis to therapeutic target. Current Atherosclerosis Reports 16 (9). https://doi.org/10.1007/s11883-014-0435-z.

Gostin, L.O., Abou-taleb, H., Roache, S.A., Alwan, A., 2016. Legal priorities for prevention of non-communicable diseases: innovations from WHO's Eastern Mediterranean region. Public Health 144 (202), 4–12. https://doi.org/10.1016/j.puhe.2016.11.001.

Gu, J., Ge, J., Li, M., Wu, F., Zhang, W., Qin, Z., 2012. Inhibition of NF-KB activation is associated with anti-inflammatory and anti-apoptotic effects of Ginkgolide B in a mouse model of cerebral ischemia/reperfusion injury. European Journal of Pharmaceutical Sciences 47 (4), 652–660. https://doi.org/10.1016/j.ejps.2012.07.016.

Gupta, S.C., Patchva, S., Aggarwal, B.B., 2013. Therapeutic roles of curcumin: lessons learned from clinical trials. The AAPS Journal 15 (1), 195–218. https://doi.org/10.1208/s12248-012-9432-8.

Ha, S.K., Moon, E., Ju, M.S., Kim, D.H., Ryu, J.H., Oh, M.S., Kim, S.Y., 2012. Neuropharmacology 6-Shogaol, a ginger product, modulates neuroinflammation: a new approach to neuroprotection. Neuropharmacology 63 (2), 211–223. https://doi.org/10.1016/j.neuropharm.2012.03.016.

Hadley, C.W., Clinton, S.K., Schwartz, S.J., 2003. The consumption of processed tomato products enhances plasma lycopene concentrations in association with a reduced lipoprotein sensitivity to oxidative damage. Journal of Nutrition 133 (3), 727–732.

Hasler, C.M., 2002. Functional Foods: benefits, concerns and challenges — a position paper from the American council on science and health. Journal of Nutrition (September), 3772–3781.

Hatcher, H., Planalp, R., Cho, J., Torti, F.M., Torti, S.V., 2008. Curcumin: from ancient medicine to current clinical trials. Cellular and Molecular Life Sciences 65, 1631–1652. https://doi.org/10.1007/s00018-008-7452-4.

He, Q., Lv, Y., Yao, K., 2006. Food Chemistry Effects of tea polyphenols on the activities of a -amylase, pepsin, trypsin and lipase. Food Chemistry 101, 1178–1182. https://doi.org/10.1016/j.foodchem.2006.03.020.

He, Z.-Y., Shi, C.-B., Wen, H., Li, F.-L., Wang, B.-L., Wang, J., 2011. Upregulation of p53 expression in patients with colorectal cancer by administration of curcumin. Cancer Investigation 29 (3), 208–213. https://doi.org/1
0.3109/07357907.2010.550592.

Heinen, M.M., Hughes, M.C., Ibiebele, T.I., Marks, G.C., Green, A.C., van der Pols, J.C., 2007. Intake of antioxidant nutrients and the risk of skin cancer. European Journal of Cancer 43, 2707–2716. https://doi.org/10.1016/j.ejca.2007.09.005.

Henry, A.G., Brooks, A.S., Piperno, D.R., 2011. Microfossils in calculus demonstrate consumption of plants and cooked foods in Neanderthal diets (Shanidar III, Iraq; Spy I and II, Belgium). Proceedings of the National Academy of Sciences of the United States of America 108 (2), 486–491. https://doi.org/10.1073/pnas.1016868108.

Jain, A.P., Aggarwal, K.K., Zhang, P.-Y., 2015. Omega-3 fatty acids and cardiovascular disease. European Review for Medical and Pharmacological Sciences 19, 441–445.

Jiang, Q., 2014. Natural forms of vitamin E: metabolism, antioxidant and anti-inflammatory activities and the role in disease prevention and therapy. Free Radical Biology and Medicine 72, 76–90. https://doi.org/10.1016/j.freeradbiomed.2014.03.035.

Kiec-wilk, B., Mykka, H., 2008. Antioxidant phytochemicals against type 2 diabetes. British Journal of Nutrition 99 (1). https://doi.org/10.1017/S000711450896579X.

Kucuk, O., Sarkar, F.H., Sakr, W., Djuric, Z., Pollak, M.N., Khachik, F., et al., August 2001. Phase II randomized clinical trial of lycopene supplementation before radical prostatectomy. Cancer Epidemiology, Biomarkers and Prevention 10, 861–868. https://doi.org/Published.

Kucuk, O., Sarkar, F.H., Djuric, Z., Sakr, W., Pollak, M.N., Khachik, F., et al., 2002. Effects of lycopene supplementation in patients with localized prostate cancer. Experimental Biology and Medicine 227 (10), 881–885.

Lastra, C.A., Villegas, I., 2005. Review Resveratrol as an anti-inflammatory and anti-aging agent: mechanisms and clinical implications. Molecular Nutrition and Food Research 49, 405–430. https://doi.org/10.1002/mnfr.200500022.

Li, Y., Yao, J., Han, C., Yang, J., Chaudhry, M.T., Wang, S., Yin, Y., 2016. Quercetin, inflammation and immunity. Nutrients 8 (167), 1–14. https://doi.org/10.3390/nu8030167.

Martínez Leo, E.E., Acevedo Fernández, J.J., Segura Campos, M.R., 2016. Biopeptides with antioxidant and anti-inflammatory potential in the prevention and treatment of diabesity disease. Biomedicine and Pharmacotherapy 83, 816–826. https://doi.org/10.1016/j.biopha.2016.07.051.

Liu, R.H., 2013. Dietary bioactive compounds and their health implications. Journal of Food Science 78 (S1), A18–A25. https://doi.org/10.1111/1750-3841.12101.

López-otín, C., Blasco, M.A., Partridge, L., Serrano, M., Kroemer, G., 2013. The hallmarks of aging. Cell 153 (6), 1194–1217. https://doi.org/10.1016/j.cell.2013.05.039.The.

Martin, C.M., 2007. Chronic disease and illness care. Canadian Family Physician 53, 2086–2091.

Meng, S., Cao, J., Feng, Q., Peng, J., Hu, Y., 2013. Roles of chlorogenic acid on regulating glucose and lipids metabolism: a review. Evidence-Based Complementary and Alternative Medicine 2013.

Mori, T.A., 2014. Omega-3 fatty acids and cardiovascular disease: epidemiology and effects on cardiometabolic risk factors. Food and Function 5, 2004–2019. https://doi.org/10.1039/C4FO00393D.

Mukherjee, S., Dudley, J.I., Das Cardiovascular, D.K., 2010. Dose-dependency of resveratrol in providing health benefits. Dose-Response 8 (4), 478–500. https://doi.org/10.2203/dose-response.09-015.Mukherjee.

Nasri, H., Baradaran, A., Shirzad, H., Rafieian-Kopaei, M., 2014. New concepts in nutraceuticals as alternative for pharmaceuticals. International Journal of Preventive Medicine 5 (12), 1487–1499.

Nasri, M., 2016. Protein Hydrolysates and Biopeptides: Production, Biological Activities, and Applications in Foods and Health Benefits. A Review. Advances in Food and Nutrition Research, first ed. Elsevier Inc. https://doi.org/10.1016/bs.afnr.2016.10.003.

Pagliaro, B., Santolamazza, C., Simonelli, F., Rubattu, S., 2015. Phytochemical compounds and protection from cardiovascular diseases: a state of the art. BioMed Research International 2015, 1–17. https://doi.org/10.1155/2015/918069.

Petrovska, B.B., 2012. Historical review of medicinal plants' usage. Pharmacognosy Reviews 6 (11), 1–6. https://doi.org/10.4103/0973-7847.95849.

Riccioni, G., Gammone, M.A., Tettamanti, G., Bergante, S., Pluchinotta, F.R., Orazio, N.D., August 2015. Resveratrol and anti-atherogenic effects. International Journal of Food Sciencies and Nutrition 1–8. https://doi.org/10.3109/09637486.2015.1077796.

Rigacci, S., Stefani, M., 2016. Nutraceutical properties of olive oil polyphenols. An itinerary from cultured cells through animal models to humans. International Journal of Mole 17 (843), 1–28. https://doi.org/10.3390/ijms17060843.

Rizzo, M., Kotur-Stevuljevic, J., Spinas, G., Rini, G.B., Jelic-ivanovic, Z., Spasojevic-kalimanovska, V., Vekic, J., 2009. Atherogenic dyslipidemia and oxidative stress: a new look. Translational Research 153 (5), 217–223. https://doi.org/10.1016/j.trsl.2009.01.008.

Ruiz, R.B., Hernández, P.S., 2014. Diet and cancer: risk factors and epidemiological evidence. Maturitas 77 (3), 202–208. https://doi.org/10.1016/j.maturitas.2013.11.010.

Sakakura, K., Nakano, M., Otsuka, F., Ladich, E., Kolodgie, F.D., Virmani, R., 2013. Pathophysiology of atherosclerosis plaque progression. Heart Lung and Circulation 22 (6), 399–411. https://doi.org/10.1016/j.hlc.2013.03.001.

Saremi, A., Arora, R., 2010. Vitamin E and cardiovascular disease. American Journal of Therapeutics 17 (3), 56–65. https://doi.org/10.1097/MJT.0b013e31819cdc9a.

Seow, S.L.S., Hong, S.L., Lee, G.S., Malek, S.N.A., Sabaratnam, V., 2017. 6-shogaol, a neuroactive compound of ginger (*Jahe gajah*) induced neuritogenic activity via NGF responsive pathways in PC-12 cells. BMC Complementary and Alternative Medicine 17 (334), 1–11. https://doi.org/10.1186/s12906-0 17-1837-6.

Si, H., Liu, D., 2014. Dietary antiaging phytochemicals and mechanisms associated with prolonged survival. The Journal of Nutritional Biochemistry 25 (6), 581–591. https://doi.org/10.1016/j.jnutbio.2014.02.001.Dietary.

Sikora, E., Scapagnini, G., Barbagallo, M., 2010. Curcumin, inflammation, ageing and age-related diseases. Immunity and Ageing 7 (1), 1–4. https://doi.org/10.1186/1742-4933-7-1.

Singh, B.N., Shankar, S., Srivastava, R.K., 2011. Green tea catechin, epigallocatechin-3-gallate (EGCG): mechanisms, perspectives and clinical applications. Biochemical Pharmacology 82 (12), 1807–1821. https://doi.org/10.1016/j.bcp.2011.07.093.Green.

Somani, S.J., Modi, K.P., Majumdar, A.S., Sadarani, B.N., 2015. Phytochemicals and their potential usefulness in inflammatory bowel disease. Phytotherapy Research 29 (3), 339–350. https://doi.org/10.1002/ptr.5271.

Stewart, B.W., Wild, C.P., 2014. World Health Organization. World cancer report 2014. https://doi.org/9283204298.

Story, E.N., Kopec, R.E., Schawartz, S.J., Harris, G.K., 2010. An update on the health effects of tomato lycopene. Annual Review of Food Science and Technology 1, 1–24. https://doi.org/10.1146/annurev.food.102308.124120.An.

Swanson, D., Block, R., Mousa, S.A., 2012. Omega-3 fatty acids EPA and DHA: health benefits throughout life. Advances in Nutrition 3, 1–7. https://doi.org/10.3945/an.111.000893.

Tan, M.-S., Yu, J.-T., Tan, C.-C., Wang, H.-F., Meng, X.-F., Wang, C., Tan, L., 2015. Efficacy and adverse effects of Ginkgo biloba for cognitive impairment and dementia: a systematic review and meta-analysis. Journal of Alzheimer's Disease 43, 589–603. https://doi.org/10.3233/JAD-140837.

Thomas, R., Butler, E., Macchi, F., Williams, M., 2015. Phytochemicals in cancer prevention and management? British Journal of Medical Practitioners 8 (2).

Tomé-carneiro, J., Larrosa, M., González-sarrías, A., Tomás-barberán, F.A., García-Conesa, M.T., Espín, J.C., 2013. Resveratrol and clinical trials: the crossroad from in vitro studies to human evidence. Current Pharmaceutical Design 19 (34), 6064–6093. https://doi.org/10.2174/13816128113199990407.

Tsao, R., 2010. Chemistry and biochemistry of dietary polyphenols. Nutrients 2, 1231–1246. https://doi.org/10.3390/nu2121231.

Tur, J.A., Bibiloni, M.M., Sureda, A., Pons, A., 2012. Dietary sources of omega 3 fatty acids: public health risks and benefits. British Journal of Nutrition 107, 23–52. https://doi.org/10.1017/S0007114512001456.

Unlu, N.Z., Bohn, T., Clinton, S.K., Schwartz, S.J., 2005. Human nutrition and metabolism carotenoid absorption from salad and salsa by humans is enhanced by the addition of avocado or avocado oil 1, 2. Journal of Nutrition 135 (3), 431–436.

Venkatesan, R., Ji, E., Kim, S.Y., 2015. Phytochemicals that regulate neurodegenerative disease by targeting neurotrophins: a comprehensive review. BioMed Research International 2015, 1–22.

Walther, B., Sieber, R., 2011. Bioactive proteins and peptides in foods. International Journal for Vitamin and Nutrition Research 81 (2–3), 181–191. https://doi.org/10.1024/0300-9831/a000054.

Wang, D.-M., Li, S.-Q., Wu, W.-L., Zhu, X.-Y., Wang, Y., Yuan, H.-Y., 2014. Effects of long-term treatment with Quercetin on cognition and mitochondrial function in a mouse model of Alzheimer's disease. Neurochemical Research (June). https://doi.org/10.1007/s11064-014-1343-x.

Wardyn, J.D., Ponsford, A.H., Sanderson, C.M., 2015. Dissecting molecular cross-talk between Nrf2 and NF- κB response pathways. Biochemical Society Transactions. 43 (3), 621–626. https://doi.org/10.1042/BST20150014.

Weissmiller, A.M., Wu, C., 2012. Current advances in using neurotrophic factors to treat neurodegenerative disorders. Translational Neurodegeneration 1 (14), 1–9.

Wu, Y., Ding, Y., Tanaka, Y., Zhang, W., 2014. Risk factors contributing to type 2 diabetes and recent advances in the treatment and prevention. International Journal of Medical Sciences 11, 1185–1200. https://doi.org/10.7150/ijms.10001.

Xiao, Q., Wang, C., Li, J., Hou, Q., Li, J., Ma, J., Wang, Z., 2010. Ginkgolide B protects hippocampal neurons from apoptosis induced by beta-amyloid 25–35 partly via up-regulation of brain-derived neurotrophic factor. European Journal of Pharmacology 647 (1–3), 48–54. https://doi.org/10.1016/j.ejphar.2010.08.002.

Yan, Y., Jiang, W., Spinetti, T., Tardivel, A., Castillo, R., Bourquin, C., et al., 2013. Omega-3 fatty acids prevent inflammation and metabolic disorder through inhibition of NLRP3 inflammasome activation. Immunity 38 (6), 1154–1163. https://doi.org/10.1016/j.immuni.2013.05.015.

Yang, C.S., Fang, M., Lambert, J.D., Yan, P., Huang, T.H.-M., 2010. Reversal of hypermethylation and reactivation of genes by dietary polyphenolic compounds. Nutrition Reviews 66 (Suppl. 1), 18–20. https://doi.org/10.1111/j.1753-4887.2008.00059.x.Reversal.

Young, J.F., Dragsted, L.O., Haraldsdóttir, J., Daneshvar, B., Kall, M.A., Loft, S., Sandström, B., 2002. Green tea extract only affects markers of oxidative status postprandially: lasting antioxidant effect of flavonoid-free diet. British Journal of Nutrition 87, 343–355. https://doi.org/10.1079/BJN2002523.

Zhang, Y., Gan, R., Li, S., Zhou, Y., Li, A., Xu, D., Li, H.-B., 2015. Antioxidant phytochemicals for the prevention and treatment of chronic diseases. Molecules 20, 21138–21156. https://doi.org/10.3390/molecules201219753.

Zhu, R., Liu, H., Liu, C., Wang, L., Ma, R., Chen, B., et al., 2017. Cinnamaldehyde in diabetes: a review of pharmacology, pharmacokinetics and safety. Pharmacological Research 122, 78–89. https://doi.org/10.1016/j.phrs.2017.05.019.

CHAPTER 14

Flavonoids: Potential Therapeutic Agents by Their Antioxidant Capacity

Patricia Hernández-Rodríguez[1,2,3], Ludy Pabón Baquero[2,3], Harold Rodríguez Larrota[4]

[1]*Researcher Professor Biology Program, Department of Basic Sciences. Universidad de La Salle, Bogotá, Colombia;* [2]*Researcher of Department of Basic Sciences, Universidad de La Salle, Bogotá, Colombia;* [3]*Biomigen Research Group, Universidad de La Salle, Bogotá, Colombia;* [4]*Student Biology Program, SIEZ Research Incubator*

14.1 Introduction

Plants, with the release of oxygen through photosynthesis and with metabolic processes such as respiration and the regulation of reactive oxygen species (ROS), seek to defend themselves and adapt to the conditions of terrestrial ecosystems (Delaux et al., 2012; Stevens et al., 2008). ROS molecules are essential in development and growth processes (Foreman et al., 2003), movement in stomata (Pei et al., 2000), and plant–microorganism interactions (Torres and Dangl, 2005). However, ROS, when they accumulate in cells, are very toxic and cause damage to DNA, proteins, and lipids, which causes alterations to cells, tissues, and organs, leading to pathologies related to reactive species such as hypoxia, cancer, dysfunction, endothelial and neurodegenerative diseases, among others (Lluis and Morales, 2008). These pathologies are related to an imbalance between oxidants and antioxidants, which leads to a situation of oxidative stress at the cellular level but which extends to tissues and systems (Boveris et al., 2008). At the intracellular level, chemical species such as hydrogen peroxide, nitric oxide, superoxide, and hydroxyl ion, etc., present differences in the steady state and changes in concentrations that affect the speed of chemical reactions, leading to the consumption of chemicals. Endogenous antioxidants alter the natural balance in biological systems, causing damage to the membranes, alterations to DNA, and damage to surface proteins (Boveris et al., 2008). These cellular effects, being cumulative, cause significant tissue changes that lead to systemic or peripheral oxidative stress, which in humans and experimental animals is mainly determined by the plasma levels of thiobarbituric acid reactive substances (TBARS) and malondialdehyde (MDA), which are considered oxidative stress markers (Abuja and Albertini, 2001). To counteract the effect of these substances, the body has antioxidant defense enzyme systems, such as superoxide dismutase (SOD) and glutathione peroxidase (GPx), which are enzymes found in the cytoplasm and mitochondria; SOD catalyzes the conversion of

Bioactive Compounds. https://doi.org/10.1016/B978-0-12-814774-0.00014-1

superoxide radical to hydrogen peroxide that is eliminated by GPx and catalase; additionally, GPx participates in the transformation of ROS and prevents the oxidation of lipoperoxides (L-OOH), which are toxic and originate ROS (Pennathur and Heinecke, 2004). Exogenous substances derived from plants such as flavonoids have been considered excellent antioxidants, for their ability to trap free radicals, thus becoming excellent candidates in the prevention of diseases associated with increased production of free radicals. Therefore, this chapter aims to establish the potential of flavonoids as free radical scavengers and as therapeutic agents for the prevention and control of diseases related to oxidative stress.

14.2 Oxidative Stress

Oxidative stress is described as an imbalance between the production and elimination of free radicals (oxidants). ROS that include hydrogen peroxide, singlet oxygen, superoxide, and hydroxyl radicals are the main oxidant molecules and arise from various sources, including oxidative metabolism of aerobic systems, metabolic processes, activity of NADPH oxidase, and environmental stressors such as radiation, ultraviolet (UV), and pollutants. The physiological concentration of ROS is essential for the preservation of cellular redox homeostasis, as well as for the regulation of cell proliferation and the particular signaling pathways. However, excess production of ROS can cause deterioration in cellular function, inflammatory changes, and adverse effects by altering cellular ingredients such as DNA, lipids, and proteins. The overproduction of ROS can lead to mutagenesis and cancer, and has been considered as a prominent factor in the pathogenesis of metabolic, neuronal, and aging diseases (Naeimi and Alizadeh, 2017).

Oxidative damage causes alterations that generate neurodegenerative, hepatic diseases, acute renal failure, cancer, endothelial dysfunction, and retinopathies. These alterations in neurodegenerative diseases, such as Alzheimer's, Parkinson's, Huntington's, and sclerosis among others, involve toxic events with biochemical and molecular changes that lead to the formation of ROS that promote inflammatory processes, oxidative stress, activation of transcription factors that trigger degeneration, apoptosis, or necrosis. In recent decades, research has focused on studying the impact of oxidative stress on the brain because this organ has several conditions that make it vulnerable mainly to the toxic action of ROS. Among these are the high content of lipids, which leads to greater peroxidation and also the high metabolic rate, associated with a greater amount of mitochondria that increases the probability of a loss of electrons through the respiratory chain; additionally, in the brain, there is limited activity of catalase, antioxidant enzyme. These reasons explain the relationship of oxidative stress to the degeneration of the nervous system (Santamaria, 2008).

In the development of cancer, ROS plays a fundamental role in the initiation, promotion, and progression of tumor growth; hyperplasia begins with a mutation, permanent modification of DNA, product of oxidative damage caused by ROS; this alteration begins to generate cascade signaling that leads to the general characteristics of the cancerous tissue, such as increased cell proliferation, changes in the expression of genes that lead to inactivation of the processes of apoptosis and changes in cellular metabolism that increase the action of the cells. ROS allows for an increase in mutations, cellular immortalization, genomic instability, loss of cellular functionality, insufficiency, and organic deterioration that lead to death (Gómez and Cuevas, 2008).

The endothelial damage related to metabolic and cardiac diseases occurs in response to an increase in the production of substances associated with oxidative stress; in this sense, the role played by the endothelium in the inflammatory process influences the expression of genes such as nitric oxide synthetase (iNOS) and cyclooxygenase 2 (COX-2) that are activated against the action of proinflammatory cytokines. Tumor necrosis factor alpha (TNF-α) and interleukin 1 (IL-1), as mediators of inflammation and immune response, act on the endothelium, increasing the production of cell adhesion molecules and cytokines that lead to cascading signals that cause leucocitary adhesion. Leukocytes and activated endothelial cells produce ROS that activates p38 protein kinase activated by the mitogen complex MAPK that leads to the phosphorylation of the heat stress protein (HSP27) that modulates oxidative stress (Keaney, 2000; Jackson and Garcia-Rojas, 2008).

Antioxidants can prevent and/or alleviate diseases related to oxidative stress by delaying or reducing such damage, which is why they are considered important nutraceuticals due to their multiple health benefits and why they are widely used in the food industry as potential inhibitors of lipid peroxidation (Deng et al., 2011). Antioxidants can be classified into two types, the enzymes that act as the first line of defense in the human body, such as superoxide dismutase, catalase, glutathione peroxidase, glutathione reductase, and glucose-6-phosphate dehydrogenase, and nonenzymatic, antioxidants such as glutathione, uric acid, dihydrolipoic acid and metallothionein.

14.3 Secondary Metabolites With Antioxidant Activity

Many properties of plant products are associated with the presence of phenolic compounds, which are found in different anatomical parts and are essential for the development and defense mechanisms of plants (Sulaiman and Balachandran, 2016). These secondary metabolites are considered the most abundant and are synthesized mainly by shikimic acid, pentose phosphate, and phenylpropanoid pathways, giving rise to more than 8000 compounds (Singh et al., 2017). Structurally they have one or more hydroxyl groups directly attached to the aromatic ring and can vary from simple molecules, such as phenolic acids, followed by flavonoids, lignans, and stilbenes; until reaching highly polymerized substances such as tannins (Amarowicz and Shahidi, 2017). Tannins occur in complexes with polysaccharides, proteins, and alkaloids and are subdivided into hydrolyzable and condensed tannins. Some of these compounds are soluble in water (phenolic acids and flavonoids), while others are insoluble (some condensed tannins). Additionally, it has been determined that flavonoids (60%) and phenolic acids (30%) mainly represent phenolic compounds in our diet (Singh et al., 2017).

For the extraction of phenolic compounds there are many available methods, however, there is no single method that can be considered as standard. The extraction of phenolic compounds is affected by several parameters (such as the particle size of the samples, the type of solvent used, the ratio of solute to solvent, speed of agitation, efficiency of mass transfer, and temperature). In this way, the finely pulverized samples significantly improve the extraction of phenolic compounds due to the increase in the surface area and the disruption of the cell walls present in the plant material used; organic solvents of high polarity (methanol, ethanol, acetone, and ethyl acetate) and their combinations with water are used to extract this type of compounds (Singh et al., 2017).

Additionally, the antioxidant potential of phenolic compounds of vegetable origin allows them to act by inhibiting, delaying, or preventing oxidation in food and in the body, because they eliminate reactive oxygen and nitrogen species, which affect lipids, proteins, and DNA (Amarowicz and Shahidi, 2017). The antioxidant activity of the phenolic compounds is directly related to their chemical structure, such as the degree of glycosylation and the number, as well as the position, of the hydroxyl groups. In turn, it has been determined that the health-promoting activities of phenolic compounds also depend on the conjugation of phenolic compounds with other compounds, together with solubility, absorption, and metabolism (Singh et al., 2017).

In this way, these compounds contribute significantly to the antioxidant activity due to the elimination of radicals and the chelation activity of metals. In this sense, phenolic compounds have the ability to donate hydrogen atoms or an electron to free radicals to form more stable intermediates (Singh et al., 2017; Riebel et al., 2017). However, its properties are not limited to antioxidant activity, but may also have antiallergic, antimicrobial, and/or antiinflammatory effects. Additionally, it has been reported that they improve the organoleptic properties of foods of vegetable origin, and can also be used as raw materials in the development of functional foods or as natural preservatives against food degradation (Ballesteros et al., 2017a,b). In this way, the phenolic compounds and the flavonoids of the plants have important functional properties and, therefore, are of interest for the chemical, pharmaceutical, and food industries (Ballesteros et al., 2017b).

14.4 Flavonoids and Mechanisms of Antioxidant Activity

Flavonoids are secondary metabolites corresponding to polyphenols, which have a varied structure, found in the form of aglycones or glycosides in many fruits and vegetables. Flavonoids have a chemical structure of 15 carbons constituted by a common skeleton of phenyl-benzo-γ-pyran (C6–C3–C6), also known as nucleus flava, composed of two phenyl rings (A and B) and a ring heterocyclic (pyran) C. Flavonoids include flavonols, flavones, flavonoids, flavanones, anthocyanidins, and isoflavones (Fig. 14.1) (Queiroz Ferreira et al., 2015; Li et al., 2017).

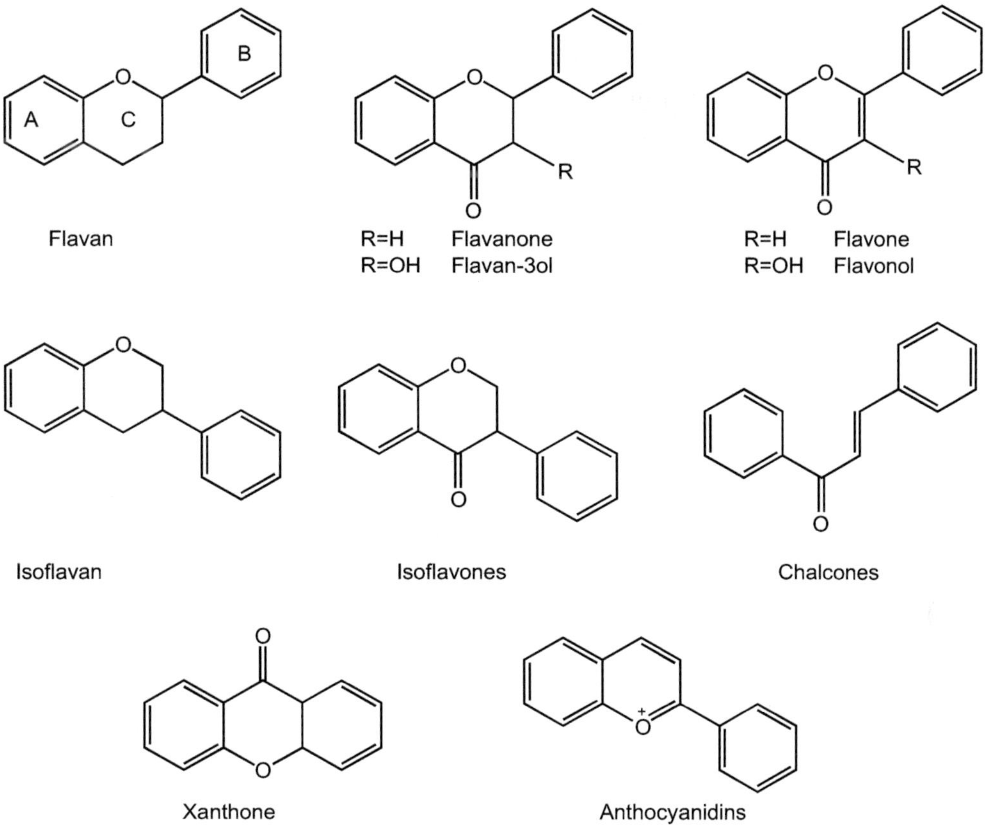

Figure 14.1
Classification of flavonoids.

These compounds are widely distributed in the leaves, seeds, bark, and flowers of plants, and correspond to pigments of different colors that perform different functions, such as protection from ultraviolet light, defense from abiotic stresses and from bacterial and fungal phytopathogens; the most recent properties found reveal that they can act as the endogenous regulator of auxin movement in plants (Li et al., 2017; Sandoval-Yañez et al., 2017).

Flavonoids have been studied due to their numerous pharmacological activities beneficial to the human body, in providing positive effects for the maintenance of health and the prevention of diseases. Some flavonoids have been used in the development of nutraceuticals and medicines, due to the wide spectrum of biological activities that present as: antioxidants, antiinflammatory, antiallergic, antimutagenic, cardioprotective, modulators of enzymatic activity, and anticancer activity, among others (Wen et al., 2017; Sandoval-Yañez et al., 2017). Flavonoids occur naturally as a glycoside; however, after they have been metabolized by intestinal enzymes, the remaining sugar is lost, resulting in the aglycone form (Chanput et al., 2016). Among the disadvantages of this type of compound is low stability against environmental stress (heat, light, oxidation, etc.) and low bioavailability after oral administration, which limits their health benefits. Therefore, attempts have been made to encapsulate flavonoids to preserve their chemical integrity and their pharmacological activity (Hu et al., 2017a,b).

The antioxidant activity of flavonoids is the result of the combination of mechanisms, such as decreased enzymatic activity of oxidases such as lipoxygenase (LO), cyclooxygenase (CO), myeloperoxidase (MPO), NADPH oxidase, and xanthine oxidase (XO) (Ferrandiz and Alcaraz, 1991). In this sense, studies conducted in liver microsomes to establish the effect of quercetin on the activity of NADPH oxidase of the cytochrome P-450 (NOX) system show how inhibition of NOX can prevent the increase in free radicals generated by the metabolism of xenobiotics (Pérez and Martínez, 2001).

It has also been shown that differences in flavonoid concentrations can inhibit oxidases selectively and more effectively by potentiating the antioxidant effect; thus at concentrations of 0.125, 0.25, and 0.5 mM for quercetin, morin, and catechin, respectively, the damage induced by oxyradicals in the endothelial cells of the porcine aorta can be avoided, but if these same concentrations in the XO are compared with the action exerted by the morin and for quercetin it is significantly higher ($P < .01$) with respect to the catechin (Zeng et al., 1997).

Flavonoids also inhibit oxidases CO and LO in a differential manner according to the concentration. In some cases the block in the activation is related to a low concentration but another oxidase may require a higher concentration. In this way, silybin inhibits 5-LO produced by human granulocytic cells with levels of 15 μM but four times higher concentrations are required to achieve half of the CO inhibition in these same cells (Pérez and Martínez, 2001).

Another antioxidant mechanism of flavonoids is related to the stimulation of enzymes such as superoxide dismutase (SOD) and catalase (CAT); in this way, the catalytic activity is increased by reducing the amount of free radicals through intervention in the reactions that involves conversion of the superoxide ion into hydrogen peroxide by the SOD, which in turn is converted into water and oxygen by CAT (Gómez and Cuevas, 2008).

Many of the pharmacological effects of flavonoids are related to their antioxidant activity, the important biological function of which is to maintain levels of oxidative stress below a critical point in the body at low toxicity (Naeimi and Alizadeh, 2017; Jabbari and Jabbari, 2016). Although several studies have been conducted to investigate the possible mechanism of action, it is still not entirely clear how flavonoids exert their beneficial effects or toxic actions on human health (Queiroz Ferreira et al., 2015). Different studies have shown that flavonoids can interact with different therapeutic targets, this ability to interact is mainly influenced by its chemical structure and redox capacity. Examples of enzymes that can be inhibited are NADH oxidases, polyphenol oxidases, peroxidases, lipoxygenase, cellulases, xylanases, pectinases, glutathione-S-transferases, glycoproteins, and kinases. Therefore, flavonoids are attractive molecules because they have many of the structural characteristics mentioned, examples of these compounds with antioxidant activity are genistein, luteolin, apigenin, and kaempferol (Sandoval-Yañez et al., 2017).

Within the antioxidant mechanisms of the flavonoids it has been noted that they act as free radical scavengers, an activity that they present because they have a stabilized structure that allows them to attenuate the highly reactive free radicals, becoming less reactive aroxyl radicals. This suppression of oxidants is produced by the donation of electrons or a hydrogen atom from free hydroxyl, thus playing a moderate role in the propagation of damage induced by radicals in biological systems, but more potent than vitamins and carotenoids, together with which synergistic effects have been demonstrated (Jabeen et al., 2017; Jabbari and Jabbari, 2016). Additionally, they are considered extremely safe and as having low toxicity, which makes them excellent chemopreventive agents. Flavonoids can also chelate metal ions and can form different complexes, whose pharmacokinetic characteristics are increased in comparison with free flavonoids. In this way, flavonoids have been shown to play a moderate role in the propagation of radicals induced by damage in biological systems, whose antioxidant capacity depends to a large extent on the number and position of hydroxyl groups, sugars, and double bonds, which also contribute to their molecular polarity (Jabeen et al., 2017; Li et al., 2017). The advances already obtained in studies of structure–activity relationship (SAR), highlight that the biochemical activity of flavonoids and their metabolites depends on the chemical structure and the relative orientation of the substituents in the molecule, with three determining factors standing out in the effective capacity to capture or sequester free radicals (Queiroz Ferreira et al., 2015) (Fig. 14.2).

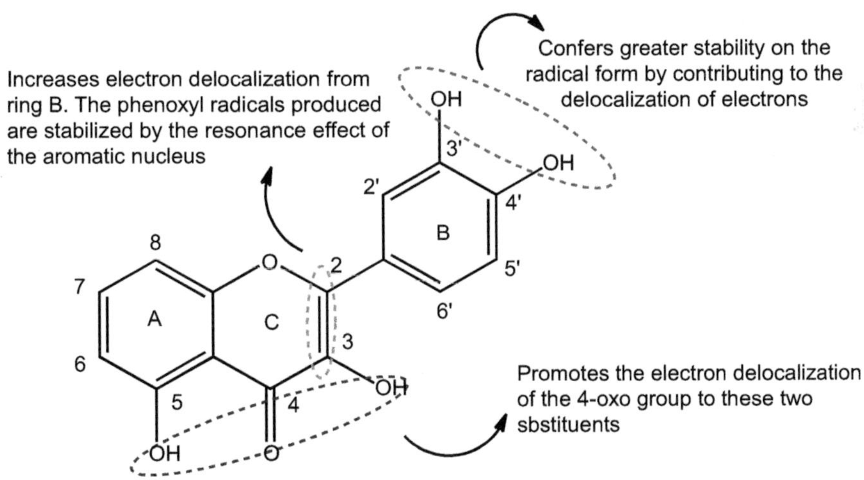

Figure 14.2
Structure of the flavonoid with three determining factors in the effective capacity for capturing/sequestering free radicals.

Despite their emphasis in the scientific community, flavonoids still have greatly unexplored potential. These gaps are mainly due to the lack of studies that quantitatively relate the mechanisms of action and the antioxidant capacity of these compounds with their chemical structure (Chanput et al., 2016).

Considering that antioxidant capacity is one of the most important biological activities, some of the main methods that determine this property in flavonoids are described below. Additionally, an exhaustive review is made of studies reported in the literature where DPPH and ABTS assays were used to evaluate antioxidant activity in flavonoids.

14.5 Methods of Evaluation of Antioxidant Activity

Total phenols and especially flavonoids are of importance, due to their high correlation with antioxidant capacity and, therefore, are generally accepted as the main parameters for the selection of species in studies that seek to evaluate the antioxidant potential of a plant material, without going so far as to affirm that these compounds are the only ones responsible for the total antioxidant activity (Li et al., 2017). In this sense it has been suggested that compounds of lipophilic and hydrophilic natures have antioxidant properties that, being present in combination in the extract, their antioxidant capacity is a function of the interactions in the mixture (Le Grandois et al., 2017). For this reason, extensive research into the natural sources of efficient radical uptake compounds is receiving great attention in health research and has led to the development of a wide variety of methods for the evaluation of antioxidants. A wide variety of methods have been developed for the evaluation of antioxidants, including the ferric reducing antioxidant power (FRAP) assay, the 2,2′-azino-bis (3-ethylbenzothiazoline-6-sulfonate) (ABTS+), the oxygen radical absorbance capacity assay (ORAC), cupric ion reducing antioxidant capacity (CUPRAC), and 1,1-diphenyl-2-picrylhydrazyl (DPPH) (Chen et al., 2013).

In general, the antioxidant compounds in plant extracts are chemically diverse and structurally complex; there is no single adequate test to accurately determine the antioxidant activity for all compounds present in a sample (Prior and Gu, 2005), and therefore the use of multiple tests is recommended. There are several methods to determine the antioxidant capacity and these differ in relation to the principle and the experimental conditions of the test. A single assay does not accurately represent all groups of antioxidant compounds, particularly in a complex system, such as fruit matrices. Therefore, it is necessary to perform different tests of antioxidants to ensure a better comparison of the data between the different fruit residues (Barros et al., 2017). Among these methods, DPPH and ABTS are two of the most commonly used to evaluate the antioxidant activity of plant extracts, food, and unique compounds, thanks to their stability, reduced costs, and easy to implement protocols (Naeimi and Alizadeh, 2017; Chen et al., 2013; Magalhães et al., 2012), which is convenient for their application and, therefore, they are the most popular, however, they are limited since they use nonphysiological radicals (Floegel et al., 2011).

DPPH (2,2-diphenyl-1-picrylhydrazyl) is one of the most commonly used methods because it is considered a simple, efficient trial with good stability, sensitivity, feasibility, and low cost (Carmona-Jiménez et al., 2014; Deng et al., 2011). The original method was developed by Blois (1955) and small modifications have been made over time. DPPH is a stable free radical that has a deep purple color and a maximum absorption wavelength in the range of 515–520 nm (Hidayat et al., 2017; Carmona-Jiménez et al., 2014). To form a more stable molecule, the DPPH radical can accept an electron or a hydrogen atom from the sequestering antioxidant molecule (Fig. 14.3) (Lee et al., 2017). This reaction has a change in coloration from violet to pale yellow that corresponds to reduced DPPH, so it is possible to determine the antioxidant activity by spectrophotometric methods. In this sense, the greater the capacity of free radical capture of an antioxidant compound, the greater the reduction of DPPH and the less purple the sample remains. The results are usually expressed as the effective concentration (EC50), which corresponds to the amount of the sample necessary to decrease the initial concentration of DPPH radicals by 50%, this unit of expression allows comparison of results independent of the concentration of the sample (Carmona-Jiménez et al., 2014; Deng et al., 2011).

Within the limitations of this method it is reported that, according to several experimental studies, the relationship between the concentration of antioxidants and the DPPH radical elimination activity in some cases is not linear, so it is necessary to study the behavior of each sample and obtain a standard curve for each one. In this way, the EC50 is determined by interpolating the plot of the DPPH percentage inhibited against the sample concentration (Deng et al., 2011).

Finally, it is important to mention that this method has been applied with many different protocols, in which there are differences regarding the solvents, the initial concentration of DPPH, the volume of the sample, the incubation time, and the presentation of the results; consequently, it is known that the results obtained according to different protocols are not

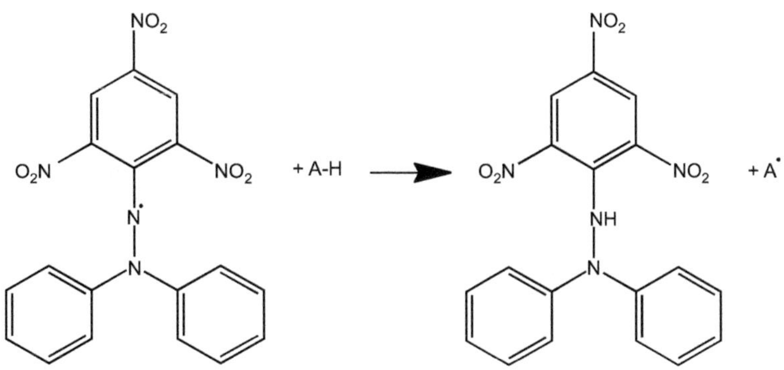

Figure 14.3
The reaction mechanism of DPPH radical with antioxidant (AH).

comparable (Carmona-Jiménez et al., 2014). Although this method is technically simple, and only requires a UV-vis spectrophotometer, if this technique is going to be used routinely, it is almost essential to use an automated spectrophotometric system that can measure kinetics in parallel (Carmona-Jiménez et al., 2014; Deng et al., 2011).

14.6 ABTS (2,2′-Azinobis-(3-Ethylbenzthiazolin-6-Sulfonic Acid))

The ABTS assay is considered one of the most sensitive techniques to identify antioxidant activity, because the response of antioxidants involves faster reaction kinetics (Chanput et al., 2016). This method was initially reported by Miller and colleagues, and is based on the ability of an antioxidant to stabilize the ABTS colored cation radical, which can be previously formed by the oxidation of ABTS (2,2-azinobis-(3-ethylbenzthiazoline-6-sulfonic acid)) by methemoglobin and hydrogen peroxide (Londoño Londoño, 2012). The modified technique for the generation of the ABTS cation radical involves direct production of the green-blue ABTS chromophore through the reaction between ABTS and potassium persulfate. This chromophore has three absorption maxima at wavelengths of 645, 734, and 815 nm. The addition of antioxidants to this previously obtained radical follows an electron transfer mechanism (Lee et al., 2017; Re et al., 1999) (Fig. 14.4), which is visualized as a discoloration corresponding to when the radical ABTS it is reduced by antioxidant (Floegel et al., 2011). In this way, the degree of discoloration makes it possible to evaluate the percentage of inhibition of the ABTS cation radical, which is determined as a function of the antioxidant concentration and the reaction time (Kuskoski et al., 2005). The results are expressed as equivalents of Trolox or TEAC (Trolox Equivalent Antioxidant Capacity) (Londoño Londoño, 2012).

Among the advantages of this method is that TEAC values are reported for a wide range of foods that allows comparisons to be made; additionally, it can be used in a wide range of pHs and ionic strengths, in addition to the fact that ABTS is soluble both in aqueous and

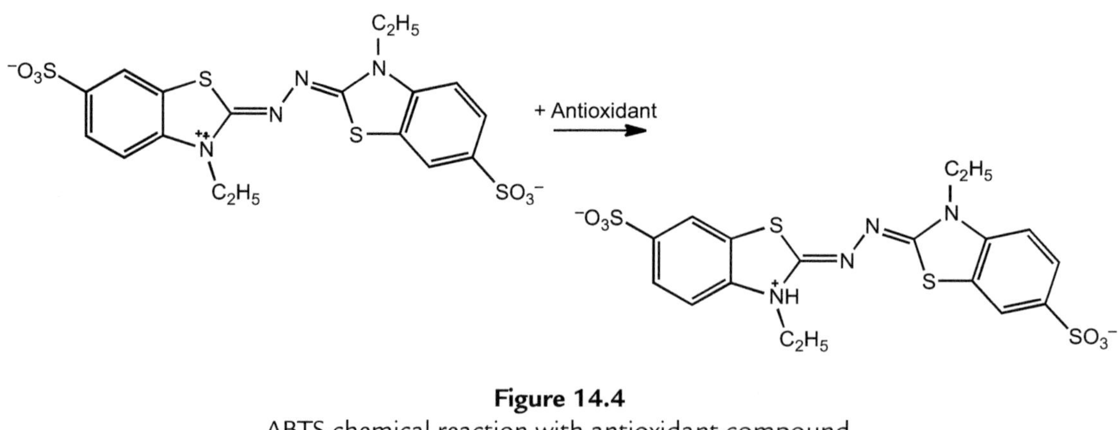

Figure 14.4
ABTS chemical reaction with antioxidant compound.

organic media, which allows the evaluation of hydrophilic and lipophilic antioxidants with high precision even in highly pigmented compounds. Among the disadvantages of this technique is the previous generation of ABTS, which must be previously generated from a chemical reaction (manganese dioxide, potassium persulfate, ABAP), enzymatic (peroxidase, myoglobin) or electrochemical; and that the reaction kinetics with some antioxidants can be slow and, therefore, can lead to errors in the determination of antioxidant capacity (Magalhaes et al., 2013; Londoño Londoño, 2012; Kuskoski et al., 2005).

Several studies have shown a linear relationship between the concentration of flavonoids and the activity of reducing the DPPH radical or ABTS radicals, finding that species with higher total flavonoid contents showed the highest antioxidant activity (Wang et al., 2017a,b; Lee et al., 2016; Cai et al., 2004). Therefore, a review of the main flavonoids to which their antioxidant capacity has been evaluated by the DPPH and ABTS methods is presented below.

14.7 Antioxidant Activity in Flavonoids by DPPH and ABTS

We identified 119 compounds of plant origin in 25 articles that reported the antioxidant activity of the DPPH and ABTS assays in the IC50 expression unit (μm), at a concentration at which 50% of the antioxidant activity was obtained (Table 14.1). These data were obtained after conducting a search in 500 scientific articles from 2008 to 2018 in the database "Science Direct" using as keywords "DPPH, ABTS, flavonoids, antioxidant, & IC50." The structures are presented in Table 14.1, where it was observed that flavones, flavonols, and flavanonols were the most abundant flavonoids.

Additionally it was identified that the substitution of functional groups by glucosides does not improve the antioxidant activity in the case of the following compounds: Naringenin 4, Epicatechin 7, Apigenin 35, Baicalin 41, Chrysin 46, and Isovitexin 54. However, in the compound Isoorientin 51 this was changed if activity was increased.

Through the compilation of the compounds in the present work, the 10 compounds that showed the best values (IC_{50}) of antioxidant activity in both trials were chosen, obtaining 15 structures (Fig. 14.5), five of them—three chalcones, one isoflavone, and one flavone—were present in both trials (Luteolin **57**, Glyurallin B **72**, Echinatin **103**, Licochalcone A **104**, and Licochalcone B **105**). In these structures their high capacity to catch free radicals can be precisely identified for the previously mentioned functional groups to which they are related (Queiroz Ferreira et al., 2015).

In addition to the reports on Glyurallin B **72**, Licochalcone A **104**, and Licochalcone B **105**, which were the structures with the best activity, they are also known for their good antiinflammatory activity due to their ability to capture free radicals and modulate mitochondrial activity (Fu et al., 2013; Furusawa et al., 2009; Kolbe et al., 2006). Furthermore, the

Table 14.1: List and Antioxidant Activity of Flavonoids Isolated and Reported From Different Species

Compound		Species	Family	Antioxidant Assay		References
				DPPH	ABTS	
Flavanone						
Isocarthamidin-7-O-glucuronide	**1**	*Scutellaria baicalensis*	Lamiaceae	>100.00	84.250	Li et al. (2017)
Japonicasins A	**2**	*Sophora japonica*	Fabaceae	35.10		He et al. (2016)
Japonicasins B	**3**			88.70		
Naringenin	**4**	*Silybum marianum*	Asteraceae	>100.00	1.03	Qin et al. (2017)
Naringenin-7-O-β-D-glucopyranoside	**5**			>100.00	2.61	
Pinocembrin	**6**	*Dodonaea viscosa*	Sapindaceae		>100.00	Muhammad et al. (2015)
Flavan 3 ol						
(−)-Epicatechin	**7**	*Litchi chinensis*	Sapindaceae	9.55		Ibrahim and Mohamed (2015)
				5.54		
Epicatechin-(7,8-bc)-4β-(4-hydroxyphenyl)-dihydro-2(3H)-pyranone	**8**	*Litchi chinensis*	Sapindaceae	20.07		
Falandioside A	**9**	*Fragaria × ananassa*	Rosaceae	>100.00	5.74	Yang et al. (2016)
Flavanonol						
2,3-*cis*-3,7,8,3,4-pentahydroxydi-hydroflavone	**10**	*Acacia confusa*	Fabaceae	3.90		Lin and Chang (2013)
2,3-*trans*-3,7,8,3,4-pentahydroxy-dihydroflavone	**11**			4.60		
4-O-Methyl-melacacidi	**12**			4.30		
Aromadendrin,(2S,3S)3,4,5,7-tetrahydroxyflavanone	**13**	*Dodonaea viscosa*	Sapindaceae		>100.00	Muhammad et al. (2015)
Dihydrokaempferol	**14**	*Silybum marianum*	Asteraceae	>100.00	1.21	Qin et al. (2017)
Dihydroquercetin-4′-methylether	**15**			>100.00	2.85	
Isomelacacidin	**16**	*Acacia confusa*	Fabaceae	4.50		Lin and Chang (2013)
Melacacidin	**17**			3.90		
Taxifolin	**18**	*Silybum marianum*	Asteraceae	11.53	1.14	Qin et al. (2017)

Flavone						
5,4′-Dihydroxy-3,7,3′-trimethoxyflavone	**19**	*Vitex negundo*	Verbenaceae	>100.00	7.57	Hu et al. (2017a,b)
3,4-Dimethoxy-5,7-dihydroxyflavone	**20**	*Dodonaea viscosa*	Sapindaceae		>100.00	Muhammad et al. (2015)
3-C-glucopyranosylapigenin	**21**	*Vitex negundo*	Verbenaceae	>100.00	2.37	
5,4′-Dihydroxy-3,6,7,3′-tetramethoxyflavone	**22**			>100.00	2.95	
5,4′-Dihydroxy-3,6,7,8,3′-pentamethoxyflavone	**23**			>100.00	2.94	
5,4′-Dihydroxy-3,6,7,8-tetramethoxyflavone	**24**			58.77	3.23	
5,4′-Dihydroxy-3,6,7-trimethoxyflavone	**25**			>100.00	3.50	
5,4′-Dihydroxy-6,7,3′-trimethoxyflavone	**26**			>100.00	5.12	
5,6,3′,4′-Tetrahydroxy-3,7-dimethoxyflavone	**27**			3.67	1.23	
5,7,4-Trihydroxy-3-(3-hydroxymethylbutyl)-3,6- dimethoxyflavone	**28**	*Dodonaea viscosa*	Sapindaceae		14.91	Muhammad et al. (2015)
5,7-Dihydroxy-3-(4-acetoxy-3-methylbutyl)-3,6,4-trimethoxy flavone	**29**				52.83	
5,7-Dihydroxy-3-(2-hydroxy-3-methyl-3-butenyl)-3,6,4-trimethoxy-flavone	**30**				33.89	
5,7-Dihydroxy-3-(3-hydroxy-methylbutyl)-3,6,4-trimethoxyflavone	**31**				40.84	
5-Hydroxy-3,6,7,8,3′,4′-hexamethoxyflavone	**32**	*Vitex negundo*	Verbenaceae	>100.00	2.85	Hu et al. (2017a,b)
6-OH-luteolin-7-O-β-D-glucoside	**33**	*Achillea millefolium*	Asteraceae	50.00	6.00	Sevindik et al. (2015)
7,4′-Dihydroxy-flavonol-3-O-α-glucoside	**34**	*Trifolium echinatum*	Fabaceae	>100.00	>100.00	Sabudak et al. (2013)

Continued

Table 14.1: List and Antioxidant Activity of Flavonoids Isolated and Reported From Different Species—cont'd

Compound		Species	Family	Antioxidant Assay		References
				DPPH	ABTS	
Apigenin	**35**	*Verbascum nigrum, Verbascum phlomoides and Verbascum thapsus*	Scrophulariaceae	>100.00		Mihailović et al. (2016)
		Dorystoechas hastata	Lamiaceae	>100.00		Erkan et al. (2011)
		Lawsonia inermis	Lythraceae		>100.00	Singh et al. (2015)
		Scutellaria baicalensis	Lamiaceae	67.37	18.22	Li et al. (2017)
Apigenin-7-O-β-D-glucoside	**36**	*Achillea millefolium*	Asteraceae	>100.00	30.00	Sevindik et al. (2015)
Apigenin-7-O-β-glucoside	**37**	*Trifolium echinatum*	Fabaceae	>100.00	45.67	Sabudak et al. (2013)
Apigenin-7-O-glucuronide	**38**	*Scutellaria baicalensis*	Lamiaceae	>100.00	32.67	Li et al. (2017)
Apiin	**39**	*Lawsonia inermis*	Lythraceae		>100.00	Singh et al. (2015)
Artogomezianone	**40**	*Artocarpus altilis*	Moraceae	>100.00	36.90	Lan et al. (2013)
Baicalein	**41**	*Scutellaria baicalensis*	Lamiaceae	66.02	17.53	Li et al. (2017)
		Oroxylum indicum	Bignoniaceae	>100.00		Dinda et al. (2015a,b)
Baicalein-7-O-gentiobioside	**42**			>100.00		
Baicalein-7-O-glucoside	**43**			>100.00		
Baicalin	**44**	*Oroxylum indicum*	Bignoniaceae	0.256		
		Scutellaria baicalensis	Lamiaceae	93.46	21.98	Li et al. (2017)
Carambolaflavone	**45**	*Averrhoa carambola*	Oxalidaceae	>100.00	6.60	Yang et al. (2015)
Chrysin	**46**	*Scutellaria baicalensis*	Lamiaceae	92.14	20.16	Li et al. (2017)
		Sida glutinosa	Malvaceae	>100.00		Dinda et al. (2015a,b)
Chrysin-7-O-glucuronide	**47**	*Scutellaria baicalensis*	Lamiaceae	>100.00	79.34	Li et al. (2017)
Cosmosiin	**48**	*Lawsonia inermis*	Lythraceae		>100.00	Singh et al. (2015)
Diosmetin-7-O-β-D-glucopyranoside	**49**	*Vitex negundo*	Verbenaceae	>100.00	9.37	Hu et al. (2017a,b)
Hydroxyartoflavone A	**50**	*Artocarpus altilis*	Moraceae	20.90		Lan et al. (2013)
Isoorientin	**51**	*Vitex negundo*	Verbenaceae	13.08	1.26	Hu et al. (2017a,b)
Isoorientin-6″-O-caffeate	**52**			3.38	0.75	
Isorhamnetin-3-(6-methylglucuronide)	**53**	*Fragaria × ananassa*	Rosaceae	>100.00	17.74	Yang et al. (2016)
Isovitexin	**54**	*Vitex negundo*	Verbenaceae	50.63	1.70	Hu et al. (2017a,b)
Isovitexin-2″-O-α-L-rhamnopyranoside	**55**	*Averrhoa carambola*	Oxalidaceae	>100.00	7.60	Yang et al. (2015)

Kaempferol-3-(6-methylglucuronide)	**56**	*Fragaria × ananassa*	Rosaceae	>100.00	4.42	Yang et al. (2016)
Luteolin	**57**	*Lawsonia inermis*	Lythraceae		>100.00	Singh et al. (2015)
		Litchi chinensis	Sapindaceae	9.98		Ibrahim and Mohamed (2015)
		Opilia amentacea	Opiliaceae	85.10		Magid et al. (2017)
		Schinus terebinthifolius	Anacardiaceae	4.57		Silva et al. (2017)
		Verbascum nigrum, Verbascum phlomoides and *Verbascum thapsus*	Scrophulariaceae	0.019		Mihailović et al. (2016)
		Vitex negundo	Verbenaceae	3.51	0.86	Hu et al. (2017a,b)
		Achillea millefolium	Asteraceae	80.00	9.00	Sevindik et al. (2015)
Luteolin-7-O-β-D-glucoside	**58**	*Achillea millefolium*	Asteraceae	60.00	6.00	
Luteolin-7-O-β-D-glucopyranoside	**59**	*Vitex negundo*	Verbenaceae	6.35	1.73	Hu et al. (2017a,b)
Orientin	**60**			12.30	0.86	
Paucatalinone C	**61**	*Paulownia catalpifolia*	Scrophulariaceae	4.82		Wang et al. (2017a,b)
Paucatalinone D	**62**			5.15		
Paucatalinone E	**63**			15.22		
Penduletin	**64**	*Dodonaea viscosa*	Sapindaceae		>100.00	Muhammad et al. (2015)
Quercetin-3-(6-methylglucuronide)	**65**	*Fragaria × ananassa*	Rosaceae	32.12	4.60	Yang et al. (2016)
Scutellarein-7-O-gentiobioside	**66**	*Oroxylum indicum*	Bignoniaceae	>100.00		Dinda et al. (2015a,b)
Scutellarein-7-O-glucoside	**67**		Bignoniaceae	>100.00		
Scutellarin	**68**	*Scutellaria baicalensis*	Lamiaceae	89.07	19.52	Li et al. (2017)
Biflavone						
5,7,4″,5″,3‴,4‴-Hexahydroxy-3″-O-β-glucosyl-3′,7″-O biflavone	**69**	*Trifolium echinatum*	Fabaceae	>100.00	82.93	Sabudak et al. (2013)
Robustaflavone	**70**	*Schinus terebinthifolius*	Anacardiaceae	19.44		Silva et al. (2017)

Continued

Table 14.1: List and Antioxidant Activity of Flavonoids Isolated and Reported From Different Species—cont'd

Compound		Species	Family	Antioxidant Assay		References
				DPPH	ABTS	
Isoflavone						
Biochanin-A-7-O-β-glucoside	**71**	*Trifolium echinatum*	Fabaceae	>100.00	>100.00	Sabudak et al. (2013)
Glyurallin B	**72**	*Glycyrrhiza glabra*	Fabaceae	0.029	0.018	Li et al. (2011)
Orobol-7- O-β-glucoside	**73**	*Trifolium echinatum*	Fabaceae	>100.00	>100.00	Sabudak et al. (2013)
Flavonol						
6,7-Dimethylkaempferol	**74**	*Dodonaea viscosa*	Sapindaceae		39.95	Muhammad et al. (2015)
Anhydrous viscosine	**75**				40.46	
Falandioside B	**76**	*Fragaria × ananassa*	Rosaceae	>100.00	8.80	Yang et al. (2016)
Hydrous santin	**77**	*Dodonaea viscosa*	Sapindaceae		38.82	Muhammad et al. (2015)
Kaempferol	**78**	*Dorystoechas hastata*	Lamiaceae	0.093		Erkan et al. (2011)
		Silybum marianum	Asteraceae	67.73	3.33	Qin et al. (2017)
Kaempferol-3-methylether	**79**	*Dodonaea viscosa*	Sapindaceae		>100.00	Muhammad et al. (2015)
Kaempferol-3-O-[α-L-rhamnopyranosyl-(1–6)]-[β-D-glucopyranosyl-(1–2)]-β-D-glucopyranoside	**80**	*Sophora japonica*	Fabaceae	25.30		He et al. (2016)
Kaempferol-3-O-D-glucopyranosyl-(1→2)--L-rhamnopyranosyl-(1→6)--L-rhamnopyranoside 2	**81**	*Hedysarum carnosum*		50.39	74.37	Ben Salah et al. (2016)
Kaempferol-3-O-α-L-rhamnopyranosyl-(1–6)-β-D-glucopyranosyl-(1–2)-β-D-glucopyranoside	**82**	*Sophora japonica*		25.50		He et al. (2016)
Kaempferol-3-O-β-glucopyranosyl(1–2)-β-galactopyranoside-7-O-α-rhamnopyranoside	**83**			26.60		
Kaempferol-3-O-a-L-rhamnopyranosyl-7- O-[β-D-glucopyranosyl-(1–2)-O-L-rhamnoside]	**84**	*Siraitia grosvenori*	Cucurbitaceae	>100.00	>100.00	Pan et al. (2012)

Kaempferol-3-O-β-D-glucose-7-O-α-Lrhamnoside	**85**			>100.00	>100.00	
Kaempferol-3-O-α-rhamnoside	**86**	*Litchi chinensis*	Sapindaceae	78.71		Ibrahim and Mohamed (2015)
Kaempferol-3-O-β-D-glucoside	**87**			>100.00		
Luteolin-7-O-β-galactoside	**88**	*Trifolium echinatum*	Fabaceae	>100.00	>100.00	Sabudak et al. (2013)
Melanoxetin	**89**	*Acacia confusa*		3.10		Lin and Chang (2013)
Myricetin	**90**	*Dorystoechas hastata*	Lamiaceae	1.37		Erkan et al. (2011)
Quercetin	**91**			0.043		
		Litchi chinensis	Sapindaceae	5.52		Ibrahim and Mohamed (2015)
		Opilia amentacea	Opiliaceae	42.10		Magid et al. (2017)
		Psidium guajava	Myrtaceae	0.036		Feng et al. (2015)
		Schinus terebinthifolius	Anacardiaceae	4.03		Silva et al. (2017)
		Vitex negundo	Verbenaceae	>100.00	2.08	Hu et al. (2017a,b)
Quercetin-3-O--D-glucopyranosyl-(1→2)--L-rhamnopyranosyl-(1→6)--L-rhamnopyranoside 1	**92**	*Hedysarum carnosum*	Fabaceae	20.27	17.19	Ben Salah et al. (2016)
Quercetin-3-O-β-D-galactopyranoside	**93**	*Psidium guajava*	Myrtaceae	58.20		Feng et al. (2015)
Quercetin-3-O-α-L-arabinofuranoside	**94**			37.13		
Quercetin-3-O-α-L-arabinopyranoside	**95**			30.92		
Quercetin-3-O-rutinoside	**96**	*Litchi chinensis*	Sapindaceae	7.40		Ibrahim and Mohamed (2015)
Quercetin-3-O-β-D-glucuronide	**97**	*Eucalyptus grandis x Eucalyptus urophylla GL9*	Myrtaceae	>100.00	>100.00	Chen et al. (2014)
Quercetrin	**98**	*Schinus terebinthifolius*	Anacardiaceae	20.43		Silva et al. (2017)
Santin	**99**	*Dodonaea viscosa*	Sapindaceae		40.44	Muhammad et al. (2015)
Transilitin	**100**	*Acacia confusa*	Fabaceae	3.10		Lin and Chang (2013)
Viscosine	**101**	*Dodonaea viscosa*	Sapindaceae		23.69	Muhammad et al. (2015)

Continued

Table 14.1: List and Antioxidant Activity of Flavonoids Isolated and Reported From Different Species—cont'd

Compound		Species	Family	Antioxidant Assay		References
				DPPH	ABTS	
			Chalcone			
5-(1,1-Dimethylallyl)-3,4,4 -trihydroxy-2-methoxychalcone	**102**	*Glycyrrhiza glabra*	Fabaceae	15.92	13.01	Li et al. (2011)
Echinatin	**103**			0.19	0.019	
Licochalcone A	**104**			0.10	0.006	
Licochalcone B	**105**			0.007	0.001	
Okanin	**106**	*Acacia confusa*	Fabaceae	3.10		Lin and Chang (2013)
			Dihydrochalcone			
Carambolasides A	**107**	*Averrhoa carambola*	Oxalidaceae	>100.00	5.90	Yang et al. (2015)
Carambolasides B	**108**			>100.00	3.90	
Carambolasides C	**109**			47.00	4.80	
Carambolasides D	**110**			>100.00	4.10	
			Xanthone			
Isocycloartobiloxanthone	**111**	*Artocarpus altilis*	Moraceae	33.90	7.20	Lan et al. (2013)
			Proanthocyanidin			
Aesculitannin A	**112**	*Litchi chinensis*	Sapindaceae	5.57		Ibrahim and Mohamed (2015)
Epicatechin-(2β-O-7,4β-8)-epiafzelechin-(4α-8)-epicatechin	**113**			9.65		
Litchitannin-A1[epicatechin-(2β-O-7,4β-6)-epicatechin-(2β-O 7,4β-8)-catechin]	**114**			5.25		
Litchitannin-A2[epicatechin-(2β-O-7,4β-6)-epicatechin-(2β-O-7,4β-6)-epicatechin]	**115**			12.61		
Proanthocyanidin A1	**116**	*Litchi chinensis*	Sapindaceae	9.53		
Proanthocyanidin A6	**117**			8.66		
Proanthocyanidin B2	**118**			5.42		
Procyanidin A2	**119**			7.17	1.66	

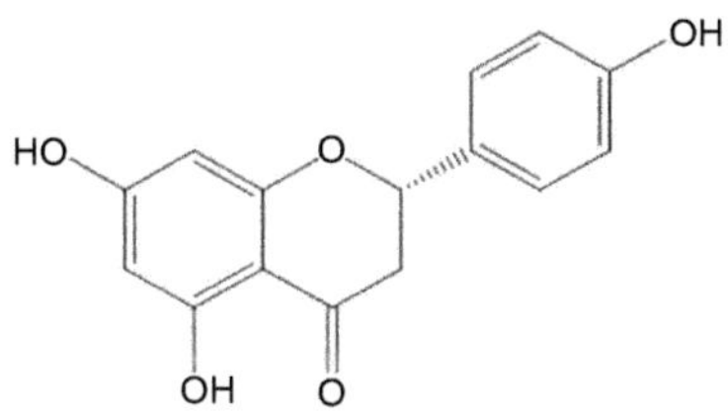

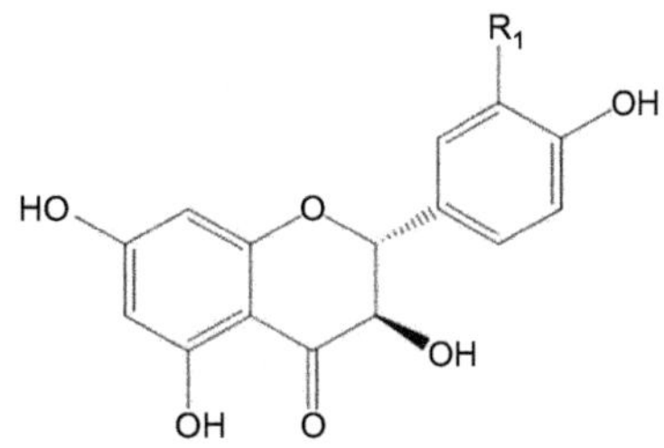

14 R_1= H
18 R_1= OH

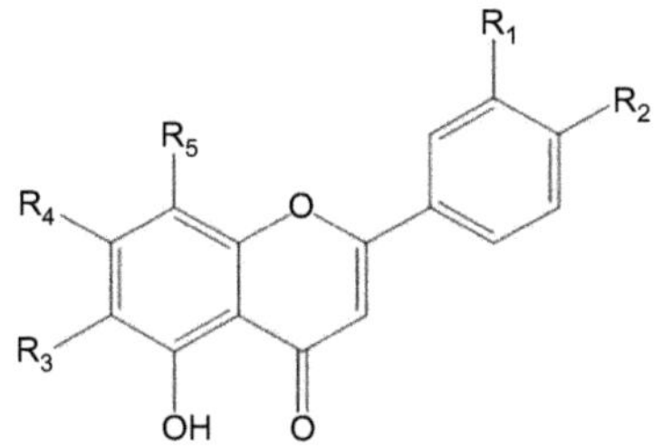

44 R_1= H R_2= H R_3= OH R_4= R_5= H

52 R_1= OH R_2= OH R_3= Glc(6″-caffeoyl) R_4= OH R_5= H
Glc= β-D-glucopyranosyl
caffeoyl=

57 R_1= OH R_2= OH R_3= H R_4= OH R_5= H
60 R_1= OH R_2= OH R_3= H R_4= OH R_5= OH

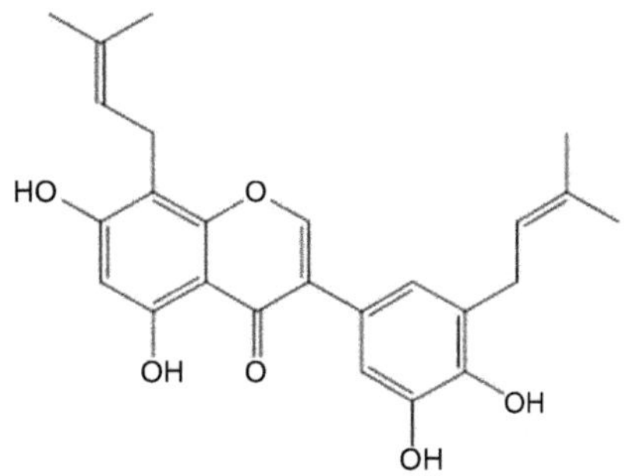

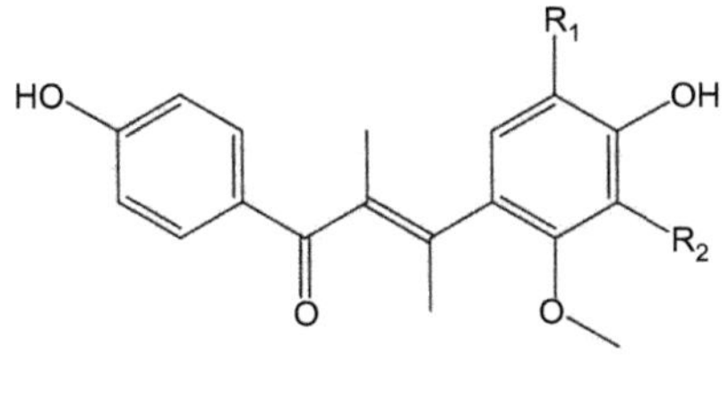

103 R_1= H; R_2= H
104 R_1= $-C(CH_3)_2CH{=}CH_2$ R_2= H
105 R_1= H; R_2= OH

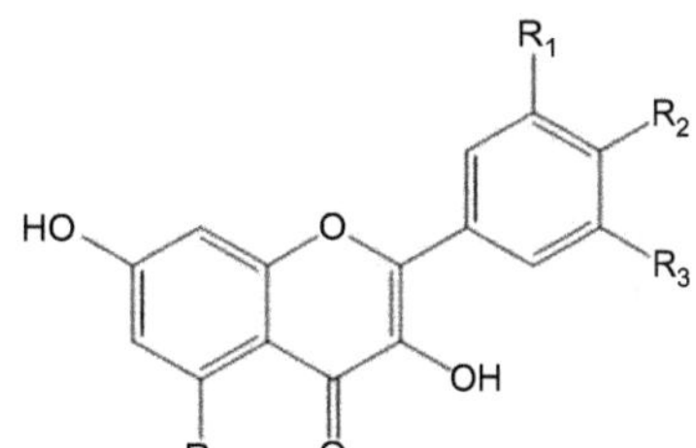

78 R_1= H R_2= OH R_3= H R_4= OH
89 R_1= OH R_2= OH R_3= H R_4= H
90 R_1= OH R_2= OH R_3= OH R_4= OH
91 R_1= OH R_2= OH R_3= H R_4= OH

Figure 14.5

Flavonoids structures which presented the best antioxidant activity in the DPPH and ABTS assays: Naringenin **4**, Dihydrokaempferol **14**, Taxifolin **18**, Baicalin **44**, Isoorientin-6″-O-caffeate **52**, Luteolin **57**, Orientin **60**, Glyurallin B **72**, kaempferol **78**, Melanoxetin **89**, Myricetin **90**, Quercetin **91**, Echinatin **103**, Licochalcone A **104**, and Licochalcone B **105**.

antioxidant activity of flavonoids is related to more than one property, such as antitumor (Raffa et al., 2017) or neuroprotection from the antiinflammatory action (Jaeger et al., 2017) making these compounds relevant for its implementation in medicine.

References

Abuja, P.M., Albertini, R., 2001. Methods for monitoring oxidative stress, lipid peroxidation and oxidation resistance of lipoproteins. Clinica Chimica Acta 306, 1–17.

Amarowicz, R., Shahidi, F., 2017. Antioxidant activity of broad bean seed extract and its phenolic composition. Journal of Functional Foods 38, 656–662.

Ballesteros, L.F., Ramirez, M.J., Orrego, C.E., Teixeira, J.A., Mussatto, S.I., 2017a. Optimization of autohydrolysis conditions to extract antioxidant phenolic compounds from spent coffee grounds. Journal of Food Engineering 199, 1–8.

Ballesteros, L.F., Ramirez, M.J., Orrego, C.E., Teixeira, J.A., Mussatto, S.I., 2017b. Encapsulation of antioxidant phenolic compounds extracted from spent coffee grounds by freeze-drying and spray-drying using different coating materials. Food Chemistry 237, 623–631.

Barros, R.G.C., Andrade, J.K.S., Denadai, M., Nunes, M.L., Narain, N., 2017. Evaluation of bioactive compounds potential and antioxidant activity in some Brazilian exotic fruit residues. Food Research International 102, 84–92.

Ben Salah, N., Casabianca, H., Essaidi, I., Chenavas, S., Fildier, A., Sanglar, C., Ben Jannet, H., Bouzouita, N., 2016. Isolation and structure elucidation of two new antioxidant flavonoid glycosides and fatty acid composition in Hedysarum carnosum Desf. Industrial Crops and Products. 81, 195–201. https://doi.org/10.1016/j.indcrop.2015.11.057.

Blois, S., 1955. A note on free radical formation in biologically occurring quinones. Biochimica et biophysica acta 18 (1), 165–165.

Boveris, A., Repetto, M., Boveris Valdez, L., 2008. Superóxido dismutase. In: Radicales libres y stress oxidative. Aplicaciones médicas. Manual Moderno, Mexico, D.F, pp. 169–182.

Cai, Y., Luo, Q., Sun, M., Corke, H., 2004. Antioxidant activity and phenolic compounds of 112 traditional Chinese medicinal plants associated with anticancer. Life Sciences 74, 2157–2184.

Carmona-Jiménez, Y., García-Moreno, M.V., Igartuburu, J.M., Barroso, C.G., 2014. Simplification of the DPPH assay for estimating the antioxidant activity of wine and wine by-products. Food Chemistry 165, 198–204.

Chanput, W., Krueyos, N., Ritthiruangdej, P., 2016. Anti-oxidative assays as markers for anti-inflammatory activity of flavonoids. International Immunopharmacology 40, 170–175.

Chen, Y., Wang, J., Ou, Y., Chen, H., Xiao, S., Liu, G., Cao, Y., Huang, Q., 2014. Cellular antioxidant activities of polyphenols isolated from Eucalyptus leaves (*Eucalyptus grandis×Eucalyptus urophylla* GL9). Journal of Functional Foods. 7, 737–745. https://doi.org/10.1016/j.jff.2013.12.003.

Chen, Z., Bertin, R., Froldi, G., 2013. EC 50 estimation of antioxidant activity in DPPH assay using several statistical programs. Food Chemistry 138 (1), 414–420.

Delaux, P., Nanda, A.K., Mathé, C., Sejalon-Delmas, N., Dunand, C., 2012. Molecular and biochemical aspects of plant terrestrialization. Perspectives in Plant Ecology, Evolution and Systematic 14 (1), 49–59.

Deng, J., Cheng, W., Yang, G., 2011. A novel antioxidant activity index (AAU) for natural products using the DPPH assay. Food Chemistry 125 (4), 1430–1435.

Dinda, B., Das, N., Dinda, S., Dinda, M., SilSarma, I., 2015a. The genus Sida L. – a traditional medicine: its ethnopharmacological, phytochemical and pharmacological data for commercial exploitation in herbal drugs industry. Journal of Ethnopharmacology. 176, 135–176. https://doi.org/10.1016/j.jep.2015.10.027.

Dinda, B., SilSarma, I., Dinda, M., Rudrapaul, P., 2015b. Oroxylum indicum (L.) Kurz, an important Asian traditional medicine: from traditional uses to scientific data for its commercial exploitation. Journal of Ethnopharmacology. 161, 255–278. https://doi.org/10.1016/j.jep.2014.12.027.

Erkan, N., Akgonen, S., Ovat, S., Goksel, G., Ayranci, E., 2011. Phenolic compounds profile and antioxidant activity of *Dorystoechas hastata* L. Boiss et Heldr. Food Research International. 44, 3013–3020. https://doi.org/10.1016/j.foodres.2011.07.015.

Feng, X., Wang, Z., Meng, D., Li, X., 2015. Cytotoxic and antioxidant constituents from the leaves of *Psidium guajava*. Bioorganic and Medicinal Chemistry Letters. 25, 2193–2198. https://doi.org/10.1016/j.bmcl.2015.03.058.

Ferrándiz, M.L., Alcaraz, M.J., 1991. Anti-inflamatory activity and inhibition of arachidonic acid metabolism by flavonoids. Agents and Actions 32, 283–288.

Foreman, J., Demidchik, V., Bothwell, J.H.F., Mylona, P., Miedema, H., Torres, M.A., Linstead, P., Costa, S., Brownlee, C., Jones, J.D.G., Davies, J.M., Dolan, L., 2003. Reactive oxygen species produced by NADPH oxidase regulate plant cell growth. Nature 422, 442–446.

Floegel, A., Kim, D.O., Chung, S.J., Koo, S.I., Chun, O.K., 2011. Comparison of ABTS/DPPH assays to measure antioxidant capacity in popular antioxidant-rich US foods. Journal of Food Composition and Analysis 24 (7), 1043–1048.

Fu, Y., Chen, J., Li, Y.-J., Zheng, Y.-F., Li, P., 2013. Antioxidant and anti-inflammatory activities of six flavonoids separated from licorice. Food Chemistry. 141, 1063–1071. https://doi.org/10.1016/j.foodchem.2013.03.089.

Furusawa, J., Funakoshi-Tago, M., Mashino, T., Tago, K., Inoue, H., Sonoda, Y., Kasahara, T., 2009. Glycyrrhiza inflata-derived chalcones, Licochalcone A, Licochalcone B and Licochalcone D, inhibit phosphorylation of NF-κB p65 in LPS signaling pathway. International Immunopharmacology. 9, 499–507. https://doi.org/10.1016/j.intimp.2009.01.031.

Gómez, L.E., Cuevas, D.B., 2008. Superóxido dismutase. In: Radicales libres y stress oxidative. Aplicaciones médicas. Manual Moderno, Mexico, D.F, pp. 169–182.

He, X., Bai, Y., Zhao, Z., Wang, X., Fang, J., Huang, L., Zeng, M., Zhang, Q., Zhang, Y., Zheng, X., 2016. Local and traditional uses, phytochemistry, and pharmacology of *Sophora japonica* L.: a review. Journal of Ethnopharmacology. 187, 160–182. https://doi.org/10.1016/j.jep.2016.04.014.

Hidayat, M.A., Fitri, A., Kuswandi, B., 2017. Scanometry as microplate reader for high throughput method based on DPPH dry reagent for antioxidant assay. Acta Pharmaceutica Sinica B 7 (3), 395–400.

Hu, P., Li, D.-H., Jia, C.-C., Liu, Q., Wang, X.-F., Li, Z.-L., Hua, H.-M., 2017a. Bioactive constituents from Vitex negundo var. heterophylla and their antioxidant and α -glucosidase inhibitory activities. Journal of Functional Foods. 35, 236–244. https://doi.org/10.1016/j.jff.2017.05.047.

Hu, Y., Li, Y., Zhang, W., Kou, G., Zhou, Z., 2017b. Physical stability and antioxidant activity of citrus flavonoids in Arabic gum-stabilized microcapsules: modulation of whey protein concentrate. Food Hydrocolloids. https://doi.org/10.1016/j.foodhyd.2017.10.037.

Ibrahim, S.R.M., Mohamed, G.A., 2015. Litchi chinensis: medicinal uses, phytochemistry, and pharmacology. Journal of Ethnopharmacology. 174, 492–513. https://doi.org/10.1016/j.jep.2015.08.054.

Jabbari, M., Jabbari, A., 2016. Antioxidant potential and DPPH radical scavenging kinetics of water-insoluble flavonoid naringenin in aqueous solution of micelles. Colloids and Surfaces A: Physicochemical and Engineering Aspects 489, 392–399.

Jabeen, E., Janjua, N.K., Ahmed, S., Murtaza, I., Ali, T., Hameed, S., 2017. Radical scavenging propensity of Cu 2+, Fe 3+ complexes of flavonoids and in-vivo radical scavenging by Fe 3+-primuletin. Spectrochimica Acta Part A: Molecular and Biomolecular Spectroscopy 171, 432–438.

Jackson, R.M., Garcia-Rojas, R., 2008. Kinase activity, heat shock protein 27 phosphorylation, and lung epithelial cell glutathione. Experimental lung research 34 (5), 245–262.

Jaeger, B.N., Parylak, S.L., Gage, F.H., 2017. Mechanisms of dietary flavonoid action in neuronal function and neuroinflammation. Molecular Aspects of Medicine. https://doi.org/10.1016/j.mam.2017.11.003.

Keaney, J.F., 2000. Atherosclerosis: from the lesión formation to plaque activation and endothelial dysfuntion. Molecular Aspect of Medicine 21, 99–166.

Kolbe, L., Immeyer, J., Batzer, J., Wensorra, U., Dieck, K., tom, Mundt, C., Wolber, R., Stäb, F., Schönrock, U., Ceilley, R.I., Wenck, H., 2006. Anti-inflammatory efficacy of Licochalcone A: correlation of clinical potency and in vitro effects. Archives of Dermatological Research. 298, 23–30. https://doi.org/10.1007/s00403-006-0654-4.

Kuskoski, E.M., Asuero, A.G., Troncoso, A.M., Mancini-Filho, J., Fett, R., 2005. Aplicación de diversos métodos químicos para determinar actividad antioxidante en pulpa de frutos. Food Science and Technology (Campinas) 25 (4), 726–732.

Lan, W.-C., Tzeng, C.-W., Lin, C.-C., Yen, F.-L., Ko, H.-H., 2013. Prenylated flavonoids from *Artocarpus altilis*: antioxidant activities and inhibitory effects on melanin production. Phytochemistry. 89, 78–88. https://doi.org/10.1016/j.phytochem.2013.01.011.

Le Grandois, J., Guffond, D., Hamon, E., Marchioni, E., Werner, D., 2017. Combined microplate-ABTS and HPLC-ABTS analysis of tomato and pepper extracts reveals synergetic and antagonist effects of their lipophilic antioxidative components. Food Chemistry 223, 62–71.

Lee, L.S., Choi, E.J., Kim, C.H., Sung, J.M., Kim, Y.B., Seo, D.H., et al., 2016. Contribution of flavonoids to the antioxidant properties of common and tartary buckwheat. Journal of Cereal Science 68, 181–186.

Li, K., Fan, H., Yin, P., Yang, L., Xue, Q., Li, X., Sun, L., Liu, Y., 2017. Structure-activity relationship of eight high content flavonoids analyzed with a preliminary assign-score method and their contribution to antioxidant ability of flavonoids-rich extract from *Scutellaria baicalensis* shoots. Arabian Journal of Chemistry. https://doi.org/10.1016/j.arabjc.2017.08.002.

Li, Y.-J., Chen, J., Li, Y., Li, Q., Zheng, Y.-F., Fu, Y., Li, P., 2011. Screening and characterization of natural antioxidants in four Glycyrrhiza species by liquid chromatography coupled with electrospray ionization quadrupole time-of-flight tandem mass spectrometry. Journal of Chromatography A. 1218, 8181–8191. https://doi.org/10.1016/j.chroma.2011.09.030.

Lin, H.Y., Chang, S.T., 2013. Antioxidant potency of phenolic phytochemicals from the root extract of *Acacia confusa*. Industrial Crops and Products. 49, 871–878. https://doi.org/10.1016/j.indcrop.2013.07.001.

Lluis, J.M., Morales, A., 2008. Hipoxia y reperfusion. In: Radicales libres y stress oxidative. Aplicaciones médicas. Manual Moderno, Mexico, D.F, pp. 169–182.

Londoño Londoño, J., 2012. Antioxidantes: importancia biológica y métodos para medir su actividad. In: Desarrollo y Transversalidad serie Lasallista Investigación y Ciencia. Corporación Universitaria Lasallista.

Magalhães, L.M., Almeida, M.I.G., Barreiros, L., Reis, S., Segundo, M.A., 2012. Automatic aluminum chloride method for routine estimation of total flavonoids in red wines and teas. Food Analytical Methods 5 (3), 530–539.

Magid, A.A., Abdellah, A., Pecher, V., Pasquier, L., Harakat, D., Voutquenne-Nazabadioko, L., 2017. Flavonol glycosides and lignans from the leaves of *Opilia amentacea*. Phytochemistry Letters. 21, 84–89. https://doi.org/10.1016/j.phytol.2017.05.023.

Mihailović, V., Kreft, S., Benković, E.T., Ivanović, N., Stanković, M.S., 2016. Chemical profile, antioxidant activity and stability in stimulated gastrointestinal tract model system of three *Verbascum* species. Industrial Crops and Products. 89, 141–151. https://doi.org/10.1016/j.indcrop.2016.04.075.

Muhammad, A., Tel-Cayan, G., Öztürk, M., Nadeem, S., Duru, M.E., Anis, I., Ng, S.W., Shah, M.R., 2015. Biologically active flavonoids from *Dodonaea viscosa* and their structure–activity relationships. Industrial Crops and Products. 78, 66–72. https://doi.org/10.1016/j.indcrop.2015.10.011.

Naeimi, A.F., Alizadeh, M., 2017. Antioxidant properties of the flavonoid fisetin: an updated review of in vivo and in vitro studies. Trends in Food Science and Technology 70, 34–44.

Pan, Y., Wei, L., Zhu, Z., Liang, Y., Huang, C., Wang, H., Wang, K., 2012. Processing of *Siraitia grosvenori*' leaves: extraction of antioxidant substances. Biomass and Bioenergy. 36, 419–426. https://doi.org/10.1016/j.biombioe.2011.11.011.

Pei, Z.M., Murata, Y., Benning, G., Thomine, S., Klusener, B., Allen, G.J., Grill, E., Schroeder, J.I., 2000. Calcium channels activated by hydrogen peroxide mediate abscisic acid signalling in guard cells. Nature 406, 731–734.

Pennathur, S., Heinecke, J.W., 2004. Mechanisms of oxidative stress in diabetes implications for the pathogenesis of vascular disease and antioxidant therapy. Frontier Times 1 (9), 565–574.

Pérez, G., Martínez, G., 2001. Flavonoides como antioxidantes naturales. Acta Farmacologica Bonaerense 20 (4), 297–306.

Prior, R.L., Gu, L., 2005. Occurrence and biological significance of proanthocyanidins in the American diet. Phytochemistry 66 (18), 2264–2280.

Qin, N., Jia, C., Xu, J., Li, D., Xu, F., Bai, J., Li, Z., Hua, H., 2017. New amides from seeds of Silybum marianum with potential antioxidant and antidiabetic activities. Fitoterapia. 119, 83–89. https://doi.org/10.1016/j.fitote.2017.04.008.

Queiroz Ferreira, R., Greco, S.J., Delarmelina, M., Weber, K.C., 2015. Electrochemical quantification of the structure/antioxidant activity relationship of flavonoids. Electrochimica Acta 163, 161–166.

Raffa, D., Maggio, B., Raimondi, M.V., Plescia, F., Daidone, G., 2017. Recent discoveries of anticancer flavonoids. European Journal of Medicinal Chemistry. 142, 213–228. https://doi.org/10.1016/j.ejmech.2017.07.034.

Re, R., Pellegrini, N., Proteggente, A., Pannala, A., Yang, M., Rice-Evans, C., 1999. Antioxidant activity applying an improved ABTS radical cation decolorization assay. Free Radical Biology and Medicine 26 (9), 1231–1237.

Riebel, M., Sabel, A., Claus, H., Xia, N., Li, H., König, H., et al., 2017. Antioxidant capacity of phenolic compounds on human cell lines as affected by grape-tyrosinase and Botrytis-laccase oxidation. Food Chemistry 229, 779–789.

Sabudak, T., Demirkiran, O., Ozturk, M., Topcu, G., 2013. Phenolic compounds from *Trifolium echinatum* Bieb. and investigation of their tyrosinase inhibitory and antioxidant activities. Phytochemistry. 96, 305–311. https://doi.org/10.1016/j.phytochem.2013.08.014.

Sandoval-Yañez, C., Mascayano, C., Martínez-Araya, J.I., 2017. A theoretical assessment of antioxidant capacity of flavonoids by means of local hyper–softness. Arabian Journal of Chemistry. https://doi.org/10.1016/j.arabjc.2017.10.011.

Santamaria, A., 2008. Daño oxidative y enfermedades neurodegenerativas. In: Radicales libres y stress oxidative. Aplicaciones médicas. Manual Moderno, Mexico, D.F, pp. 169–182.

Sevindik, H.G., Güvenalp, Z., Yerdelen, K.Ö., Yuca, H., Demirezer, L.Ö., 2015. The discovery of potential anticholinesterase compounds from *Achillea millefolium* L. Industrial Crops and Products. 76, 873–879. https://doi.org/10.1016/j.indcrop.2015.05.088.

Silva, M.M. da, Iriguchi, E.K.K., Kassuya, C.A.L., Vieira, M. do C., Foglio, M.A., Carvalho, J.E. de, Ruiz, A.L.T.G., Souza, K.deP., Formagio, A.S.N., 2017. *Schinus terebinthifolius*: phenolic constituents and in vitro antioxidant, antiproliferative and in vivo anti-inflammatory activities. Revista Brasileira de Farmacognosia. 27, 445–452. https://doi.org/10.1016/j.bjp.2016.12.007.

Singh, D.K., Luqman, S., Mathur, A.K., 2015. Botros -Lawsonia inermis L. – a commercially important primaeval dying and medicinal plant with diverse pharmacological activity: a review. Industrial Crops and Products. 65, 269–286. https://doi.org/10.1016/j.indcrop.2014.11.025.

Singh, B., Singh, J.P., Kaur, A., Singh, N., 2017. Phenolic composition and antioxidant potential of grain legume seeds: a review. Food Research International 101, 1–16.

Stevens, C.L., Taylor, D.I., Delaux, S., Smith, M.J., Schiel, D.R., 2008. Characterisation of wave-influenced macroalgal propagule settlement. Journal of Marine Systems 74 (1–2), 96–107.

Sulaiman, C.T., Balachandran, I., 2016. LC/MS characterization of antioxidant flavonoids from *Tragia involucrata* L. Beni-Suef University Journal of Basic and Applied Sciences 5 (3), 231–235.

Torres, M.A., Dangl, J.L., 2005. Functions of the respiratory burst oxidase in biotic interactions, abiotic stress and development. Current Opinion in Plant Biology 8, 397–403.

Wang, Y., Xue, J., Jia, X., Du, C., Tang, W., Wang, X., 2017b. New antioxidant C-geranylated flavonoids from the fruit peels of Paulownia catalpifolia T. Gong ex D.Y. Hong. Phytochemistry Letters. 21, 169–173. https://doi.org/10.1016/j.phytol.2017.06.025.

Wang, X., Wang, M., Cao, J., Wu, Y., Xiao, J., Wang, Q., 2017a. Analysis of flavonoids and antioxidants in extracts of ferns from Tianmu Mountain in Zhejiang Province (China). Industrial Crops and Products 97, 137–145.

Wen, L., Zhao, Y., Jiang, Y., Yu, L., Zeng, X., Yang, J., et al., 2017. Identification of a flavonoid C-glycoside as potent antioxidant. Free Radical Biology and Medicine 110, 92–101.

Yang, D., Xie, H., Jia, X., Wei, X., 2015. Flavonoid C-glycosides from star fruit and their antioxidant activity. Journal of Functional Foods. 16, 204–210. https://doi.org/10.1016/j.jff.2015.04.048.

Yang, D., Xie, H., Jiang, Y., Wei, X., 2016. Phenolics from strawberry cv. Falandi and their antioxidant and α-glucosidase inhibitory activities. Food Chemistry. 194, 857–863. https://doi.org/10.1016/j.foodchem.2015.08.091.

Zeng, L.H., Wu, J., Fung, B., Tong, J.H., Mic-kle, D., Wu, T.W., 1997. Comparative protection against oxyradicals by three flavonoids on cultured endothelial cells. Biochemistry and Cell Biology 75, 717–720.

Further Reading

Alara, O.R., Abdurahman, N.H., Olalere, O.A., 2017a. Optimization of microwave-assisted extraction of total flavonoids and antioxidants from *Vernonia amygdalina* leaf using response surface methodology. Food and Bioproducts Processing 107, 36–48.

Alara, O.R., Abdurahman, N.H., Olalere, O.A., 2017b. Ethanolic extraction of flavonoids, phenolics and antioxidants from *Vernonia amygdalina* leaf using two-level factorial design. Journal of King Saud University science. https://doi.org/10.1016/j.jksus.2017.08.001.

Gómez, L.E., 2008. Especies reactivas de oxígeno y cáncer. In: Radicales libres y stress oxidative. Aplicaciones médicas. Manual Moderno, Mexico, D.F, pp. 169–182.

Kump, L.R., 2008. The rise of atmospheric oxygen. Nature 451, 277–278.

Index

'*Note*: Page numbers followed by "f" indicate figures and "t" indicate tables.'

CPI Antony Rowe
Chippenham, UK
2018-12-08 10:56